Sustainable Innovation

This book explores the convergence of cutting-edge engineering techniques, innovative design approaches, and fundamental ecological principles to address the challenges associated with sustainable infrastructure, including green construction, resilient urban planning, integration of smart systems, low-carbon emissions, and waste reduction. Each chapter focuses on a specific aspect of sustainable infrastructure, such as eco-friendly building materials, energy-efficient systems, and waste management practices, supported by case studies, illustrated processes, and detailed technical explanations.

Key features

- Alignment with the central theme of eco-smart infrastructure, emphasizing sustainability, innovative technologies, and global best practices
- Discussion of critical aspects of climate-resilient infrastructure
- Coverage of essential topics such as eco-friendly materials, waste management, and life cycle assessment to address key challenges in sustainable infrastructure
- Exploration of smart waste management systems
- Review of real-world case studies supported by detailed technical explanations

This book is intended for graduate students and researchers in civil engineering, infrastructure development, sustainability, and urban planning.

Sustainable Innovation

Smart Strategies for Resilient Infrastructure

Edited by Peerzada Danish, Anil Sagar Srinivasa,
Jasir Mushtaq, Mohan Ganesh G., and
A.S. Santhi

CRC Press
Taylor & Francis Group
Boca Raton London New York

CRC Press is an imprint of the
Taylor & Francis Group, an **informa** business

First edition published 2027
by CRC Press
2385 NW Executive Center Drive, Suite 320, Boca Raton FL 33431

and by CRC Press
4 Park Square, Milton Park, Abingdon, Oxon, OX14 4RN

CRC Press is an imprint of Taylor & Francis Group, LLC

ISBN: 978-1-041-12317-0 (hbk)
ISBN: 978-1-041-12344-6 (pbk)
ISBN: 978-1-003-66428-4 (ebk)

DOI: 10.1201/9781003664284

Typeset in Times
by Apex CoVantage, LLC

Contents

Editors' Biographies

Peerzada Danish received his PhD in structural engineering (civil engineering) from Vellore Institute of Technology (VIT), Vellore, with research focused on self-compacting concrete incorporating mineral admixtures as supplementary materials. He is currently Postdoctoral Research Associate III at the Indian Institute of Science (IISc), Bengaluru, working on the development of low-carbon alternative materials for dam construction and rehabilitation under projects supported by the Ministry of Jal Shakti (MoJS), Central Water Commission, Government of India, and the World Bank. He has authored more than 20 SCI and Scopus-indexed publications presented at international conferences, along with books and book chapters. Dr Danish previously served as Assistant Professor at several universities and institutes, teaching undergraduate and postgraduate courses and supervising student research. His research interests include sustainable construction materials, alkali-activated binders, and durable infrastructure systems.

Anil Sagar Srinivasa is Assistant Professor in the Department of Civil Engineering at NIT Calicut. He completed his PhD in civil engineering from NITK Surathkal. His research focuses on one-part geopolymer pastes and mortars, eco-friendly alkali-activated concrete, and sustainable construction materials. He applies response surface methods for optimization of geopolymer binders and studies the influence of solid activators on fly ash/slag-based systems. He has authored SCI and Scopus-indexed papers and presented at international conferences. He is guiding PhD, MTech, and BTech students in the areas of geopolymer composites, alkali-activated binders, and durable infrastructure systems.

Jasir Mushtaq is Saudi Aramco Approved Environmental Engineer with over six years of professional experience in environmental management across the Gulf region and India. He holds a PhD in civil and environmental engineering from the National Institute of Technology Srinagar, India. His expertise includes solid waste management, environmental monitoring, carbon footprint assessment, and sustainable infrastructure development. Dr Mushtaq has authored several international journal publications and book chapters focusing on climate change, waste management, and environmental sustainability. He is actively involved in implementing international environmental standards and promoting sustainable engineering practices.

Mohan Ganesh G. obtained his PhD in structural engineering from Indian Institute of Technology Roorkee, India. He is Professor with higher academic grade for the Department of Structural and Geotechnical Engineering at VIT, Vellore, India. He has more than 60 publications in the Scopus-indexed journals. His area of research interest lies in high-volume fly ash concrete, self-compacting concrete, fibre-reinforced concrete, green corrosion inhibitors, steel concrete composite structures, artificial neural network, and genetic algorithm.

A.S. Santhi is Professor with a higher academic grade in the School of Civil Engineering, VIT, Vellore. She holds a PhD in structural engineering from IIT Roorkee (2006). She has supervised 11 PhD theses, one MS student, and 76 MTech theses and has published more than 40 research papers in referred and peer-reviewed Scopus-indexed journals. Her research interest includes high-volume fly ash in concrete, utilizations of mineral admixtures in concrete, corrosion studies, and impact studies on concrete. She has successfully completed two R&D projects sponsored by DST, Government of India and NLC, Neyveli.

Contributors

Awodele Imoleayo Abraham
Department of Construction Management and Quantity Surveying
Durban University of Technology
KwaZulu-Natal, South Africa

Dibyendu Adak
Department of Civil Engineering
NIT Karnataka (NITK)
Surathkal, Karnataka, India

Rajib Ahmed
Department of Civil Engineering, Assam Don Bosco
University, Azara, Assam, India

Wisal Ahmed
Department of Architecture and Civil Engineering
City University of Hong Kong
Kowloon Tong
Hong Kong SAR, P.R. China

Arslan Akbar
Department of Civil and Environmental Engineering
Pennsylvania State University
University Park, Pennsylvania, USA

Akhil Avchar
Department of Mining Engineering
NITK Surathkal, Mangalore
Karnataka, India

Samreen Bano
CSIR-Central Building Research Institute
Roorkee, Uttarakhand, India

Salim Barbhuiya
Department of Engineering and Construction
University of East London, London, UK

Mehvish Bilal
Department of Civil and Environmental Engineering
King Fahd University of Petroleum & Minerals, Dhahran
Saudi Arabia

Shreekanth Birgonda
Department of Civil Engineering
National Institute of Technology – Tiruchirappalli
Tiruchirappalli, Tamil Nadu

Prithiviraj Chidambaram
Department of Civil Engineering
SRM Madurai College for Engineering and Technology
Sivagangai, Tamil Nadu, India

S.C. Chin
Centre for Research in Advanced Fluid & Processes
Universiti Malaysia Pahang Al-Sultan
Abdullah, Gambang, Pahang
Malaysia

Halton Dan
Department of Civil Engineering, Assam Don Bosco
University, Azara, Assam
India

Bibhuti Bhusan Das
Department of Civil Engineering
NITK Surathkal, Karnataka, India

Deepana G.
Department of Civil Engineering, S.A. Engineering College
Chennai, Tamil Nadu
India

Jitong Ding
Department of Architecture and Civil Engineering
City University of Hong Kong, Tat Chee Avenue, Kowloon Tong
Hong Kong SAR, P.R. China

Muneeb Farooq
University of Regina, Regina
Canada

Jolius Gimbun
Centre for Research in Advanced Fluid & Processes
Universiti Malaysia Pahang Al-Sultan Abdullah, Gambang, Pahang, Malaysia

Shweta Goyal
Civil Engineering Department
Thapar Institute of Engineering and Technology
Patiala, Punjab, India

Himanshu Guleria
Civil Engineering Department
Thapar Institute of Engineering and Technology
Patiala, Punjab, India

Mohammed Jafer Sathik Ali Abdul Hameed
Department of Civil Engineering
PSG Institute of Technology and Applied Research
Coimbatore, Tamil Nadu, India

Harshani R.
Department of Civil Engineering, National Engineering College
Kovilpatti, Tamil Nadu, India

Noor Md. Sadiqul Hasan
Department of Civil Engineering,
International University of Business Agriculture and Technology, Dhaka
Bangladesh

Shahriar Hasan
Department of Civil Engineering
International University of Business Agriculture and Technology
Dhaka, Bangladesh

Md. Maruf Hossain
Department of Civil Engineering
International University of Business Agriculture and Technology
Dhaka, Bangladesh

M. Shariful Islam
Department of Engineering
South Carolina State University
Orangeburg, South Carolina
USA

Jeya Arthi A.J.
Department of Civil Engineering
Rajalakshmi Engineering College
Chennai, Tamil Nadu, India

Rajvardhan Kallimani
Dr. Vishwanath Karad MIT World Peace University
Pune, Maharashtra, India

Muhammad Basit Khan
Department of Architecture and Civil Engineering
City University of Hong Kong, Tat Chee Avenue, Kowloon Tong
Hong Kong SAR, P.R. China

Zishan Raza Khan
Department of Civil Engineering
Integral University, Lucknow
Uttar Pradesh, India

Sheh Ching, Khong
Faculty of Civil Engineering Technology
Universiti Malaysia Pahang Al-Sultan Abdullah
Gambang, Pahang, Malaysia

Abhishek Kumar
Department of Civil Engineering
Integral University, Lucknow
Uttar Pradesh, India

Mritunjay Kumar
Department of Mining Engineering
NITK Surathkal, Mangalore
Karnataka, India

Babitdor Marwein
Department of Civil Engineering, Assam Don Bosco
University, Azara, Assam, India

Pravin Minde
Dr. Vishwanath Karad MIT World Peace University
Pune, Maharashtra, India

Saidu Mohammed
Department of Quantity Surveying
Ahmadu Bello University
Zaria, Nigeria

Arun Murugesan
Department of Civil Engineering
PSG Institute of Technology and Applied Research
Coimbatore, Tamil Nadu, India

Jasim Nisar
Indian Institute of Technology Kanpur, Kanpur
Uttar Pradesh, India

Alice Ogunlolu
Department of Quantity Surveying and Commercial Management
Coventry University, Coventry
UK

Aryamanki Pala
Department of Civil Engineering, Assam Don Bosco
University, Azara, Assam, India

Jagruti Patil
Dr. Vishwanath Karad MIT World Peace University
Pune, Maharashtra, India
Ghaziabad, Uttar Pradesh
India

Suvash Chandra Paul
Department of Civil Engineering
International University of Business Agriculture and Technology, Dhaka
Bangladesh
and
Singapore Centre for 3D Printing
Nanyang Technological University
Singapore, Singapore

K. Srinivasa Rao
Department of Civil Engineering
College of Engineering, Andhra University, Visakhapatnam
Andhra Pradesh, India

Nidhya Rathinavel
Department of Civil Engineering
PSG Institute of Technology and Applied Research
Coimbatore, Tamil Nadu
India

Ubaid Rehman
National Institute of Technology
Srinagar, Jammu and Kashmir (J&K)
India

Rafiq Gymbad Rymbai
Department of Civil Engineering, Assam Don Bosco
University, Azara, Assam, India

Mrinal Kanti Sen
Department of Civil Engineering, Assam Don Bosco
University, Azara, Assam, India

Senthilkumar
Department of Civil Engineering
National Institute of Technology – Tiruchirappalli
Tiruchirappalli, Tamil Nadu

Mohammad Yousuf Shah
National Institute of Technology
Srinagar, J&K, India

Sivamalar Shijumon
Department of Civil Engineering
University of Greater Manchester, Ras Al Khaimah
United Arab Emirates

V. Shijumon
Department of Civil Engineering
Projects & Operations Department, ElectroRAK Group, Ras Al Khaimah
United Arab Emirates

Vikash Singh
Department of Civil Engineering
Integral University, Lucknow
Uttar Pradesh, India

Shambhavi Sinha
Department of Mining Engineering
NITK Surathkal, Mangalore
India

Ming Jen Tan
Singapore Centre for 3D Printing
School of Mechanical & Aerospace Engineering
Nanyang Technological University
Singapore, Singapore

Yi Wei Daniel Tay
Singapore Centre for 3D Printing
School of Mechanical & Aerospace Engineering
Nanyang Technological University
Singapore, Singapore

Kong Fah, Tee
Department of Civil and Environmental Engineering
King Fahd University of Petroleum and Minerals, Dhahran, Saudi Arabia
and
Interdisciplinary Research Center for Construction and Building Materials, King Fahd
University of Petroleum and Minerals, Dhahran, Saudi Arabia

Chockalingam T.
Department of Civil Engineering, Ramco Institute of Technology
Rajapalayam, Tamil Nadu, India

Girija T.R.
Department of Civil Engineering, Assam Don Bosco
University, Azara, Assam, India

Vidya Balagam
Department of Civil Engineering
GVP College for Degree and PG Courses (A)
Visakhapatnam, Andhra Pradesh, India

Chandramouli Viswanathan
Department of Construction Science and Organizational
Purdue University Northwest, Indiana, USA

Jia Jun, Yee
Faculty of Civil Engineering Technology
Universiti Malaysia Pahang Al-Sultan Abdullah
Gambang, Pahang, Malaysia

Iaemlangi Ymbon
Department of Civil Engineering, Assam Don Bosco
University, Azara, Assam, India

Mohammad Younis
National Institute of Technology
Srinagar, J&K, India

Preface

The rapid pace of urbanization and infrastructure development across the world has brought with it significant environmental and societal challenges. Increasing carbon emissions, depletion of natural resources, and growing volumes of construction and industrial waste have made it clear that conventional approaches to infrastructure development are no longer sufficient. There is a pressing need to rethink how we design, construct, and maintain infrastructure systems so that they are not only durable and efficient but also environmentally responsible and socially inclusive.

This book, *Sustainable Innovation: Smart Strategies for Resilient Infrastructure*, has been developed with this objective in mind. It brings together contributions from researchers and practitioners working across diverse areas of civil and environmental engineering, with a shared focus on advancing sustainable infrastructure solutions. The volume explores the convergence of innovative engineering techniques, eco-friendly materials, and fundamental ecological principles to address current and emerging challenges in the built environment.

The chapters in this book cover a wide range of topics, including the use of biochar in cementitious systems, alkali-activated and geopolymer materials, self-compacting concrete, waste-derived aggregates, natural fibres, and advanced functional materials. Several contributions focus on the valorization of industrial and agricultural waste such as ceramic waste, coconut ash, bagasse, and plastic, highlighting their potential in reducing environmental impact while maintaining or enhancing performance. Other chapters address important aspects such as rheology, durability under thermal loading, adsorption mechanisms for environmental remediation, and the behavior of sustainable binders and asphalt systems. In addition to material innovations, the book also considers broader sustainability perspectives. Topics such as life cycle assessment, waste management strategies, and the integration of circular economy principles are discussed to provide a more comprehensive understanding of sustainable infrastructure development. The aim is to connect fundamental scientific understanding with practical applications, enabling readers to translate research outcomes into real-world solutions.

This volume is intended for researchers, engineers, academicians, industry professionals, and policymakers who are engaged in the field of sustainable construction and infrastructure. It is our hope that the insights presented here will support informed decision-making and encourage the adoption of more sustainable practices across different stages of infrastructure development. The editors would like to express their sincere gratitude to all the authors for their valuable contributions and timely efforts. We also acknowledge the support of the publishing team for their guidance throughout the process. We hope that this book will serve not only as a technical reference but also as a step towards fostering a more sustainable and resilient built environment for the future.

Acknowledgement

The corresponding editor, Dr Peerzada Danish, gratefully acknowledges the Indian Institute of Science (IISc) Bengaluru for providing institutional support and infrastructure facilities, and the editor also thanks the Central Water Commission (CWC), Ministry of Jal Shakti (MoJS), Government of India, for providing research funding and salary support at IISc Bengaluru.

1 Biochar as a Sustainable Supplement in Cement Mortar and Concrete

Salim Barbhuiya, Bibhuti Bhusan Das and Dibyendu Adak

1.1 INTRODUCTION

Biochar is a stable, carbon-rich material produced through the pyrolysis of biomass – such as crop residues, wood waste, or manure – under limited oxygen conditions. This thermochemical process sequesters carbon that would otherwise be released into the atmosphere as carbon dioxide, thereby contributing to climate change mitigation. When incorporated into soil, biochar enhances soil fertility, water retention, and microbial activity, supporting improved plant growth and yield. Its porous structure also reduces nutrient leaching and enhances cation exchange capacity (CEC), making it especially effective in degraded or nutrient-poor soils. Moreover, biochar can lower nitrous oxide and methane emissions from agricultural soils, further contributing to greenhouse gas (GHG) reduction. Biochar also presents a sustainable waste management solution, converting agricultural residues into a valuable soil amendment rather than letting them decompose or burn. While its performance depends on factors such as feedstock, pyrolysis conditions, and application methods, biochar is increasingly recognized as a promising tool for sustainable agriculture and climate resilience.

Sustainable materials play a crucial role in modern construction, offering significant environmental, economic, and social benefits. As the global population grows and urban development accelerates, traditional building methods and materials contribute heavily to resource depletion, carbon emissions, and waste. Sustainable materials – such as bamboo, recycled steel, reclaimed wood, and low-carbon concrete – help mitigate these impacts by reducing energy consumption, lowering GHG, and minimizing ecological footprints.

Using sustainable materials promotes energy efficiency and often enhances the durability and lifespan of buildings, leading to reduced maintenance and operational costs. They also support circular economy principles, enabling materials to be reused, repurposed, or recycled at the end of a structure's life. This approach conserves natural resources and reduces landfill burden. Moreover, sustainable construction materials contribute to healthier indoor environments by limiting the use of toxic chemicals and improving air quality. These materials often carry certifications ensuring ethical sourcing and minimal environmental harm. As awareness grows,

DOI: 10.1201/9781003664284-1

the construction industry increasingly adopts green building practices, driven by regulations, market demand, and environmental responsibility. In essence, integrating sustainable materials is not only a strategic response to climate change but also a commitment to building resilient, efficient, and future-ready infrastructure. It's a vital step towards a more sustainable and liveable planet.

The motivation behind writing this chapter stems from the urgent need to reduce the environmental impact of the construction industry. Cement production significantly contributes to global CO_2 emissions, prompting the search for eco-friendly alternatives. Biochar, a carbon-rich by-product of biomass pyrolysis, offers promising potential as a sustainable additive due to its ability to improve mechanical properties, reduce carbon footprint, and enhance durability of cementitious materials. This chapter aims to explore biochar's benefits, challenges, and practical applications, contributing to the advancement of green construction technologies and circular economy principles.

1.2 BIOCHAR: PROPERTIES AND PRODUCTION

1.2.1 Feedstocks for Biochar Production

Biochar is produced through the pyrolysis of various biomass feedstocks under limited oxygen conditions. The selection of feedstock plays a critical role in determining the physical, chemical, and structural properties of the resulting biochar, thereby influencing its performance when used in applications such as cement mortar and concrete. Common feedstocks include agricultural residues (e.g. rice husks, corn stalks, wheat straw), forestry waste (e.g. sawdust, bark, wood chips), animal manure, and organic municipal solid waste. These materials are often abundant, renewable, and low-cost, making biochar production economically and environmentally attractive.

Each feedstock has a unique composition of cellulose, hemicellulose, lignin, and ash content, which affects the yield and characteristics of the biochar. For instance, woody biomass generally produces biochar with higher carbon content and stability, while manure-based feedstocks may offer greater nutrient retention but higher ash content. Pre-treatment methods like drying, grinding, or pelletizing can further enhance the efficiency and consistency of biochar production. Utilizing diverse biomass feedstocks not only supports waste valorization and resource recovery but also contributes to sustainable waste management practices. The choice of feedstock must align with the intended application of biochar, ensuring optimal performance and environmental benefit in construction and other industries. Table 1.1 presents a comparative analysis of various biochar feedstocks, highlighting essential properties such as carbon content, ash content, porosity, and availability. These characteristics play a critical role in determining the suitability and performance of biochar as a sustainable additive in cement mortar and concrete for enhanced durability and reduced environmental impact.

1.2.2 Pyrolysis and Other Production Techniques

Biochar is primarily produced through pyrolysis, a thermal decomposition process carried out in the absence or limited presence of oxygen. Pyrolysis typically occurs

TABLE 1.1
Properties 0f Common Biochar Feedstocks for Use in Cement Mortar and Concrete

Feedstock type	Examples	Carbon content (% by weight)	Ash content (% by weight)	Porosity (m^2/g)	Availability	Impact on concrete properties	Remarks
Agricultural Waste	Rice husk, wheat straw, corn stalk	55–75%	10–20%	150–350	High (seasonal)	Enhances workability, internal curing	May require grinding/ sieving for uniformity
Forestry Residue	Sawdust, bark, wood chips	70–85%	1–5%	100–300	High (consistent)	Improves strength and durability	Produces high-quality, stable biochar
Animal Manure	Poultry litter, cow dung	40–60%	30–50%	20–100	Moderate (regional)	May reduce compressive strength	Rich in nutrients; may increase water demand
Municipal Solid Waste	Organic kitchen waste, green waste	45–65%	15–30%	50–200	Moderate to High	Variable; needs pre-treatment	Risk of contaminants; biochar quality inconsistent
Industrial by-products	Paper sludge, brewery waste	50–70%	10–25%	80–250	Low to moderate	Mixed effects; depends on consistency	Sustainable reuse potential with processing

at temperatures ranging from 300°C to 700°C and can be classified into slow, fast, and flash pyrolysis. Slow pyrolysis, with lower temperatures and longer residence times, yields a higher proportion of biochar and is well suited for use in cementitious applications. Fast and flash pyrolyses, operating at higher temperatures and shorter durations, produce more bio-oil and gas, with less biochar output. Other techniques include gasification, where biomass undergoes partial oxidation at temperatures above 700°C, resulting in syngas as the main product and biochar as a secondary by-product with lower carbon content. Hydrothermal carbonization (HTC) processes wet biomass in water under pressure at moderate temperatures, generating a carbon-rich solid known as hydrochar. This method is effective for high-moisture feedstocks. Torrefaction is a milder thermal process performed at 200°C to 300°C in low oxygen, producing a brittle, energy-dense char that may have limited utility in concrete. Each production method affects the yield, structure, and chemical properties of the biochar, influencing its performance when used as a supplementary material in cement mortar and concrete.

Figure 1.1 illustrates the conversion of urban (food waste) and agro wastes (rice husk, grape waste) into biochar through slow pyrolysis at temperatures of 400, 500, and 600°C under CO_2 and N_2 atmospheres. The process produces biochar, which is then characterized for agro-environmental applications. The CO_2 atmosphere results in biochar with higher yield, greater surface area, and enhanced porosity compared to the N_2 atmosphere. This highlights the potential of CO_2-assisted pyrolysis in improving biochar quality. The schematic emphasizes sustainable waste management and the production of value-added materials for soil enhancement and environmental benefits.

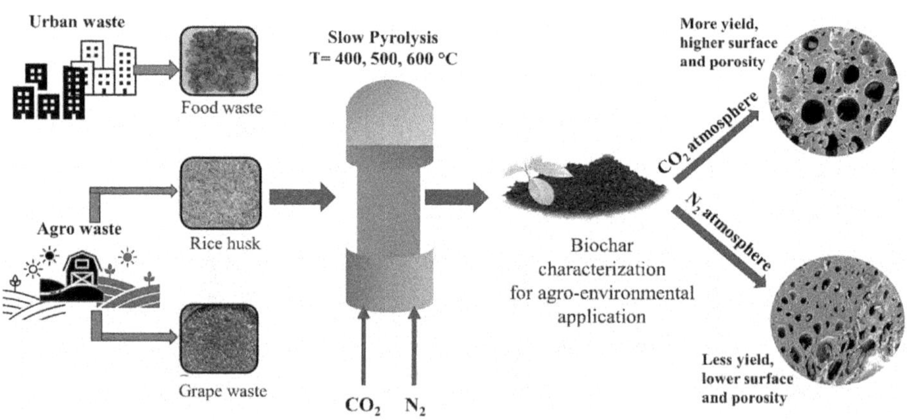

FIGURE 1.1 Schematic overview of biochar production from urban (food) and agro-industrial (rice husk, grape) wastes via slow pyrolysis.

Source: Premchand et al., (2023)

1.2.3 Environmental Benefits of Biochar

Biochar, a stable form of carbon-rich charcoal produced through pyrolysis of organic materials, offers significant environmental benefits. One of its primary advantages is its ability to sequester carbon. When biomass is converted into biochar, the carbon within is stabilized, preventing it from re-entering the atmosphere as carbon dioxide. This makes biochar an effective tool in combating climate change by reducing GHG emissions. In addition to carbon sequestration, biochar improves soil health. It enhances soil structure, increases water retention, and promotes microbial activity, all of which contribute to more resilient agricultural systems. By improving nutrient retention, biochar reduces the need for chemical fertilizers, thus minimizing nutrient runoff into water bodies – a key contributor to water pollution and eutrophication. Biochar can also help mitigate methane and nitrous oxide emissions from soils, particularly in agricultural contexts. Its porous nature supports beneficial microbial populations that suppress the formation of these potent GHGs. Furthermore, biochar production often utilizes agricultural or forestry waste, turning potential pollutants into a valuable resource. This waste-to-resource approach supports sustainable waste management and reduces landfill use.

1.2.4 Physical, Chemical, and Mechanical Properties of Biochar

Biochar exhibits a unique combination of physical, chemical, and mechanical properties that contribute to its wide range of applications in agriculture, environmental remediation, and material science.

Biochar is highly porous, with a large specific surface area that varies depending on the feedstock and pyrolysis conditions. This porosity enhances its water-holding capacity and enables the retention of nutrients and microbial populations in soil. Its low bulk density makes it lightweight, facilitating easier handling and application in agricultural systems. The colour of biochar is typically black or dark brown, reflecting its carbon-rich composition, and its texture ranges from fine powder to coarse granules.

Chemically, biochar is primarily composed of carbon, but it also contains essential nutrients such as nitrogen, phosphorus, potassium, calcium, and magnesium, depending on the feedstock used. It often has a high pH, making it suitable for ameliorating acidic soils. Biochar also possesses CEC, which allows it to adsorb and retain positively charged ions, improving nutrient availability in soil. Its surface functional groups (e.g. carboxyl, hydroxyl) contribute to its reactivity and interaction with soil and pollutants.

Mechanically, biochar is relatively brittle and friable, which means it can easily be broken down into smaller particles. This property is beneficial for soil integration but limits its structural use without modification. When used as a filler or additive in construction materials like concrete or polymers, biochar can influence mechanical strength, thermal stability, and insulation properties. Overall, the physical porosity, chemical functionality, and mechanical behaviour of biochar are highly dependent on feedstock type and pyrolysis temperature, making its properties tenable for specific applications. This versatility positions biochar as a promising material in both environmental and industrial domains.

When used in concrete, biochar's physical, chemical, and mechanical properties play critical roles. Its high porosity and surface area improve water retention and internal curing, enhancing concrete durability. Chemically, biochar contains reactive surface groups that can interact with cement hydration products, potentially increasing bonding strength. Its alkaline nature may also help stabilize concrete pH. Mechanically, although biochar is brittle, it acts as a lightweight filler that can reduce the overall weight of concrete without significantly compromising strength. Depending on the feedstock and production conditions, biochar can improve thermal insulation, reduce shrinkage, and contribute to more sustainable, carbon-negative concrete formulations. Table 1.2 summarizes the key physical, chemical, and mechanical properties of biochar, which influence its performance in concrete. It

TABLE 1.2
Key Physical, Chemical, and Mechanical Properties of Biochar Relevant to Concrete Applications

Property type	Specific property	Description	Relevance to concrete
Physical	Porosity	High internal pore structure	Enhances internal curing and moisture retention
	Surface area	Large specific surface area (up to 1,000 m^2/g)	Improves bonding with cement matrix
	Particle size and shape	Fine to coarse, angular or irregular	Affects workability and mechanical interlocking
	Density	Low bulk density	Reduces overall weight of concrete
Chemical	Surface functional groups	Carboxyl, hydroxyl, carbonyl, etc.	Promotes interaction with cement hydration products
	pH	Typically alkaline (pH 8–10)	Helps maintain stable pH in concrete
	Carbon content	High fixed carbon	Contributes to carbon sequestration and sustainability
	Mineral content	Contains Ca, K, Mg, P, depending on feedstock	May influence setting time and mechanical properties
Mechanical	Compressive strength (filler)	Brittle, but can act as a functional additive	Slightly reduces strength but improves thermal and shrinkage performance
	Thermal stability	Stable up to high temperatures (~500–700°C)	Enhances fire resistance and thermal insulation
	Elastic modulus	Lower than traditional aggregates	May reduce brittleness, allows slight flexibility under stress

highlights how characteristics like porosity, surface chemistry, and thermal stability contribute to improved durability, sustainability, and functionality in concrete composites.

1.3 BIOCHAR IN CEMENTITIOUS SYSTEMS

1.3.1 Mechanisms of Interaction with Cement Matrix

Biochar interacts with the cement matrix through a combination of physical and chemical mechanisms which influence hydration, microstructure development, and overall performance. Its high porosity and water retention capacity enable internal curing, which promotes more complete hydration of cement particles. This leads to improved strength gain and reduced shrinkage. Chemically, biochar contains surface functional groups (e.g. carboxyl, hydroxyl) which can interact with calcium ions and hydration products like calcium silicate hydrate (C-S-H), enhancing the bond at the interfacial transition zone (ITZ). Additionally, biochar acts as a micro-filler, reducing pore connectivity and refining the pore structure, which improves durability by lowering permeability and enhancing resistance to harmful ingress. However, excessive amounts of biochar or poor compatibility with the cement matrix can negatively affect workability and strength. The carbon content may also play a role in influencing alkali-silica reactivity and thermal conductivity. Proper characterization and proportioning are essential for effective integration into concrete.

Figure 1.2 compares cement hydration without (a) and with (b) biochar particles. In (a), cement particles hydrate individually, forming a metastable barrier and leaving voids, which can weaken the structure. In contrast, (b) shows that biochar particles act as nucleation sites, promoting clustering of cement particles. This clustering enhances hydration, leading to more uniform and compact hydrate formation. The presence of biochar reduces voids and improves the microstructure of the hydrated cement. Overall, biochar enhances the hydration process and results in a denser, potentially stronger cement matrix by facilitating more effective nucleation and hydration cluster development.

1.3.2 Influence on Fresh Properties

Biochar incorporation significantly influences the fresh properties of concrete, particularly workability, consistency, and setting time. According to Yang and Wang (2021), biochar's high surface area and porosity increase water demand, leading to reduced workability in cement blends. Similarly, Gupta and Kua (2019) observed that finer biochar particles and higher dosages decreased the flow of cement mortar, attributed to greater water absorption and interaction with cement paste. Wang et al. (2025a)- highlighted that biochar used as a partial sand replacement can promote internal curing, enhancing moisture retention during hydration but potentially impacting slump and compaction. Sirico et al. (2021) also reported a reduction in workability due to the irregular particle shape and high absorptivity of wood-derived biochar. These studies collectively suggest that while biochar can improve internal curing and sustainability, its influence on fresh concrete properties must be carefully managed through mix design optimization and water-to-binder ratio adjustments.

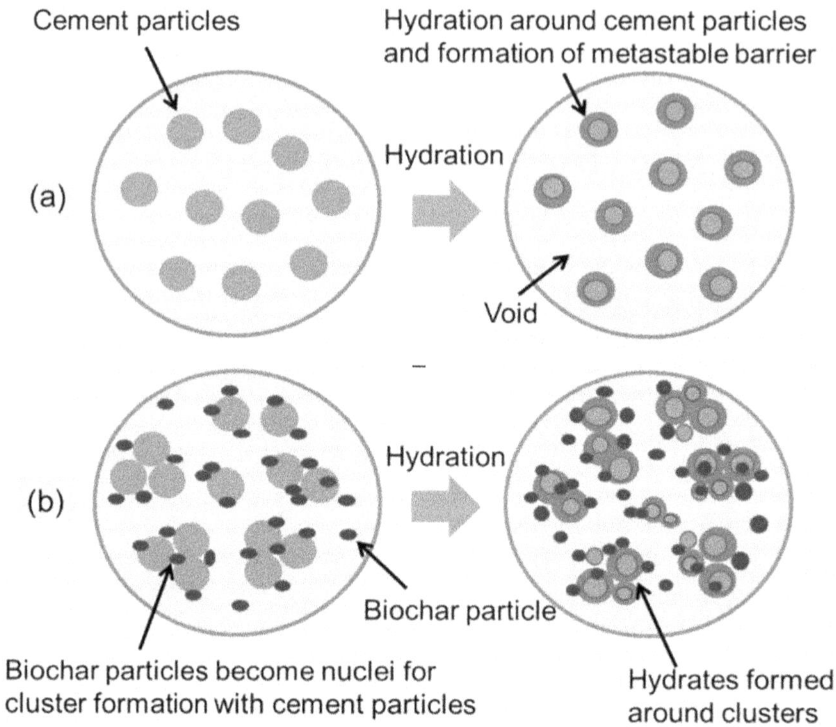

FIGURE 1.2 Schematics showing hydration product formation in (a) cement-only system and (b) system containing cement and biochar.

Source: Mensah et al. (2021)

1.3.3 Effects on Mechanical Properties

The incorporation of biochar into concrete has been shown to influence its mechanical properties in both beneficial and adverse ways, largely depending on biochar's characteristics and dosage. Akhtar and Sarmah (2018) demonstrated that optimal amounts of biochar can improve compressive strength due to enhanced particle packing and reduced porosity. Qing et al. (2023) reported similar findings, noting that biochar contributes to increased compressive and fracture strength by promoting internal curing and refining the concrete microstructure. Murali and Wong (2024) emphasized that the porous structure and surface chemistry of biochar can influence hydration and bonding within the cement matrix, with excessive dosages leading to strength reductions due to increased void content.

Li et al. (2025) further explored the role of particle size, finding that finer biochar enhances strength more effectively by accelerating hydration and improving matrix homogeneity. Javed et al. (2022) also observed that the mechanical response is closely tied to the physical and chemical nature of the biochar used. Sirico et al. (2023) highlighted that combining biochar with recycled plastic aggregates can yield significant strength benefits when balanced properly. Additionally, Wang et al.

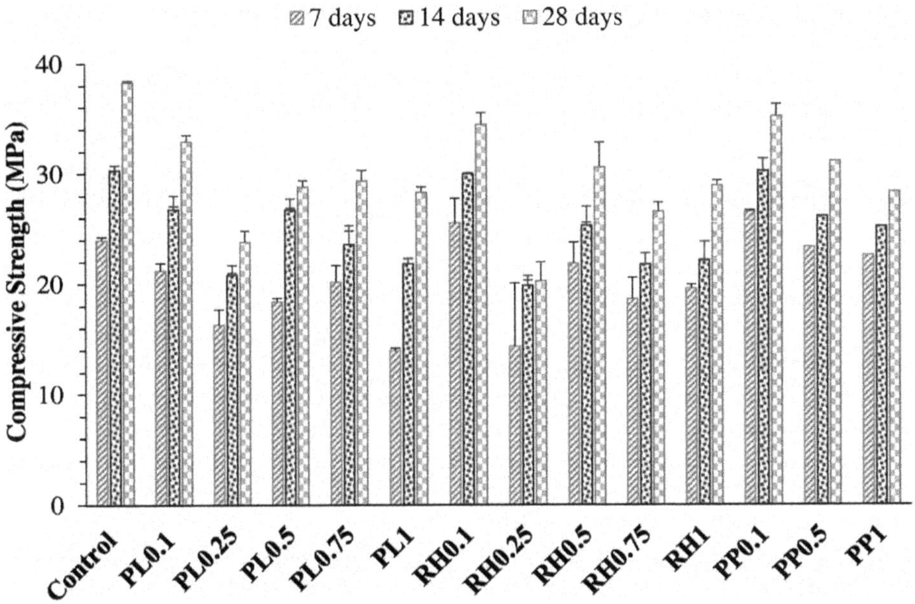

FIGURE 1.3 Compressive strength of biochar-concrete composites after 7, 14, and 28 days of curing.

Source: Akhtar and Sarmah (2018)

(2025b) demonstrated that biochar-modified Engineered Cementitious Composites exhibited improved ductility and crack resistance, suggesting potential for structural applications requiring toughness.

Figure 1.3 presents the compressive strength of concrete samples containing different types and dosages of biochar (PL, RH, and PP) at 7, 14, and 28 days of curing. The control mix consistently shows the highest compressive strength across all curing periods, peaking near 38 MPa at 28 days. Among the biochar-modified mixes, lower dosages (0.1% and 0.25%) of PL (pine lumber), RH (rice husk), and PP (plastic pyrolysis) generally yield better strength compared to higher dosages (0.75% and 1%). All biochar mixes exhibit strength gain over time, with 28-day values significantly higher than 7-day results. PL0.1, RH0.1, and PP0.1 closely approach the control's strength, indicating the potential of biochar to partially replace cement without compromising performance. However, higher dosages, particularly RH0.75 and PL1, show considerable reductions in compressive strength, likely due to increased porosity and interference with hydration. These results suggest that optimal biochar content is critical to maintain mechanical integrity.

1.3.4 Impact on Microstructure and Porosity

The incorporation of biochar into concrete significantly influences its microstructure and porosity, with effects varying based on biochar type, dosage, and processing. Biochar can enhance microstructural refinement by acting as nucleation sites for hydration products, leading to denser matrices and improved ITZs (Jafari & Sadeghian, 2023; Liu et al., 2025). Studies show that at optimal levels, biochar reduces microcracking and fosters a more homogeneous distribution of hydration products (Ali et al., 2023). However, biochar's intrinsic porosity can also increase total pore volume, particularly at higher dosages or with poorly processed materials. Javed et al. (2022) and Sobuz et al. (2024) found that while certain biochars refine pore structure and improve performance, others can increase porosity and compromise strength. Ying et al. (2025) noted that rice straw biochar improved carbon sequestration but introduced higher void content. Thus, biochar's impact is dual-faceted: it can not only enhance microstructure and sustainability but also increase porosity if not optimized. Effective performance depends on controlling biochar properties and replacement levels. Tailored biochar selection and mix design are essential to balance mechanical performance with environmental benefits. Ongoing research is needed to establish standardized guidelines for biochar use in cementitious composites.

The Scanning Electron Microscopy (SEM) micrographs in Figure 1.4 illustrate the microstructural evolution of hydrated cementitious composites containing bamboo biochar (BBC). Panel (a) shows BBC particles embedded in a matrix of hydration products. In (b), BBC is surrounded by key hydration phases such as calcium silicate hydrate (CSH), ettringite (AFt), and monosulphate (AFm), indicating effective pozzolanic activity. Panel (c) highlights the presence of crystalline calcium hydroxide (CH), while (d) shows intergrowths of CH, CSH, and amorphous calcium carbonate (ACC). These images confirm that BBC participates in hydration reactions,

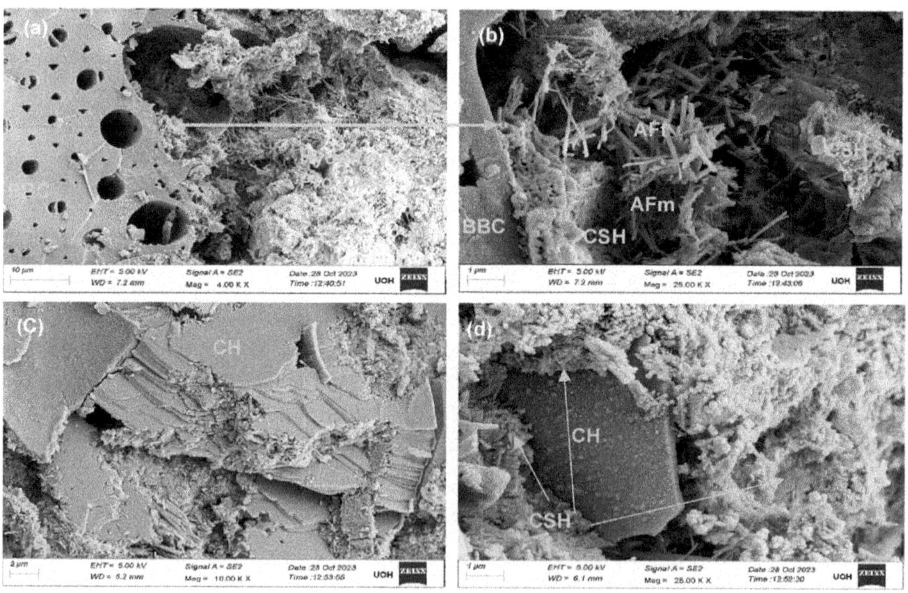

FIGURE 1.4 SEM micrographs of water-cured bamboo biochar cement paste.

Source: Bhagat and Pancharathi (2025)

contributing to the development of a dense microstructure, enhancing the composite's strength and durability.

1.4 OPTIMIZATION STRATEGIES

1.4.1 Selection of Biochar Type and Characteristics

The selection of biochar type and its characteristics plays a critical role in determining its effectiveness as a concrete additive. Biochar is derived from various biomass sources through pyrolysis, and its properties – such as surface area, porosity, particle size, and chemical composition – depend heavily on feedstock and production conditions. For use in concrete, biochars with high surface area and porous structure are desirable, as they can enhance internal curing, reduce shrinkage, and improve mechanical strength. Low ash content and appropriate pH are also important to avoid adverse chemical reactions with cement. Woody biochars, produced at moderate-to-high temperatures (500–700°C), tend to exhibit favourable characteristics, including stability and compatibility with cementitious matrices. Additionally, proper particle size distribution ensures good dispersion within the mix and minimizes negative impacts on workability. Ultimately, selecting the right type of biochar requires balancing its physical and chemical traits with the desired performance outcomes in concrete applications.

1.4.2 Proportioning and Mixing Techniques

Proportioning and mixing techniques are crucial for the effective integration of biochar into concrete. Typically, biochar is added as a partial replacement for cement or fine aggregate, with replacement levels ranging from 1% to 10% by weight, depending on desired performance outcomes. Lower percentages (1–3%) often enhance mechanical strength and durability, while higher levels may negatively impact workability and setting time due to increased water demand. Proper dispersion of biochar in the mix is essential to prevent agglomeration and ensure uniform performance. Pre-soaking biochar can help mitigate its high absorptive capacity, reducing water loss from the mix. Mechanical mixing, ideally in a high-shear mixer, ensures even distribution. In some cases, superplasticizers or other admixtures are used to counteract reduced workability. Trial batching is recommended to optimize the biochar dosage and mixing method for specific applications, ensuring compatibility with other mix components and achieving consistent, high-quality concrete performance.

1.4.3 Surface Modification of Biochar for Enhanced Performance

Surface modification of biochar is a promising approach to enhance its performance in concrete applications. Due to its naturally porous and irregular surface, untreated biochar can sometimes negatively affect concrete's workability and bonding. Surface treatments aim to improve biochar's compatibility with cementitious materials and optimize its reactivity. Common modification techniques include acid or alkali treatments, which alter surface functional groups to increase hydrophilicity and improve

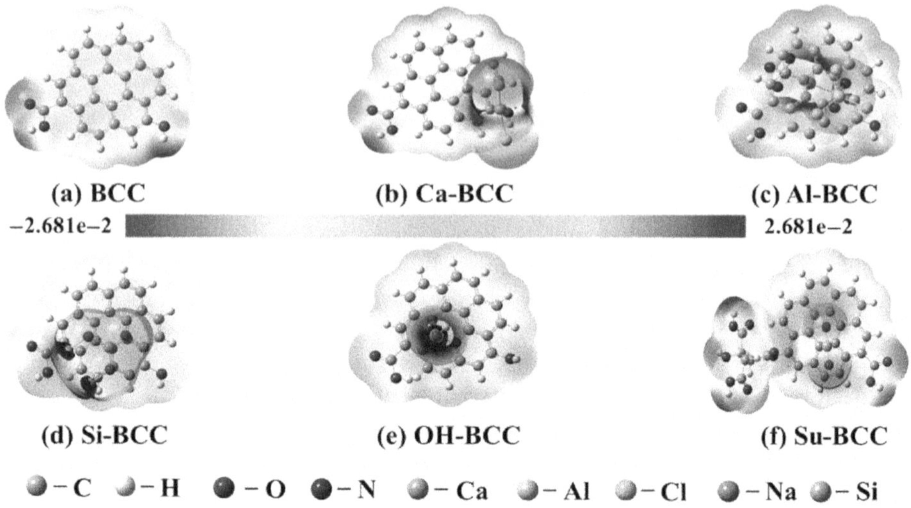

FIGURE 1.5 Electrostatic potential maps of carbon surfaces for biochar with and without modification.

Source: Ye et al. (2024)

bonding with cement paste. Silane or polymer coatings can further enhance interfacial adhesion and reduce water absorption. Oxidative treatments (e.g. using hydrogen peroxide) can introduce additional reactive sites, promoting better integration within the cement matrix. These modifications can lead to improved compressive strength, reduced porosity, and enhanced durability of the resulting concrete. However, the type of modification must be carefully selected on the basis of the biochar feedstock and desired concrete properties. Overall, surface-modified biochar offers a tailored and more efficient additive for sustainable, high-performance concrete.

Figure 1.5 illustrates electrostatic potential maps of various modified biochar-based carbon composites (BCCs), labelled (a) to (f). The colour scale ranges from red (negative potential, −2.681e-2) to blue (positive potential, 2.681e-2), indicating charge distribution. (a) is pristine BCC, while (b–f) are BCCs modified with Ca, Al, Si, OH, and Su (sucrose or sulphate group). Each modification introduces distinct electronic environments, impacting potential distributions. For instance Ca-BCC and Al-BCC exhibit strong blue regions near the metal atoms, indicating high positive potential. These variations impact the adsorptive and catalytic properties of the materials, relevant for applications in environmental remediation and material science.

1.5 DURABILITY AND LONG-TERM PERFORMANCE

1.5.1 Resistance to Water Penetration and Freeze–Thaw Cycles

Biochar concrete exhibits enhanced resistance to water penetration and freeze–thaw cycles, though performance varies with biochar type and dosage. Jia et al. (2023)

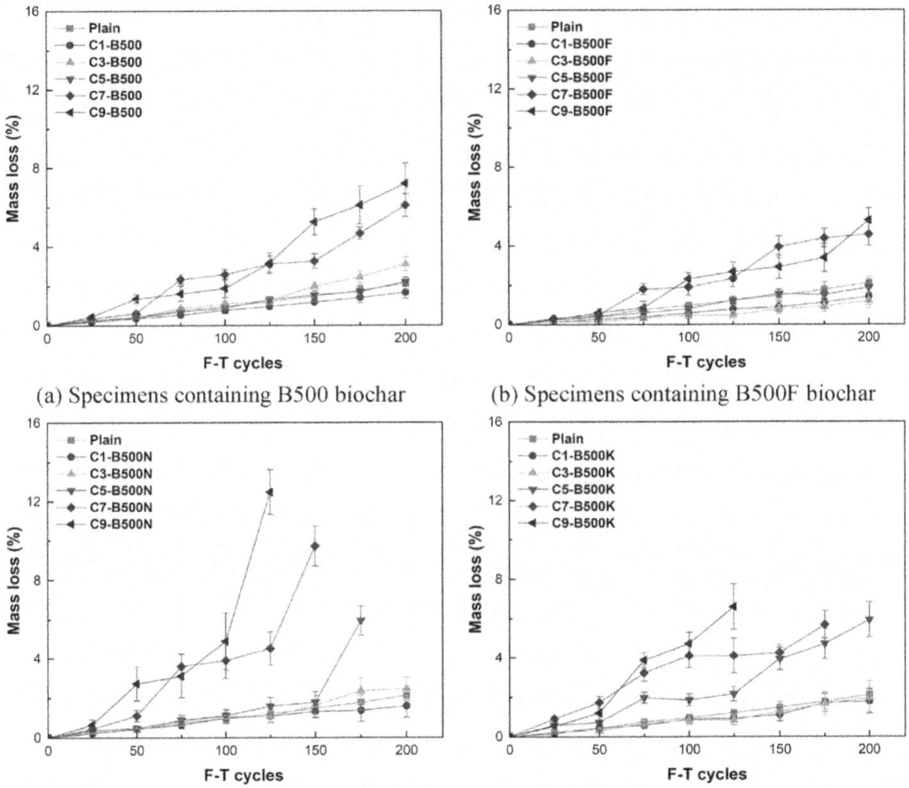

(a) Specimens containing B500 biochar

(b) Specimens containing B500F biochar

(c) Specimens containing B500N biochar

(d) Specimens containing B500K biochar

FIGURE 1.6 Mass loss of specimens during freeze–thaw cycles: (a) specimens containing B500 biochar; (b) specimens containing B500F biochar; (c) specimens containing B500N biochar; and (d) specimens containing B500K biochar. Error bars represent the standard deviation of the measurements.

Source: Chen et al. (2025)

found that incorporating biochar derived from municipal solid waste improved concrete's freeze–thaw durability by refining pore structure and reducing permeability. However, excessive biochar content led to internal cracking due to increased porosity. Complementing this, Kalpokaitė-Dičkuvienė and Zabulionis (2025) reported that biochar-silica fume blends mitigated strength loss during freeze–thaw and salt wet–dry cycles, owing to pozzolanic reactions and densified microstructure. Together, these studies suggest that optimized biochar incorporation enhances durability but requires careful material balancing. Figure 1.6 illustrates mass loss (%) over freeze–thaw (F-T) cycles for specimens with varying contents of different B500 biochars. Generally, higher biochar content (C9) results in greater mass loss, especially in B500 and B500N. Specimens with lower biochar contents (C1–C3) exhibit better durability, indicating optimal performance at lower biochar dosages.

1.5.2 Chemical Resistance

Biochar concrete demonstrates promising chemical resistance against sulphates, chlorides, and acids; yet performance varies with biochar type and environmental exposure. Gupta et al. (2021) showed that silica-rich biochar improved cement mortar resistance to chloride and sulphate ingress due to its pozzolanic activity and refined pore structure, unlike inert biochar. Xu et al. (2024) further confirmed that in marine environments, biochar enhanced durability and micromechanical integrity, primarily by mitigating ion diffusion and microcrack formation. Lin et al. (2023) emphasized biochar's role in densifying the matrix and buffering against acidic degradation, contributing to overall chemical stability. However, chemical resistance is not uniform; Zanotto et al. (2024) observed increased corrosion in reinforcing bars under chloride exposure, suggesting that biochar's effectiveness depends on exposure conditions and dosage. These findings collectively highlight the nuanced role of biochar in enhancing chemical durability, requiring a careful selection of biochar characteristics to optimize performance in aggressive environments.

Figure 1.7 illustrates mass change (%) of various cement mortar mixes over 160 days of sodium sulphate exposure. Control mix CO (black squares) shows the highest mass gain, indicating greater sulphate ingress and possible expansive reaction.

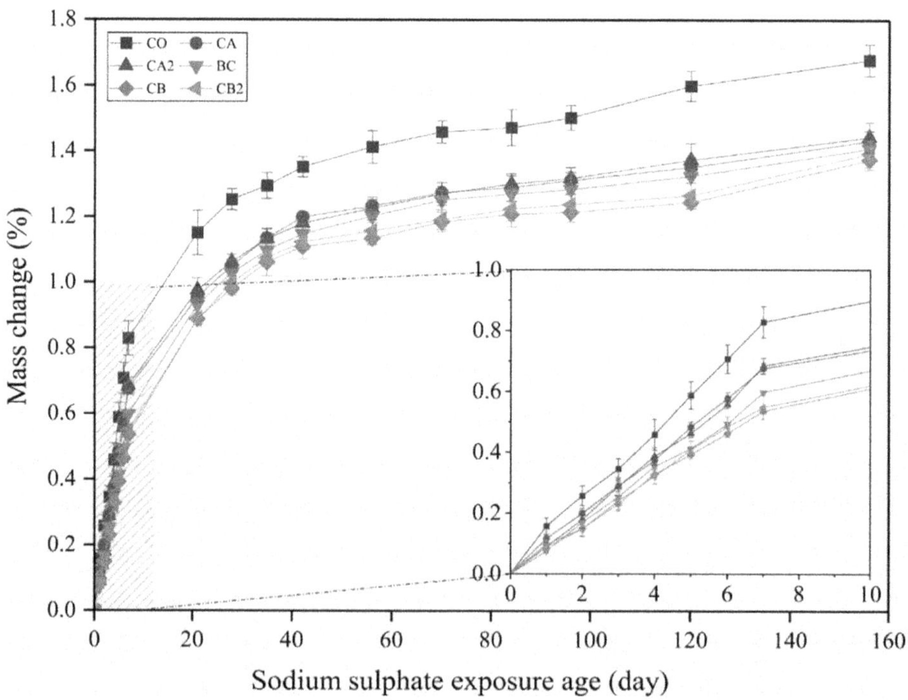

FIGURE 1.7 Mass change of specimens in 5% sodium sulphate immersion.

Source: Lin et al. (2025)

Mixes containing biochar (CA, CA2, CB, CB2, BC) exhibit consistently lower mass changes, demonstrating improved resistance to sulphate attack. The inset magnifies the initial 10 days, where all mixes show rapid mass gain, but biochar-containing samples show delayed and reduced uptake. This suggests that biochar effectively mitigates sulphate-induced expansion by refining pore structure and reducing permeability, corroborating improved durability in sulphate-rich environments.

1.5.3 Carbonation and Corrosion Mitigation

Biochar concrete offers dual benefits of carbon sequestration and improved durability, but its effectiveness in mitigating carbonation-induced corrosion depends on multiple factors. Studies like Praneeth et al. (2020) and Tang and Qiu (2024) showed that biochar enhances CO_2 uptake and strengthens the matrix through accelerated carbonation. However, excessive carbonation may lower pH, increasing steel corrosion risk. Hakuzweyezu et al. (2025) demonstrated that supercritical CO_2-modified biochar improves corrosion resistance by refining pore structure and stabilizing hydration products. Similarly, Xiao et al. (2025) and Yosri et al. (2024) emphasized the need for engineered biochar to optimize both mechanical performance and chemical stability. While some biochars improve freeze–thaw resistance (Chen et al., 2025), others may worsen durability if poorly integrated. Effective corrosion mitigation in biochar concrete requires tailored biochar properties, controlled curing, and matrix compatibility. Overall, biochar's success in carbonation and corrosion control hinges on material design and long-term performance validation.

Figure 1.8 schematically illustrates how biochar influences freeze–thaw damage in a cementitious matrix. In panel (a), biochar particles are embedded in the cement matrix, forming an interfacial zone where surface functional groups interact with Ca^{2+} and Al^{3+}, improving bonding and ion adsorption. In panel (b), the porous structure of biochar traps air bubbles and absorbs water (H_2O), acting as internal reservoirs and pressure-relief sites. In panel (c), during freezing, absorbed water expands to form ice, generating stress. The surrounding pores and interfacial buffering help

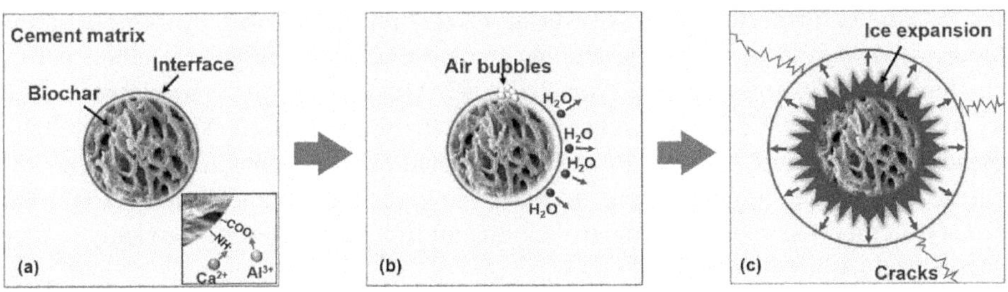

FIGURE 1.8 Mechanism of absorbing water in biochar during freeze–thaw of cement paste: (a) Biochar particle and its interface with the cement matrix. (b) Generation of air voids around the biochar particle. (c) Relief of ice-expansion stress by air voids, reducing cracking in the cement matrix.

Source: Chen et al. (2025)

dissipate expansion pressure, thereby reducing crack initiation and propagation within the cement matrix.

1.6 SUSTAINABILITY AND LIFE CYCLE ASSESSMENT

1.6.1 Carbon Sequestration Potential

Biochar incorporation into concrete exhibits significant potential for carbon sequestration, both through physical entrapment and enhanced carbonation. Zou et al. (2025) demonstrated that encapsulated biochar can synergistically boost CO_2 uptake via carbonation, supported by simulation and experimental data. Similarly, Mishra et al. (2023) highlighted that biochar-enriched cementitious systems could permanently store atmospheric carbon while maintaining mechanical integrity. Tan et al. (2022) further emphasized the dual benefit of enhanced permeability and sequestration capacity in pervious concrete modified with biochar. Song et al. (2023) reviewed applications confirming biochar's viability in cement matrices, while Liu et al. (2022) outlined its high surface area and porosity as key factors in CO_2 adsorption. Ying et al. (2025) presented evidence of effective sequestration using rice straw biochar in recycled concrete, with improved durability. Chen et al. (2022) validated the concept of carbon-negative concrete using biochar, underscoring its environmental advantages. Collectively, these studies affirm biochar's strategic role in sustainable construction.

1.6.2 Energy Efficiency and Embodied Carbon

The integration of biochar in concrete presents significant opportunities for enhancing energy efficiency and reducing embodied carbon in construction. Biochar, a carbon-rich by-product of biomass pyrolysis, serves as both a supplementary cementitious material and a means of carbon sequestration. Zhang et al. (2022) highlighted biochar's potential in construction materials to aid carbon neutrality by stabilizing carbon within concrete matrices. Winters et al. (2022) critically reviewed biochar–cement–$CaCO_3$ composites, indicating notable carbon reductions but emphasizing variability in performance. Lin et al. (2023) reinforced this by demonstrating biochar's dual role in mechanical enhancement and CO_2 mitigation, albeit noting challenges in mix optimization. Wen et al. (2023) further underscored biochar's ability to reduce cement demand and energy use during production. Yosri et al. (2024) explored engineered biochar in ultra-high-performance concrete, achieving low emissions without compromising strength. However, Room and Bahadori-Jahromi (2024) stressed the need for standardization and lifecycle assessment to fully realize energy and carbon benefits.

1.6.3 Comparative Life Cycle Analysis with Traditional Materials

Life Cycle Analysis (LCA) of biochar in concrete presents a nuanced perspective on its environmental and economic viability. Mekky et al. (2024) demonstrated that incorporating biochar into concrete reduces carbon emissions and improves sustainability across the supply chain when assessed through LCA and multi-criteria decision-making tools. Their findings indicate significant environmental and cost benefits, particularly in scenarios involving carbon taxation or green policy incentives. Onuaguluchi et al. (2024) focussed on sludge-derived biochar, revealing notable mechanical strength enhancements in cement mortar while maintaining a favourable environmental profile. Their study emphasizes the potential of waste-derived biochar as a circular economy solution. Meanwhile, Labianca et al. (2024) adopted a holistic approach, evaluating biochar's role not only in concrete but across multiple applications, and highlight its broader systemic benefits through technical, environmental, and economic lenses. Collectively, these studies not only confirm biochar's promise but also call for context-specific assessments to account for variability in feedstock, production, and application.

A Comparative LCA of biochar in concrete versus traditional materials reveals significant environmental advantages. Biochar, a carbon-rich by-product of biomass pyrolysis, serves as a partial cement or sand replacement. Its inclusion in concrete reduces GHG emissions due to the lower energy input required for its production compared to Portland cement. Traditional concrete materials, particularly cement, are responsible for substantial CO_2 emissions during manufacturing. In contrast, biochar can sequester carbon, potentially turning concrete into a carbon-negative material. Additionally, biochar improves insulation and durability, extending the lifespan of structures and reducing maintenance needs. However, challenges include sourcing consistent biochar quality and potential trade-offs in mechanical strength if not properly formulated. Overall, while traditional materials offer proven performance and scalability, integrating biochar in concrete supports a more sustainable construction industry by lowering embodied carbon and promoting circular economy principles. Further research and standardization can enhance its practical adoption. Table 1.3 compares traditional and biochar-modified concrete across life cycle stages, including emissions, energy use, strength, and durability. It highlights biochar's environmental benefits – lower CO_2 emissions, energy demand, and enhanced insulation – while noting trade-offs in strength and cost. Overall, biochar offers a more sustainable alternative with promising long-term advantages.

1.7 CHALLENGES AND LIMITATIONS

Variability in biochar quality significantly affects its performance in concrete applications. Differences in feedstock type, pyrolysis temperature, and production methods influence properties such as surface area, porosity, carbon content, and ash composition. These factors impact biochar's interaction with cementitious materials, potentially altering workability, strength, and durability of concrete. High variability can lead to inconsistent performance, posing challenges in standardization and large-scale use. To harness biochar's benefits – like improved sustainability and carbon sequestration – consistent production and characterization methods are essential. Standardized testing protocols and quality control measures are crucial to ensure that biochar-enhanced concrete meets structural and environmental performance criteria reliably.

Standardization and certification issues pose major challenges to the widespread adoption of biochar in concrete. Currently, there is a lack of universally accepted

TABLE 1.3
Comparative Life Cycle Metrics of Traditional Concrete and Biochar-Modified Concrete Across Key Environmental and Performance Indicators

LCA category	Traditional concrete	Biochar-modified concrete
CO_2 emissions (kg CO_2 /ton)	800–900	500–700 *(can be <0 if carbon sequestration considered)*
Energy use (MJ/ton)	4,000–5,000	2,000–3,500
Water use (L/ton)	150–200	100–150 *(biochar improves moisture retention)*
Compressive strength (MPa)	30–50	25–45 *(with 5–10% biochar replacement)*
Thermal conductivity (W/m·K)	1.4–1.8	0.7–1.2 *(better insulation)*
Service life (years)	50–100	60–120 *(enhanced durability observed)*
Carbon sequestration (kg CO_2)	0	Up to -250 per ton *(biochar stores stable carbon)*
Material cost ($/ton)	$100–$120	$110–$140 *(biochar cost depends on local availability)*
Recyclability potential	Low–moderate	Moderate–High *(biochar beneficial in reuse)*
Scalability	Very high	Moderate *(needs industrial standardization)*

standards governing biochar's properties, testing methods, and performance benchmarks. This absence leads to inconsistent product quality and unpredictable outcomes in concrete applications. Without certification protocols, engineers and manufacturers face difficulty in ensuring reliability, safety, and regulatory compliance. Developing standardized classifications based on feedstock, production conditions, and intended use is essential. Certification frameworks would promote trust, support quality assurance, and enable broader commercial use. Addressing these gaps is critical to integrating biochar into mainstream construction practices effectively.

Economic and scalability considerations are key factors influencing the adoption of biochar in concrete. While biochar offers environmental benefits, its production can be costly due to energy-intensive pyrolysis processes and limited economies of scale. Feedstock availability, transportation logistics, and processing infrastructure further impact cost-effectiveness. For large-scale concrete applications, consistent supply and quality are critical, but current production levels often fall short. Integrating biochar into existing construction supply chains requires investment and optimization to reduce costs. Economies of scale, technological advancements, and supportive policies can enhance affordability and scalability, making biochar a viable, sustainable additive for widespread use in the construction industry.

1.8 FUTURE PERSPECTIVES

Integrating biochar with other sustainable technologies enhances its potential in green construction. When combined with supplementary cementitious materials like fly ash or slag, biochar can improve concrete's environmental performance and mechanical properties. Pairing biochar with carbon capture and utilization (CCU) strategies further boosts its role in reducing carbon footprints. Additionally, using renewable energy in biochar production minimizes emissions, increasing overall sustainability. Biochar-infused concrete can also complement green building certifications, such as Leadership in Energy and Environmental Design (LEED). Synergies with water-saving admixtures or self-healing materials may enhance durability and resource efficiency. Such integration supports circular economy goals, promoting resilient, low-carbon infrastructure for sustainable urban development.

Despite growing interest, significant research gaps remain in understanding biochar's role in concrete. Key uncertainties include its long-term durability, effects on hydration chemistry, and optimal dosage for various applications. Limited data on the interaction between different biochar types and cementitious matrices hampers standardization efforts. Opportunities exist to explore biochar's role in enhancing thermal insulation, reducing shrinkage, and improving carbon sequestration in concrete. Interdisciplinary studies involving materials science, civil engineering, and environmental science are needed to unlock its full potential. Advancing characterization techniques, lifecycle assessments, and pilot-scale implementations will help bridge these gaps and pave the way for sustainable construction innovations.

Policy and regulatory support is essential to accelerate biochar adoption in construction. Government incentives – such as subsidies, tax credits, or carbon pricing mechanisms – can offset production costs and encourage investment in biochar infrastructure. Establishing clear regulations and standards for biochar quality and use in concrete provides certainty for builders and material suppliers. Inclusion of biochar in green building codes and low-carbon procurement policies can drive demand. Regulatory recognition within carbon accounting frameworks supports its role in sequestering carbon over the long term. Public funding for research, pilot projects, and demonstration sites can further validate performance. Holistic policy measures will catalyse market development and sustainable integration.

1.9 CONCLUDING REMARKS

This study highlights the potential of biochar – a carbon-rich by-product of biomass pyrolysis – as a sustainable supplement in cement mortar and concrete. Incorporating biochar enhances concrete performance by improving mechanical strength, reducing porosity, increasing durability, and enabling carbon sequestration. The effectiveness of biochar depends on variables such as feedstock type, pyrolysis conditions, and particle size. Optimal dosages and particle characteristics promote better hydration, internal curing, and microstructural refinement. Additionally, biochar helps reduce water penetration, freeze–thaw damage, and chemical ingress, making concrete more durable in harsh environments. However, excessive biochar can impair workability and compressive strength due to increased porosity. Life cycle assessments confirm that biochar integration lowers embodied carbon and energy use compared to traditional concrete. Challenges remain in quality consistency, standardization, and scalability. Future opportunities lie in combining biochar with other green technologies, surface modifications to enhance compatibility, and supportive policy frameworks to accelerate adoption. Ultimately, biochar offers a multifunctional, eco-friendly solution for building resilient and low-carbon infrastructure, aligning with circular economy goals and climate mitigation strategies.

The integration of biochar into cementitious materials paves the way for a more sustainable construction industry. By utilizing waste-derived biochar as a partial replacement for cement or aggregates, the construction sector can significantly reduce its carbon footprint, enhance material efficiency, and promote circular economy practices. Future construction envisions materials that not only meet performance standards but also contribute to climate resilience through carbon sequestration and enhanced durability. Standardization, innovation in biochar processing, and supportive regulatory frameworks will be crucial in scaling its use. Combining biochar with other sustainable technologies – such as renewable energy, carbon capture, and recycled materials – can further amplify environmental benefits. Research-driven optimization and industry-wide adoption will lead to buildings that are more energy-efficient, longer-lasting, and environmentally regenerative. This vision supports the transition towards low-carbon, resource-efficient infrastructure essential for a liveable, climate-safe future.

REFERENCES

Akhtar, A. and Sarmah, A.K., 2018. Novel biochar-concrete composites: Manufacturing, characterization and evaluation of the mechanical properties. *Science of the Total Environment, 616*, pp.408–416.

Ali, D., Agarwal, R., Hanifa, M., Rawat, P., Paswan, R., Rai, D., Tyagi, I., Naik, B.S. and Pippal, A., 2023. Thermo-physical properties and microstructural behaviour of biochar-incorporated cementitious material. *Journal of Building Engineering*, *64*, p.105695.

Bhagat, T.S. and Pancharathi, R.K., 2025. Performance, microstructure and carbon sequestration potential of agro biochar based cement mortars. *Cement and Concrete Composites*, *156*, p.105867.

Chen, T., Yang, Z., Liu, H., Li, L., Qin, L. and Gao, X., 2025. Effect of biochar characteristics on freeze-thaw durability of biochar-cement composites. *Journal of Building Engineering*, p.111959.

Chen, T., Zhao, L., Gao, X., Li, L. and Qin, L., 2022. Modification of carbonation-cured cement mortar using biochar and its environmental evaluation. *Cement and Concrete Composites*, *134*, p.104764.

Gupta, S. and Kua, H.W., 2019. Carbonaceous micro-filler for cement: Effect of particle size and dosage of biochar on fresh and hardened properties of cement mortar. *Science of the Total Environment*, *662*, pp.952–962.

Gupta, S., Muthukrishnan, S. and Kua, H.W., 2021. Comparing influence of inert biochar and silica rich biochar on cement mortar–Hydration kinetics and durability under chloride and sulfate environment. *Construction and Building Materials*, *268*, p.121142.

Hakuzweyezu, T., Gan, M., Zhang, L., Wang, Y. and Onyekwena, C.C., 2025. Supercritical CO_2-modified biochar as a novel additive for enhancing CO_2 corrosion resistance in wellbore cement: Characterization and mechanisms. *Energy & Fuels*, *39*(15), pp.7409–7419.

Jafari, A. and Sadeghian, P., 2023. Influence of biochar and recycled gypsum on the strength and microstructure of conventional and sustainable cementitious composites. *Construction and Building Materials*, *408*, p.133715.

Javed, M.H., Sikandar, M.A., Ahmad, W., Bashir, M.T., Alrowais, R. and Wadud, M.B., 2022. Effect of various biochars on physical, mechanical, and microstructural characteristics of cement pastes and mortars. *Journal of Building Engineering*, *57*, p.104850.

Jia, Y., Li, H., He, X., Li, P. and Wang, Z., 2023. Effect of biochar from municipal solid waste on mechanical and freeze–thaw properties of concrete. *Construction and Building Materials*, *368*, p.130374.

Kalpokaitė-Dičkuvienė, R. and Zabulionis, D., 2025. Effect of freeze-thawing and wet-dry cycling in salt solution on strength of biochar-silica fume blended composites. *Construction and Building Materials*, *479*, p.141496.

Labianca, C., Zhu, X., Ferrara, C., Zhang, Y., De Feo, G., Hsu, S.C. and Tsang, D.C., 2024. A holistic framework of biochar-augmented cementitious products and general applications: Technical, environmental, and economic evaluation. *Environmental Research*, *245*, p.118026.

Li, S., Jin, L., Chen, X., Hu, M., Geng, Y., Li, K., Liu, Y., Hu, Q. and Zhang, S., 2025. Effect and mechanisms of biochar particle size and dosage on the hydration process and mechanical properties of cementitious materials. *Construction and Building Materials*, *472*, p.140931.

Lin, X., Li, W., Guo, Y., Dong, W., Castel, A. and Wang, K., 2023. Biochar-cement concrete toward decarbonisation and sustainability for construction: Characteristic, performance and perspective. *Journal of Cleaner Production*, *419*, p.138219.

Lin, X., Nguyen, Q.D., Castel, A., Pang, Y., Deng, Z., Shi, T., Li, W. and Tam, V.W., 2025. Durability of biochar-cementitious composites incorporating crystalline admixture in chloride and sulphate environments. *Construction and Building Materials*, *458*, p.139554.

Liu, J., Liu, G., Zhang, W., Li, Z., Xing, F. and Tang, L., 2022. Application potential analysis of biochar as a carbon capture material in cementitious composites: A review. *Construction and Building Materials*, *350*, p.128715.

Mekky, K.M., Ibrahim, M.G., Sharobim, K., Fujii, M. and Nasr, M., 2024. Evaluating environmental and economic benefits of using biochar in concrete: A life cycle assessment and multi-criteria decision-making framework. *Case Studies in Construction Materials*, *21*, p.e03712.

Mensah, R.A., Shanmugam, V., Narayanan, S., Razavi, N., Ulfberg, A., Blanksvärd, T., Sayahi, F., Simonsson, P., Reinke, B., Försth, M. and Sas, G., 2021. Biochar-Added cementitious materials – A review on mechanical, thermal, and environmental properties. *Sustainability*, *13*(16), p.9336.

Mishra, G., Danoglidis, P.A., Shah, S.P. and Konsta-Gdoutos, M.S., 2023. Carbon capture and storage potential of biochar-enriched cementitious systems. *Cement and Concrete Composites*, *140*, p.105078.

Murali, G. and Wong, L.S., 2024. A comprehensive review of biochar-modified concrete: Mechanical performance and microstructural insights. *Construction and Building Materials*, *425*, p.135986.

Onuaguluchi, O., Mohamed, B., Adwan, A., Li, L. and Banthia, N., 2024. Sludge-derived biochar as an additive in cement mortar: Mechanical strength and life cycle assessment (LCA). *Construction and Building Materials*, *425*, p.135959.

Praneeth, S., Guo, R., Wang, T., Dubey, B.K. and Sarmah, A.K., 2020. Accelerated carbonation of biochar reinforced cement-fly ash composites: Enhancing and sequestering CO_2 in building materials. *Construction and Building Materials*, *244*, p.118363.

Premchand, P., Demichelis, F., Chiaramonti, D., Bensaid, S. and Fino, D., 2023. Study on the effects of carbon dioxide atmosphere on the production of biochar derived from slow pyrolysis of organic agro-urban waste. *Waste Management*, *172*, pp.308–319.

Qing, L., Zhang, H. and Zhang, Z., 2023. Effect of biochar on compressive strength and fracture performance of concrete. *Journal of Building Engineering*, *78*, p.107587.

Room, S. and Bahadori-Jahromi, A., 2024. Biochar-enhanced carbon-negative and sustainable cement composites: A scientometric review. *Sustainability*, *16*(23), p.10162.

Sirico, A., Bernardi, P., Sciancalepore, C., Belletti, B., Milanese, D. and Malcevschi, A., 2023. Combined effects of biochar and recycled plastic aggregates on mechanical behavior of concrete. *Structural Concrete*, *24*(5), pp.6721–6737.

Sirico, A., Bernardi, P., Sciancalepore, C., Vecchi, F., Malcevschi, A., Belletti, B. and Milanese, D., 2021. Biochar from wood waste as additive for structural concrete. *Construction and Building Materials*, *303*, p.124500.

Sobuz, M.H.R., Khan, M.H., Kabbo, M.K.I., Alhamami, A.H., Aditto, F.S., Sajib, M.S., Alengaram, U.J., Mansour, W., Hasan, N.M.S., Datta, S.D. and Alam, A., 2024. Assessment of mechanical properties with machine learning modeling and durability, and microstructural characteristics of a biochar-cement mortar composite. *Construction and Building Materials*, *411*, p.134281.

Song, S., Liu, Z., Liu, G., Cui, X. and Sun, J., 2023. Application of biochar cement-based materials for carbon sequestration. *Construction and Building Materials*, *405*, p.133373.

Tan, K., Qin, Y. and Wang, J., 2022. Evaluation of the properties and carbon sequestration potential of biochar-modified pervious concrete. *Construction and Building Materials*, *314*, p.125648.

Tang, Y. and Qiu, J., 2024. CO2-sequestering ability of lightweight concrete based on reactive magnesia cement and high-dosage biochar aggregate. *Journal of Cleaner Production*, *451*, p.141922.

Wang, D., Jantwal, A., Kaynak, E., Sas, G. and Das, O., 2025a. Promoting internal curing in concrete by replacing sand with sustainable biochar. *Case Studies in Construction Materials*, *22*, p.e04542.

Wang, T., Tang, Y., Qin, S., Li, G., Wu, H. and Leung, C.K., 2025b. Sustainable and mechanical properties of engineered cementitious composites with biochar: Integrating micro- and macro-mechanical insight. *Cement and Concrete Composites*, *155*, p.105813.

Wen, J., Wang, B., Dai, Z., Shi, X., Jin, Z., Wang, H. and Jiang, X., 2023. New insights into the green cement composites with low carbon footprint: The role of biochar as cement additive/alternative. *Resources, Conservation and Recycling*, *197*, p.107081.

Winters, D., Boakye, K. and Simske, S., 2022. Toward carbon-neutral concrete through biochar–cement–calcium carbonate composites: A critical review. *Sustainability*, *14*(8), p.4633.

Xiao, S., Li, Y., Xu, C., Deng, J., Cheng, L., Zhang, H., Zhou, Y. and Hong, J., 2025. The influence of biochar addition on the mechanical performance, hydration mechanism, and carbonation capacity of Super Sulfated Cement. *Construction and Building Materials*, *463*, p.140073.

Xu, W., Zhang, Y., Li, M., Qu, F., Poon, C.S., Zhu, X. and Tsang, D.C., 2024. Durability and micromechanical properties of biochar in biochar-cement composites under marine environment. *Journal of Cleaner Production*, *450*, p.141842.

Yang, X. and Wang, X.Y., 2021. Hydration-strength-durability-workability of biochar-cement binary blends. *Journal of Building Engineering*, *42*, p.103064.

Ye, P., Guo, B., Qin, H., Wang, C., Li, J., Chen, Y., Lu, D., Wang, L., Gao, P., Ma, P. and Zhan, B., 2024. Investigation of the properties and sustainability of modified biochar-doped cement-based composite. *Cement and Concrete Composites*, *153*, p.105684.

Ying, X., Zhao, X., Ye, M., Wang, C., Zhan, B., Zhao, J., He, Z. and Nie, X., 2025. Waste rice straw biochar recycled concrete: Carbon sequestration, durability and microstructure. *Journal of Cleaner Production*, p.145690.

Yosri, A.M., Zaid, O. and Hamad, M., 2024. Valorization of engineered biochar to develop ultra-high-performance fiber-reinforced concrete with low carbon emission. *Journal of Sustainable Cement-Based Materials*, *13*(6), pp.865–887.

Zanotto, F., Sirico, A., Balbo, A., Bernardi, P., Merchiori, S., Grassi, V., Belletti, B., Malcevschi, A. and Monticelli, C., 2024. Study of the corrosion behaviour of reinforcing bars in biochar-added concrete under wet and dry exposure to calcium chloride solutions. *Construction and Building Materials*, *420*, p.135509.

Zhang, Y., He, M., Wang, L., Yan, J., Ma, B., Zhu, X., Ok, Y.S., Mechtcherine, V. and Tsang, D.C., 2022. Biochar as construction materials for achieving carbon neutrality. *Biochar*, *4*(1), p.59.

Zou, S., Chen, X., Sham, M.L., Lu, J.X. and Poon, C.S., 2025. Carbon sequestration in aggregate and concrete by encapsulated biochar and carbonation: Experiment and simulation. *Cement and Concrete Composites*, *159*, p.105990.

2 Examining the Use of Biochar as a Sustainable Green Supplement in Cement Mortar and Concrete Blend

A Comprehensive Review

Rajvardhan Kallimani, Pravin Minde and Jagruti Patil

2.1 INTRODUCTION

2.1.1 Background and Importance of Sustainability in Construction

The construction industry plays a prominent contribution in shaping social and economic progress, but it also has a big effect on the environment [1]. The climate crisis caused by too much carbon dioxide being released into the atmosphere by human activities has hurt Earth's ecosystems in the last few years. This has caused things like global warming and loss of biodiversity [2]. The growing focus on sustainable building around the world shows how important it is to deal with environmental and resource issues. Sustainable construction has the potential to meet societal requirements by improving well-being of individuals, but the industry also has to deal with a lot of problems with management, strategy, and operations [3]. Portland that is normal cement is a key part of concrete. Making it requires a lot of energy, which has a big effect on the amount of carbon dioxide (CO_2) that is released into the atmosphere. The construction sector contributes nearly 38% to worldwide CO_2 emissions, with cement production by itself accounting for approximately 7% [4]. We need sustainable materials right away because making regular cement uses 3.2 GJ of energy and releases 900 kg of CO_2 for every ton of clinker [5]. These figures emphasize the importance of the construction sector to start implementing environmentally friendly building practices to lessen its environmental impacts. To meet international carbon reduction goals and address future environmental sustainability issues, the construction sector needs to adopt greener ways of making cement and concrete [6]. Furthermore, concrete is by far the least sustainable building material

DOI: 10.1201/9781003664284-2

as it is composed of the greatest availability of unprocessed resources [7]. There has continued to be pressure each year for economic or urban development while both population growth and economic growth happen on the same timelines. To begin to tackle these issues, the sector must change its practices to begin using new, more sustainable materials instead of conventional materials, especially cement, to mitigate carbon and reduce the consumption of resources. If the construction sector starts to look for more sustainable materials and practices, then it may be able to establish a sustainable pathway to safeguard global ecosystems and humanity.

2.1.2 Role of Cement and Concrete in the Built Environment

Concrete, composed of cement, aggregates, and water, is the most widely used engineered material in the built environment, forming the backbone of infrastructure such as buildings, transportation systems, energy networks, and water management structures. Cement, which functions as the binder, typically accounts for about 10% of the mix by mass but governs the performance and environmental impact of concrete production. With growing sustainability concerns, the industry has shifted from exclusive reliance on ordinary Portland cement (OPC) toward blended formulations incorporating supplementary cementitious materials (SCMs) and industrial by-products. Extensive research demonstrates that optimum replacement levels of SCMs can improve both durability and eco-efficiency. High-volume fly ash (HVFA) concretes, incorporating up to 60% cement replacement, have been shown to achieve long-term strength comparable to OPC, although early-age strength is significantly reduced, requiring extended curing. Class C fly ash replacement levels of 20–40% often surpass the 28-day compressive strength of control mixes, while class F fly ash requires longer curing to achieve similar performance [8]. Metakaolin (MK), a highly reactive pozzolan containing 50–55% SiO_2 and 40–46% Al_2O_3, performs best when used at 10–15% replacement of cement, where it enhances early strength, refines pore structure, and improves resistance to chloride and sulfate attack [9, 10]. At higher dosages (above ~20%), MK reduces workability due to its fineness, necessitating increased superplasticizer dosages [10]. Similarly, the use of waste marble powder (WMP) as a partial replacement has been widely investigated, with around 10% cement substitution yielding the best balance between strength and sustainability, and up to 20% fine aggregate replacement improving cohesiveness and reducing sorptivity in self-compacting concretes [9, 10]. Beyond these materials, bottom ash (BA) has gained prominence as a potential sand replacement. Researchers [11] reported that at 30% BA replacement, concretes achieved compressive strengths of 40–50 MPa with durability superior to control mixes due to reduced chloride permeability. However, higher replacement levels compromised strength, with 60% BA showing a 10–17% reduction and 100% BA leading to significant losses, though still demonstrating enhanced durability relative to conventional concrete. Parallel findings by [12] emphasized that self-compacting concretes integrating mineral admixtures such as fly ash, silica fume, and MK require careful optimization. For instance, silica fume replacement of 12–17% by binder weight significantly reduced plastic viscosity and improved flowability, while high binder contents of 450–525 kg/m^3

and lower water-to-binder ratios were essential for achieving adequate filling ability and segregation resistance. Moreover, fiber additions up to 1% by volume enhanced tensile and flexural strength, although excessive dosages impaired workability. Taken together, these results demonstrate that cement and concrete are not static construction materials but evolving composites. By strategically integrating SCMs and waste-derived fillers at optimized replacement levels 20–40% for fly ash, 10–15% for MK, 10–20% for waste marble powder, and 30% for BA, modern concretes can deliver high performance and durability while minimizing environmental impact. Moreover, the incorporation of regression and statistically based predictive tools further advances the field, enabling accurate forecasting of strength development and providing a scientific basis for the sustainable design of future infrastructure [8–12].

2.1.3 The Concept of Biochar as a Green Supplement

Biochar is a promising carbon-neutral material that offers a carbon-negative solution by converting agricultural and forestry waste into a useful additive for soil and construction. Produced through biomass pyrolysis, it features high porosity, large surface area, and strong carbon sequestration potential. In concrete, biochar can improve mechanical performance, lower the carbon footprint, and promote waste valorization. Used as a supplementary cement replacement, it presents an innovative approach to reducing emissions in construction. The use of biochar can help civil engineering shift toward more sustainable carbon-neutral practices while providing a solution to waste management practices and green initiatives [2]. Biochar certainly can help manage carbon emissions associated with building materials, but adding biochar directly into concrete could negatively affect its mechanical performance, especially at higher biochar content. In order to avoid potential structural issues, whilst maintaining the engineering properties of the material, it will be essential to optimize both the amount of biochar and the processing steps [13]. By adding biochar into cement-based materials would increase their durability, sound absorption, thermal insulation, and pollution mitigation properties [14]. In conclusion, biochar represents an innovative alternative to sustainable construction materials and a nuanced approach to managing carbon footprint reductions for civil engineering. It appears to provide a good carbon-negative (or as close as possible) additive, but significant effort still needs to take place to find the appropriate trade-off that both achieves environmental benefits and maintains mechanical performance. The future of building materials lies in similar initiatives that attempt to meld innovation, technological advancement, waste valorization, and being environmentally responsible.

2.1.4 Purpose and Outline of the Chapter

The chapter delves into the potential of biochar as a sustainable additive in cement and concrete, addressing a primary challenge facing our built environment – the environmental impacts of the construction industry, in particular the CO_2 emissions attributed to cement production. The chapter discusses the use of biochar, a form of carbon-rich material created from the pyrolysis of organic biomass, which possesses the ability to reduce carbon emissions going forward, promote greener practices, and generally advance the circular economy initiatives noted in prior chapters, specifically relating to the bioremediation and re-use of multiple construction materials. The chapter begins with detailing the production of biochar and the general properties that make biochar a potentially viable component of cement-based materials. Next, the chapter examines how biochar can be used in cement and concrete formulations, while considering how biochar affects the strength, durability, and overall environmental profile of these materials. Biochar's ability to retain and maintain carbon within concrete forms one of the most appealing outcomes addressed in the chapter, illustrating biochar's potential value for carbon sequestration. The chapter then compliments its earlier findings with some of the challenges and limitations presented when considering the scale-up of biochar usage in the construction sector, with a closing call for future research directions and actionable measures to facilitate a wider adoption of biochar use within our built environment. It works as a useful guide for academics and practitioners interested in more sustainable building solutions.

2.2 OVERVIEW OF BIOCHAR

2.2.1 Definition and Characteristics of Biochar

Biochar is a carbon-rich product generated through the pyrolysis of organic matter like wood, crop residues, and animal waste, a process that is conducted in low oxygen conditions [15]. Biochar has received considerable investigation (and utilization in farming as a soil amendment) and is being researched in other fields, including construction. In the construction industry, biochar is also being considered for non-structural applications such as insulation, humidity control, and as an absorbent for electromagnetic radiation and pollutants. Additionally, biochar has been demonstrated to be useful in the production of green cementitious materials. Biochar, as a chemically inert carbonaceous material, can be used as an additive in paste, mortar, and concrete, with improved physical and mechanical properties. Research indicates that it can densify mixes and improve water retention. Due to the fine particle size, biochar can be used as a micro-aggregate, which will displace volume in the pore structure, allowing for additional hydration (especially useful in aggressive curing scenarios) based on the biochar's ability to uptake water during mixing and then release moisture in the hydration process. Biochar has some exceptional qualities that make it exceedingly versatile for a wide variety of applications. Biochar is also characterized by high porosity, high surface area, and reactive functional groups. Biochar is also characterized by high cation exchange capacity and high stability, enhancing its performance in different applications. Biochar's specific qualities like carbon content and ash content are affected by factors like the source of biomass, manufacturing operations, and reactor designs. Key parameters like processing temperature, processing time, and heating rate also affect its properties [16].

2.2.2 Biochar Production Methods

2.2.2.1 Fast Pyrolysis

It is a thermally initiated chemical process that uses rapid heating of biomass in an anaerobic atmosphere within temperature thresholds from 400°C to 600°C to modify it into bio-oil, gas, and a few dense products, like charcoal. The process involves a short vapor residence time of about 0.5 to 3 seconds. Before pyrolysis, the biomass must be dried and ground to reduce moisture content and particle size, which guarantees improved purity and output of bio-oil, along with enhanced thermal flow efficiency. Several reactor configurations, such as fluidized bed, ablative, and circulating fluid bed reactors, are used for fast pyrolysis. The biomass is raised in temperature using a heat carrier, like sand, to produce bio-oil vapor, which is then instantly condensed to minimize secondary reactions. The resulting char is used to generate heat, and off gases can be refurbished for drying or fluidizing the biomass. During fast pyrolysis, biomass decomposes into liquid bio-oil, char, and gases. The process requires very intense heating rates (over 1,000°C/s) and short vapor retention time, typically under 2 seconds. The pyrolysis vapors undergo rapid breaking down and disintegration of lignocellulosic biomass. The vapors are then cooled rapidly so as to create bio-oil and to avoid subsequent reactions that would modify the liquid product. Feedstock type, particle size, and reactor configuration used determine the nature of the bio-oil, char, and gas. Fast pyrolysis is generally conducted at intermediate temperatures (400–600°C) where small-scale biomass has to be fed into order to overcome the mass/heat transfer constraints. Besides the high liquid yields, the process is dependent on fast cooling to save the bio-oil and limit any other undesired secondary reactions. Popular reactor choices in this process are fluidized systems and other unique arrangements that provide maximum heat exchange and minimum vapor retention period [17–19].

2.2.2.2 Slow Pyrolysis

Slow pyrolysis is a thermochemical technique with a relatively long process time of typically a few hours over a range of temperatures of 300 to 600 °C, heating rate of 10 °C/min, and retention time of 60–120 min. The result of this process of slower decomposition of biomass is a larger quantity of biochar compared to fast pyrolysis and smaller quantities of bio-oil and syngas and a reduced quantity of condensable liquids. Continuous high-capacity slow-pyrolysis at large scale is traditionally employed industrially in the form of drum pyrolyzers and rotary kilns. Otherwise, fixed-bed reactors like earth pits and metal kilns are used to catalyze the thermal conversion of biomass into biochar. Once produced, the biochar is removed out of the combustion chamber, and a metal sheet is placed on top of it in order to avoid oxidative reaction occurring on the hot surface of the char. Lastly, they grind the material by hand to smaller size in preparation to use it in the mortar or other purposes as well [20–24].

2.2.2.3 Intermediate Pyrolysis

Intermediate pyrolysis narrows the disparity connecting fast and slow pyrolysis in terms of both speed and efficiency. Recent technological advancements have addressed the challenges associated with fast pyrolysis, leading to improvements in intermediate pyrolysis. The key dissimilarity between intermediate and fast pyrolysis lies in the method of heat transfer to the feedstock. In intermediate pyrolysis, heat is applied at significantly lower rates, typically ranging from 100 to 500°C/min. While slow pyrolysis operates at a heating rate of approximately 100°C/s and fast pyrolysis at around 1,000°C/s, intermediate pyrolysis falls among these ranges. This process is conducted at low-to-moderate temperatures and involves a moderate residence time for the hot vapors. It yields approximately 50% liquid bio-oil (including 50% water content), 25% solid biochar, and 25% syngas [25, 26].

2.2.2.4 Torrefaction

Torrefaction is a new thermal pre-treatment method that improves the energy properties of biochar. The process happens in a setting environment with little oxygen and at temperatures between 200 and 300 degrees Celsius for 20 to 180 minutes with low heating rates (less than 20 degrees Celsius per minute). It makes the biomass's physicochemical and thermochemical properties better. The parameters change according to the type of biomass. Three types of torrefaction exist: wet, dry, and steam. It raises the $C/H/O_2$ ratio and makes the material more uniform and water-resistant. People often use nitrogen gas to stop oxidation, but it costs a lot of money, so people are looking into oxidative torrefaction as a cheaper way to do the same thing. Biochar quality is highly dependent on the temperature at which it is produced and how long it stays in one place. Changes in the biomass's chemistry, which are often seen through changes in color, conversion levels, and the presence of compounds like carbonyls and phenolics, are caused by changes in the lignocellulosic content and surface properties [22, 27–29]. At the same time, studies [30] showed that torrefied biochar has better hydrophobic properties, which could help keep concrete from absorbing too much moisture. These results suggest that customizing the ways biochar is made can make it work best in cement composites.

2.2.2.5 Gasification

Gasification is a thermochemical method that changes biomass, including carbonaceous materials and lignocellulosic substances, as well as organic fossil fuels into various biofuels. It can also convert hydrocarbons such as coal and organic waste such as municipal solid waste (MSW) to a gaseous fuel named syngas. In contrast with slow pyrolysis, the gasification process occurs at notably higher temperatures that can typically range between 700°C and 1,000°C. Under this process, a certain quantity of air or oxygen is injected inside the reactor and reacted with the biomass under elevated temperatures forming syngas. The syngas has a large composition of hydrogen and carbon monoxide, and there is less carbon dioxide and methane. Parameters of pyrolysis that influence biochar, for example, ambient temperature, thermal increase rate, and retention time, significantly influence the reactivity of biochar during gasification. Gasification offers a distinct advantage by yielding two valuable products: biochar and syngas. The syngas is employed for energy production or in another purposes. Additionally, there is considerable potential for recycling the biochar derived from gasification back into the process as fuel. This is a consequence of biochar's high carbon content and calorific value, thus rendering an efficient fuel for gasification. By using gasification biochar as a feedstock, the leftover carbon from a previous gasification cycle can be converted into more gaseous fuel. In a study, it was shown that biochar produced through gasification can be incorporated in cement paste for 3D printing, with dry material content of up to 10% biochar without impacting printability or dimensional stability. However, when biochar content exceeded 10%, the mechanical stability of the samples was negatively impacted [31–33].

2.2.3 Types of Feedstock Used for Biochar Production

Biochar can be produced from a wide range of feedstocks, each influencing its final properties. Common sources include woody biomass like maple wood, bamboo, almond shells, and sawdust, valued for their high carbon content and low moisture. Agricultural residues such as rice husk, rice straw, groundnut and peanut shells, tomato plant waste, and castor residues are widely used for their rich organic matter. Other renewable feedstocks include vetiver grass, bagasse, and corn cobs. Waste materials like municipal solid waste, sewage sludge, and food waste offer waste valorization opportunities while producing biochar with varied chemical profiles. Nutrient-rich materials such as distillers' grains, peanut hulls, cow dung, and poultry manure are also used, especially for biochar aimed at soil improvement [14, 34–37]. Table 2.1 shows various types of feedstock, including but not limited to different feedstock categories.

2.2.4 Compatibility of Biochar with Cementitious Materials

The compatibility of biochar with cement-based materials is an increasingly explored topic due to its potential to enhance the properties of concrete and mortar. When biochar is added to cement composites, it can improve performance by modifying pore structure, decreasing permeability, and boosting resistance to chemical degradation. Its significant surface area, porosity, and functional groups allow for stronger interaction with the cement matrix, leading to enhanced mechanical and durability properties. However, the impact of biochar on cement-based materials is influenced by factors like its particle size, quantity, production method, and surface characteristics. While controlled amounts of biochar can support hydration and reduce shrinkage, excessive amounts can negatively affect workability and compressive strength because of higher water demand. Thus, fine-tuning biochar's properties and dosage is essential to ensure compatibility and fully leverage its benefits in cement systems.

TABLE 2.1
Different Types of Feedstocks

Feedstock	Category
Wood (e.g., maple wood, bamboo, almond shell, sawdust)	Woody biomass
Rice husk, rice straw	Agricultural residues
Groundnut shell, peanut shell	Agricultural residues
Tomato plant waste, castor plant	Agricultural residues
Vetiver grass, bagasse (sugar cane), corn cob	Energy crops and agricultural residues
Municipal solid waste, sewage sludge, food waste	Waste biomass
Distillers' grains, peanut hull	Agricultural by-products
Cow dung, poultry manure	Animal manure

2.2.5 Characterization of Biochar

It is important to understand the properties of biochar and optimize the biochar for a sustainable and multiuse product. Biochar is created from different types of biomass resources and uses many types of pyrolysis methods. Understanding its properties is important for consistency and quality in production to promote its use. Complete characterization allows scientists to change the production strategies to optimize biochar properties for use in specific applications. Additionally, the regulatory authority may need composition and qualities of biochar to be regulated for particular uses, and characterization allows for it to be compliant with those qualities, promoting its acceptability and industry viability.

2.2.5.1 X-Ray Diffraction (XRD)

XRD characterization of biochar helps to identify components with regards to both mineral presence and crystal structure. Biochar will be analyzed using computerized XRD machine, with (amongst other components) a monochromator, stepping motor, and the radiation source. XRD characterizes the mineral phases (i.e., minerals mineralogy) in feed and alteration of such minerals though pyrolysis. The information gained in this process will assist in writing conclusions about implication of feedstock composition on biochar properties. XRD characterization may also determine crystallinity properties, representing the degree of carbonization and heat decomposition during pyrolysis. The crystallinity variations also will give some indication of thermal stability and reactivity of any biochar produced. The XRD information will also assist in determining the level of carbonization or possible graphitization from peak broadening and intensity analysis of each peak and provide useful information about the thermal treat condition obtained during the hybridization of biochar.

2.2.5.2 Nuclear Magnetic Resonance (NMR)

NMR spectroscopy is a valuable method for investigating the chemical makeup and structural characteristics of biochar. An NMR spectrometer comprises a needle coil, magnet, and electronic controller. NMR spectroscopy provides details on the chemical structure of biochar by identifying the types and orientation of hydrogen and carbon atoms in addition to the chemical context that surrounds them. NMR is able to quantify the proportion of carbon atoms that occur in an aromatic ring, termed "aromaticity," which contributes to the stable and adsorptive ability of biochar. Biochar, which has moderate-to-higher proportions of carbon that occur in a structure consisting of aromatic rings, is more stable because of its presumed resistance to microbial breakdown and likely enhanced durability. NMR spectroscopy follows the chemical structure of biochar during the carbonization and thermal degradation process. Researchers can compare the NMR spectra of biochar samples derived from different pyrolysis conditions to determine if the molecular structures must be carbonized, quantify the relative amount of aromatic carbon to assess biochar formation and decide whether volatile organic compounds have also been generated. Although NMR spectroscopy had traditionally been used to evaluate organic compounds in solution, NMR spectroscopy can be applied to understand and evaluate how biochar may impact cementitious materials when combined into biocomposites. Utilizing the capability of NMR will allow researchers to further develop sustainable and efficient biochar-based construction materials and encourage environmentally based innovation in the building industry [38, 39].

2.2.5.3 Scanning Electron Microscopy (SEM)

SEM serves an important role in examining the microstructure of biochar by providing information on biochar surface morphology, particle size distribution, and structural features. SEM provides researchers with high magnification images of biochar particles or clusters and facilitates the study of particle shape, degree of porosity, and surface features. The typical microstructural assessment using SEM involves two phases – data collection and image processing and analysis. As the SEM system provides actual physical data, the structure of biochar can be assessed both externally and internally. The collective information provided by SEM concerning the morphology of biochar is multidimensional as it can examine the shape, size, and irregularities on the surface of particles at different magnifications and angles. Furthermore, combining SEM with energy-dispersive X-ray spectroscopy (EDS) can provide elemental analysis of biochar, and elemental mapping can be conducted to assess the chemical composition of an area or feature within biochar. This is beneficial to assess the elemental spatial distribution of components within biochar and identify metal impurities or minerals [40, 41].

2.2.5.4 Raman Spectroscopy

Raman spectroscopy is a highly adaptable and dependable spectroscopic method that provides valuable benefits for characterizing chemical composition. For example, in carbon-based materials such as biochar, Raman spectroscopy provides access to more complete bonding and structural information. Briefly, Raman spectroscopy uses a light source (normally, a laser) emitting monochromatic light to illuminate a specimen and collects scattered light in response to molecular vibrations within the specimen material. The Raman spectrum of biochar should have peaks relevant to characteristics of the structure such as graphitic layers, defects, or functional groups. The position, overall width, and intensity of peaks contribute to determining varying levels of graphitization, degrees of aromaticity, or surface functionalizations of biochar. The ratio of D-band (which indicates structural disorder and defects) to G-band (which indicates graphitic carbon) also provides information about graphitization or structural disorder of biochar. While the D/G ratio will increase with increasing structural disorder (i.e., higher levels of more disordered carbon structures), it may also represent microstructural variability in the biochar [42].

2.2.5.5 Fourier Transform Infrared Spectrometer (FTIR)

FTIR is a sophisticated analytical tool that is widely employed to study the molecular composition of many different substrates, including biochar. It provides critical information about the molecular bonds and functional groups, as well as the chemistry overall. A typical configuration of the instrument contains an infrared light source, a monochromater or interferometer to leave only certain wavelengths of light, a sample holder, and a detector, all of which work together to provide the spectra. In preparing biochar for analysis, the biochar is ground very well and combined with a suitable matrix (typically potassium bromide (KBr)) to form a pellet, which will attempt to improve the consistency and accuracy of the measurement. The pellet is then placed in the sample holder of the FT-IR spectrometer. The FT-IR spectrum for biochar typically displays a complex pattern of absorption bands, illustrating the diverse chemical constituents and structural properties of biochar. In response to the presentation of a pattern of bands, accepted bands can be matched, with the accepted bands on a prior reference spectrum, to identify functional groups and molecular bonds. Furthermore, the intensity and location of these bands can be used to derive information about the sample's chemical composition, carbonization, and degree of surface functionalization. Functional groups typically detected and identified from biochar spectra include hydroxyl (OH), carbonyl (C = O), carboxyl (COOH), aromatic (C = C), and aliphatic (C - H) compounds. However, a significant challenge for biochar FT-IR analyses is that biochar is generally opaque, which can decrease reflectivity and has been shown to reduce the signal-to-noise ratio for reflectance contact measurements [43, 44].

2.3 METHODS OF INCORPORATING BIOCHAR

To utilize biochar as a supplementary material in cement-based composites, it could be integrated through several techniques. One common method involves directly blending biochar with cement during the dry mixing stage to ensure uniform distribution. Alternatively, biochar can be pretreated or activated, often by soaking in water or chemical solutions, to enhance its surface properties and compatibility with the cement matrix. Another approach includes substituting a portion of the fine aggregate with biochar, which can contribute to improved pore structure and durability. Additionally, the particle size of biochar can be optimized through grinding to ensure better dispersion and reactivity. These methods aim to leverage biochar's high surface area and porous structure to improve concrete performance while promoting sustainability by reducing cement usage [45].

2.3.1 Compressive Strength of Biochar Concrete

Biochar's impact on concrete strength varies based on its type, size, amount, and treatment. At optimal levels, it can improve compressive and flexural strength by refining pores and enhancing load transfer. However, studies show mixed results, indicating a need for further research and optimization. Researchers [46] reported that replacing 2% of cement with biochar originating from rice husk improved compressive strength by 8%, whereas coconut shell biochar caused a 5% reduction due to increased porosity. In a study by others [47], biochar produced at 500°C exhibited a 15% strength gain, whereas biochar pyrolyzed at 700°C led to a 10% strength loss due to carbonization-induced hydrophobicity, which hindered cement hydration. While numerous studies highlight biochar's benefits in improving concrete's mechanical performance, significant inconsistencies exist in reported outcomes. Researchers [48] found that 1–3% biochar replacement enhanced compressive strength by 5–10%, whereas others [49] observed a 7% reduction at similar dosages, attributing it to poor dispersion. Inconsistencies in biochar performance stem from variations in particle size, porosity, and surface chemistry. Fine particles ($< 50\ \mu m$) enhance cement hydration, while coarser ones ($> 100\ \mu m$) increase porosity and weaken the matrix.

Pyrolysis temperature (300–700°C) also affects reactivity and performance. Excess biochar can reduce strength due to higher porosity and lower density. However, pre-treatments like activation can improve compatibility and durability. Overall, optimizing biochar properties is key to enhancing concrete's mechanical performance. Table 2.2 presents the effects of biochar at various dosages and feedstocks as investigated by previous studies.

TABLE 2.2
Summary of Mechanical Observations by Different Researchers on Biochar Concrete with Various Dosages, Percentages, and Curing Age

Feedstock	Dosage (%)	Curing age (days)	Observation	Ref
Water treatment sludge	1, 2, 5	7, 14, 28	After a 28-day curing period, incorporating 1% to 5% biochar into the composite resulted in improved compressive strength compared to the conventional material. Notably, even with a 10% biochar addition, the compressive strength was on par with that of the reference material.	[50]
Mixed wood sawdust	2, 5, 8	1, 7, 28	With a 2% by-weight cement replacement, F2 (fine biochar), M2 (medium biochar), and C2 (coarse biochar) demonstrated strength similar to the convectional mix (R1) after 1 day. However, by 7 days, a slight reduction in strength was observed, with decreases of 5.20%, 8.80%, and 9.60% for F2, M2, and C2, respectively, compared to the control. By 28 days, the strength decline became more pronounced, with reductions of 10.67% for F2, 8% for M2, and 11.80% for C2 relative to R1.	[51]
Bamboo biochar	0.2, 1, 2, 3, 4	3, 7, 14, 28, 40	A 1 wt.% substitution of bamboo biochar for cement yielded the most effective enhancement in the compressive strength of biochar mortar.	[52]
Sawdust	1, 3, 5	14, 28	The compressive strength improved with a partial replacement of 3% and 5% biochar. At 7, 14, and 28 days, the cement paste containing varying biochar levels initially showed higher strength than the control up to a 3% replacement. However, as the replacement increased from 3% to 5%, a decrease in strength was noted.	[53]

(Continued)

TABLE 2.2 (*Continued*)
Summary of Mechanical Observations by Different Researchers on Biochar Concrete with Various Dosages, Percentages, and Curing Age

Feedstock	Dosage (%)	Curing age (days)	Observation	Ref
Coconut shell	5	7, 28	Compared to ordinary mortar, the partial substitution with coconut shell biochar resulted in 30% and 21% reductions in 1-day and 7-day strengths, respectively.	[54]
Mixed wood sawdust	1, 3, 5, 8	7, 28, 56	Incorporating biochar at a dosage of 1–2 wt.% enhanced compressive strength, particularly at early ages. However, as the biochar content exceeded 2 wt.%, a decline in strength was observed.	[55]
Corn straw	1, 3, 5, 7, 9, 15	1, 3, 7, 14, 28	After one day, biochar-modified concrete exhibited greater compressive strength than regular concrete. The highest compressive strength at 28 days was achieved with a 1 wt.% replacement of cement with biochar, resulting in a 5.3% increase compared to ordinary concrete.	[56]
Food waste digestate	20	7, 28	Cement composites incorporating food waste digestate, produced through pyrolysis at significantly higher temperatures of 650°C and 750°C, demonstrated superior strength compared to biochar blocks manufactured at 550°C.	[57]
Corn stover	2, 4, 6, 8	3, 28	Carbonation led to enhanced compressive strength across all mix designs. Notably, a 4% addition of biochar resulted in a strength increase, but further increases in biochar content caused a decline in strength.	[58]
Rice husk, coconut shell, bamboo	2.5, 5, 7.5, 10	28	The results revealed that adding bamboo particles enhanced the compressive strength of the specimens compared to red clay. Strength evaluations indicated that while the compressive strength generally increased with the addition of coconut shell and bamboo biochar, it decreased with the incorporation of rice husk biochar	[59]

2.3.2 Durability of Biochar Concrete

Research on the durability of cement mortar and concrete with partial biochar replacement, particularly its resistance to challenges such as acid attack, chloride penetration, exposure to seawater, sulfates, shrinkage, and water absorption, remains relatively sparse. Developing optimal mix designs through systematic testing and trials is crucial to ensure that biochar improves properties like strength, permeability, and resistance to both chemical and physical degradation. Biochar-modified concrete exhibits enhanced durability under various environmental conditions. Researchers [49] found that biochar-reinforced cement composites demonstrated reduced chloride penetration, making them more resistant to marine exposure. Others [23] reported that biochar incorporation decreases shrinkage by up to 16.4%, improving long-term durability. Additionally, others [60] reported that biochar-modified cementitious materials demonstrated 68% lower water absorption, enhancing their resistance to moisture damage. These results indicate that biochar can serve as an effective approach to extending the service life of concrete structures. Long-term observations and evaluations are central to understanding how biochar-enhanced concrete functions under real-world environmental conditions. Such studies are vital for gaining insights into its durability and for advancing construction practices and design strategies. Table 2.3 shows the effect of biochar on durability in concrete performed by pervious researchers.

2.4 CARBON SEQUESTRATION POTENTIAL IN BIOCHAR

The integration of biochar to cement-based materials noticeably enhances their carbon capture potential, helping limit the environmental impact of cement production. Biochar supports carbon storage through carbonation – where co_2 reacts with cement to form stable carbonates and through physical adsorption due to its high surface area and porous structure. This not only lowers co_2 levels but also boosts the strength and durability of concrete. Research suggests that biochar can sequester up to 250 kg of CO_2 per ton. Researchers [70] demonstrated that pervious concrete with pulverized biochar absorbed significantly more atmospheric CO_2 during curing, further reducing its carbon footprint. Additionally, the existence of biochar can affect the hydration process within the cement matrix, modifying its porosity and surface area, which in turn accelerates carbonation. When biochar is introduced into cementitious materials, the carbonation process can be enhanced using methods like accelerated carbonation curing (ACC), in which materials are exposed to high CO_2 concentrations. This method boosts both carbonation rate and CO_2 sequestration by increasing surface area and reactive sites, thanks to biochar. It not only improves carbon uptake but also enhances mechanical properties like compressive strength. The long-term carbon storage depends on biochar's durability. Biochar made at higher temperatures (e.g., 700°C) is more stable and less prone to leaching than that made at lower temperatures. Additionally, a higher biochar-to-cement ratio can increase porosity, offering more surface area for CO_2 absorption. Researchers [71] demonstrated that biochar-modified pervious concrete could absorb up to 250 kg of CO_2 per ton of biochar. Additionally, others [72] found that carbonation in biochar-enhanced

TABLE 2.3
An Overview Summary of the Biochar-Concrete's Endurance from Several Researchers Using Varying Dosages

Feedstock	Dosage (%)	Curing age (days)	Observation	Reference
Coconut shell	5%	7, 28	Adding 5% treated coconut biochar and 5% commercial wood biochar to plain mortar reduced water absorption per unit area (g/cm^2) by 12% to 17%	[61]
Mixed wood sawdust	1, 3, 5, 8	7, 28, 56	Incorporating 1% biochar incinerated at 300°C reduced penetration depth by up to 60%. All mortar specimens containing biochar exhibited lower water permeability compared to the reference samples. The results indicated that biochar pyrolyzed at 300°C reduced water penetration depth by 12% to 60% after 56 days of water curing.	[62]
Rice husk (RHB)	10, 40	7, 28, 120	The incorporation of 2% and 8% rice husk biochar, along with 18% and 12% industrial rice husk ash, respectively, delayed the onset of shrinkage until after the eight-day mark. When combined with 18% industrial rice husk ash, 2% rice husk biochar reduced the initial absorption rate by 36% compared to the control concrete	[63]
Unwashed hull of processed peanut	1, 3	7, 28	The addition of 1% and 3% biochar had no impact on initial absorption; however, 1% biochar increased total absorption by 11% compared to the reference. In contrast, adding 3% biochar decreased total absorption by 16% compared to plain mortar after 28 days of water curing	[64]
Olive stone, rice husk, wood chip	0.5, 1, 2, 4	7, 28	The addition of 0.5% olive stone biochar and 4% forest wood biochar resulted in water absorption levels comparable to plain mortar. Incorporating biochar from different feedstocks into cement produced varying absorption rates depending on the dosage. Notably, forest wood biochar exhibited the lowest absorption rate as its dosage increased from 0.5% to 4%, outperforming other biochar types	[65]

(Continued)

TABLE 2.3 (*Continued*)
An Overview Summary of the Biochar-Concrete's Endurance from Several Researchers Using Varying Dosages

Feedstock	Dosage (%)	Curing age (days)	Observation	Reference
Weed tree	2	7, 28, 90	Adding 2% biochar by weight reduced the autogenous shrinkage of cement paste to 485 microstrains, representing a 16.4% reduction compared to the control.	[66]
Mixture of wood chips	1, 2.5	7, 28	Adding biochar up to 2.5% by weight to the admixture showed no significant change in hardened density compared to ordinary mortar after 7 and 28 days of curing.	[67]
Wood chips	2.5, 5, 10 15	7	Relative to the control specimen, the sample containing 15% super-hydrophobic carbonaceous powder (SHCP) absorbed 68% less water in total.	[68]
Waste synthetic eucalyptus	0.65, 3.2, 6.5, 9.5, 13.5	7, 28	Samples with biochar demonstrated a significant increase in water absorption, particularly those with biochar content below 6.5%. Samples with 13.5% biochar had approximately 0.8% higher water absorption compared to samples without biochar.	[69]

cement promotes calcium carbonate formation, improving both strength and durability. Studies show that biochar not only assists in in carbon sequestration but also boosts the durability and strength of cement-based materials by enhancing carbonation resistance and reducing moisture penetration. With its dual benefits of carbon capture and structural improvement, biochar establishes itself as a promising eco-friendly additive. Its high surface area, role in promoting carbonation, and long-term stability make it ideal for sustainable concrete. Continued research and optimization are essential to reduce carbon emissions while enhancing concrete durability and performance [72–77].

2.5 CHALLENGES AND LIMITATIONS

2.5.1 Potential Drawbacks of Biochar in Cementitious Systems

While biochar offers sustainability benefits in cement-based materials, several challenges must be addressed for its effective use. One significant concern is the decrease in early-age compressive strength; biochar is inherently less reactive compared to cement, and it can weaken matrix bonding particularly at higher substitution rates. Its highly porous structure increases total moisture demand, impacting workability, reducing concrete density, and ultimately lowering compressive strength. Also, as biochar is hydrophobic, it can impact hydration processes, leading to inconsistencies in performance, both in fresh and hardened properties. Durability is also a concern, as concrete modified with biochar may be less resistant to moisture, temperature changes, and chemical attacks, although the addition of biochar may improve freeze–thaw resistance. A critical problem is that biochar is inconsistently available, and it is important to note that biochar varies depending on low-cost feedstock and pyrolysis conditions, attributing to its unpredictable performance. Researchers [78] noted that biochar produced at lower temperatures (~≤ 500°C) may contain remnants of organic material that may produce inconsistent hydration reactions. Others [79] also reported that untreated biochar increased total moisture demand, causing poor workability. These problems must be addressed with optimized processing and modification techniques in order for biochar to be effectively used in cementitious materials. Furthermore, environmental complexities for biochar arise from a transparent renewable energy process for biochar production, which may neutralize its benefits. Additionally, there could be undesired consequences of impurities and toxicants in biochar derived from agricultural waste that could potentially reduce the structure properties of concrete. These issues must be overcome to help facilitate further opportunities for biochar to be used beyond its broader applications toward cementitious applications. Upcoming studies should prioritize improving processing techniques, harmonizing material properties, and evaluating long-term performance to enhance the viability of biochar in sustainable construction [80, 81].

2.5.2 Compatibility with Standards and Codes

OPC and standard aggregates are regulated nationally and internationally. These rules ensure safety and performance. Biochar is largely framework-free and consequently commonly unaccepted in buildings whose defining parameters have not been effectively regulated. Despite its carbon-storing and thermal-insulating properties, biochar is rarely described as a concrete alternative. The major issue is that biochar's strength and durability in concrete are still unknown due to its different features (porosity, particle, size, and chemical components). Biochar's qualities vary; thus, thorough testing is needed to determine its cost per project or ensure consistency with the permitted concrete mix. Biochar, like all admixtures and modifiers, must function under saline water, continuous freeze–thaw cycles, and harsh chemical exposure before accuracy comments can be made regarding its durability. As ACI and CEN do not recognize biochar as an acceptable additive, alternative material, or modifier, building experts are not as eager to proceed; hence, biochar standards must be developed. To successfully use biochar as an approved alternative, several durability studies must be conducted to demonstrate its safety, durability, and efficacy as a concrete additive. Additionally, others [83] emphasized the need for performance benchmarks to ensure consistent mechanical and durability properties. Coordinated initiatives among industry stakeholders, researchers, and regulatory authorities will be crucial in establishing standards that support the incorporation of biochar into mainstream construction practices. Despite its benefits, biochar lacks standardized regulations in cementitious applications. Researchers [84] argued that the absence of guidelines limits its adoption in structural concrete. Additionally, it is suggested that 3D-printing technologies incorporating biochar could pave the way for new standardization efforts in digital construction. These studies highlight the need for developing regulations to promote the broader adoption of biochar within the construction sector.

2.6 RECOMMENDATIONS AND FUTURE DIRECTIONS

The incorporation of biochar into cement-based systems presents significant potential for advancing eco-friendly construction methods. However, to fully harness its benefits, further investigation and development are necessary to overcome the issues mentioned earlier. Drawing from current research and emerging trends, several suggestions and future avenues can be proposed for the ongoing exploration of biochar in cementitious composites.

1. Standards and Regulations: Develop clear guidelines and testing methods for biochar in concrete. Collaboration among researchers, manufacturers, and regulatory bodies is crucial for safe and consistent use.
2. Property Optimization: Refine biochar characteristics – such as particle size, porosity, and chemistry – for better compatibility with cement. Pretreatment and modification can improve performance.
3. Long-Term Performance: Conduct extended studies to assess durability under real-world conditions, including harsh weather and chemical exposure, to ensure structural reliability.
4. Environmental Assessment: Perform life cycle assessments to evaluate the sustainability of biochar concrete, focusing on carbon footprint, energy use, and circular economy potential.

2.7 CONCLUSION

Incorporating biochar into cementitious materials provides an eco-friendly alternative in construction, offering benefits such as improved mechanical strength and carbon sequestration. Research indicates that at replacement levels of 3% or less, biochar has the potential to improve compressive strength via enhanced particle packing and interactions within the cement matrix. Its role in carbonation and carbon capture supports its potential as a green additive. However, challenges remain due to variability in feedstock and production methods. While some consistency can be achieved through producer assurances, concerns around long-term durability, production energy, availability, and acceptance persist. Although potential durability benefits exist, additional studies are required to assess its resistance to external degradation and chemical interactions. Significant study is required to address optimization, standardization, and large-scale application. Future work should focus on refining production techniques, validating performance, and integrating biochar into construction standards. Overcoming these barriers could elevate biochar from being an experimental material to a mainstream solution for lowering the carbon footprint and conserving resources.

REFERENCES

1. G. Sotorrío Ortega, S. Otero Seseña, A. Cobo Escamilla, and J. A. Tenorio Ríos, "Assessment of industrialised construction through sustainability tools," *Results in Engineering*, vol. 24, Dec. 2024, https://doi.org/10.1016/j.rineng.2024.103663.
2. H. Liu and Q. Li, "Preparation of green concrete from bamboo biochar (BB) and concrete slurry waste (CSW): Preparation method and performance evaluation," *Constr Build Mater*, vol. 462, p. 139964, Feb. 2025, https://doi.org/10.1016/j.conbuildmat.2025.139964.
3. L. Lima, E. Trindade, L. Alencar, M. Alencar, and L. Silva, "Sustainability in the construction industry: A systematic review of the literature," Mar. 20, 2021, *Elsevier Ltd*, https://doi.org/10.1016/j.jclepro.2020.125730.
4. "Global Carbon Atlas." Accessed: Mar. 05, 2025. [Online], https://globalcarbonatlas.org/.
5. "World Energy Outlook 2024 – Analysis – IEA." Accessed: Mar. 05, 2025. [Online], https://www.iea.org/reports/world-energy-outlook-2024.
6. Y. Li, X. Mu, Y. Li, S. Zhang, and W. Ni, "Substitution of blast furnace slag by melting furnace slag as active component in green concrete application," *Constr Build Mater*, vol. 449, Oct. 2024, https://doi.org/10.1016/j.conbuildmat.2024.138509.
7. M. Behera, S. K. Bhattacharyya, A. K. Minocha, R. Deoliya, and S. Maiti, "Recycled aggregate from C&D waste & its use in concrete – a breakthrough towards sustainability in construction sector: A review," Oct. 15, 2014, *Elsevier Ltd*, https://doi.org/10.1016/j.conbuildmat.2014.07.003.
8. S. B. Murugan, G. M. Ganesh, and A. S. Santhi, "Regression models for prediction of compressive strength of high volume fly ash (HVFA) concrete," *Arab J Sci Eng*, vol. 39, no. 3, pp. 1659–1669, Jan. 2012, https://doi.org/10.1007/s13369-013-0783-y.
9. P. Danish and M. G. Ganesh, "Behavıour of self-compactıng concrete usıng dıfferent mıneral powders addıtıons ın ternary blends," *Romanian Journal of Materials*, vol. 50, no. 2, pp. 232–239, 2020.
10. M. Ganesh, P. Danish, and M. G. Ganesh, "Durabılıty propertıes of self-compactıng concrete usıng dıfferent mıneral powders addıtıons ın ternary blends," *Romanian Journal of Materials*, vol. 50, no. 3, pp. 369–378, 2020.
11. A. A. Kumar, A. S. Santhi, and G. M. Ganesh, "Strength and durability of green concrete," *International Journal of Structural Engineering*, vol. 6, no. 1, pp. 23–32, Jan. 2015, https://doi.org/10.1504/IJSTRUCTE.2015.067967.
12. V. Athiyamaan and G. M. Ganesh, "Statistical and detailed analysis on fiber reinforced self-compacting concrete containing admixtures – a state of art of review," in *IOP Conference Series: Materials Science and Engineering*, Institute of Physics Publishing, Dec. 2017, https://doi.org/10.1088/1757-899X/263/3/032037.
13. S. Zou, M. L. Sham, J. Xiao, L. M. Leung, J. X. Lu, and C. S. Poon, "Biochar-enabled carbon negative aggregate designed by core-shell structure: A novel biochar utilizing method in concrete," *Constr Build Mater*, vol. 449, Oct. 2024, https://doi.org/10.1016/j.conbuildmat.2024.138507.
14. L. Chen *et al.*, "A review on the roles of biochar incorporated into cementitious materials: Mechanisms, application and perspectives," Dec. 15, 2023, *Elsevier Ltd*, https://doi.org/10.1016/j.conbuildmat.2023.134204.

15. L. Wang *et al.*, "Biochar as green additives in cement-based composites with carbon dioxide curing," *J Clean Prod*, vol. 258, Jun. 2020, https://doi.org/10.1016/j.jclepro.2020.120678.
16. A. Sirico *et al.*, "Effects of biochar addition on long-term behavior of concrete," *Theoretical and Applied Fracture Mechanics*, vol. 122, Dec. 2022, https://doi.org/10.1016/j.tafmec.2022.103626.
17. Y. Hu, A. Bassi, and C. C. Xu, "Energy from biomass," in *Future Energy: Improved, Sustainable and Clean Options for Our Planet*, Elsevier, 2020, pp. 447–471, https://doi.org/10.1016/B978-0-08-102886-5.00021-9.
18. A. Pattiya, "Fast pyrolysis," in *Direct Thermochemical Liquefaction for Energy Applications*, Elsevier, 2017, pp. 3–28, https://doi.org/10.1016/B978-0-08-101029-7.00001-1.
19. V. K. Guda, P. H. Steele, V. K. Penmetsa, and Q. Li, "Fast pyrolysis of biomass: Recent advances in fast pyrolysis technology," in *Recent Advances in Thermochemical Conversion of Biomass*, Elsevier Inc., 2015, pp. 177–211, https://doi.org/10.1016/B978-0-444-63289-0.00007-7.
20. D. Cuthbertson, U. Berardi, C. Briens, and F. Berruti, "Biochar from residual biomass as a concrete filler for improved thermal and acoustic properties," *Biomass Bioenergy*, vol. 120, pp. 77–83, Jan. 2019, https://doi.org/10.1016/j.biombioe.2018.11.007.
21. O. Awogbemi and D. V. Von Kallon, "Application of biochar derived from crops residues for biofuel production," *Fuel Communications*, vol. 15, p. 100088, Jun. 2023, https://doi.org/10.1016/j.jfueco.2023.100088.
22. S. Safarian, "Performance analysis of sustainable technologies for biochar production: A comprehensive review," Dec. 01, 2023, *Elsevier Ltd*, https://doi.org/10.1016/j.egyr.2023.03.111.
23. H. Maljaee, R. Madadi, H. Paiva, L. Tarelho, and V. M. Ferreira, "Incorporation of biochar in cementitious materials: A roadmap of biochar selection," May 10, 2021, *Elsevier Ltd*, https://doi.org/10.1016/j.conbuildmat.2021.122757.
24] A. Danish, M. Ali Mosaberpanah, M. Usama Salim, N. Ahmad, F. Ahmad, and A. Ahmad, "Reusing biochar as a filler or cement replacement material in cementitious composites: A review," Sep. 20, 2021, *Elsevier Ltd*, https://doi.org/10.1016/j.conbuildmat.2021.124295.
25. A. Hornung, "Intermediate pyrolysis of biomass," in *Biomass Combustion Science, Technology and Engineering*, Elsevier Inc., 2013, pp. 172–186, https://doi.org/10.1533/9780857097439.2.172.
26. A. A. Boateng, "Introduction," in *Pyrolysis of Biomass for Fuels and Chemicals*, Elsevier, 2020, pp. 1–21, https://doi.org/10.1016/b978-0-12-818213-0.00001-1.
27. M. Khairy, M. Amer, M. Ibrahim, S. Ookawara, H. Sekiguchi, and A. Elwardany, "The influence of torrefaction on the biochar characteristics produced from sesame stalks and bean husk," *Biomass Convers Biorefin*, vol. 14, no. 15, pp. 17127–17148, Aug. 2024, https://doi.org/10.1007/s13399-023-03822-9.
28. C. Lokmit, K. Nakason, S. Kuboon, A. Jiratanachotikul, and B. Panyapinyopol, "A comparison of char fuel properties derived from dry and wet torrefaction of oil palm leaf and its techno-economic feasibility," *Mater Sci Energy Technol*, vol. 6, pp. 192–204, Jan. 2023, https://doi.org/10.1016/j.mset.2022.12.010.
29. S. Pambudi, J. S. Jongyingcharoen, and W. Saechua, "Thermochemical treatment of spent coffee grounds via torrefaction: A statistical evidence of biochar properties similarity between inert and oxidative conditions," *Results in Engineering*, vol. 21, Mar. 2024, https://doi.org/10.1016/j.rineng.2024.102012.
30. C. Lokmit, K. Nakason, S. Kuboon, A. Jiratanachotikul, and B. Panyapinyopol, "A comparison of char fuel properties derived from dry and wet torrefaction of oil palm leaf and its techno-economic feasibility," *Mater Sci Energy Technol*, vol. 6, pp. 192–204, Jan. 2023, https://doi.org/10.1016/J.MSET.2022.12.010.

31. A. Akgül, S. Ş. Akın, F. Güleç, and F. Kazanç, "Biochar gasification: Insights from pyrolysis atmospheres and gasification heating rates," *Fuel*, vol. 360, Mar. 2024, https://doi.org/10.1016/j.fuel.2023.130469.
32. L. A. Vergara, J. F. Perez, and H. A. Colorado, "3D printing of ordinary Portland cement with waste wood derived biochar obtained from gasification," *Case Studies in Construction Materials*, vol. 18, Jul. 2023, https://doi.org/10.1016/j.cscm.2023.e02117.
33. S. You *et al.*, "A critical review on sustainable biochar system through gasification: Energy and environmental applications," Dec. 01, 2017, *Elsevier Ltd*, https://doi.org/10.1016/j.biortech.2017.06.177.
34. J. Liu, G. Liu, W. Zhang, Z. Li, F. Xing, and L. Tang, "Application potential analysis of biochar as a carbon capture material in cementitious composites: A review," Oct. 03, 2022, *Elsevier Ltd*, https://doi.org/10.1016/j.conbuildmat.2022.128715.
35. P. R. Yaashikaa, P. S. Kumar, S. Varjani, and A. Saravanan, "A critical review on the biochar production techniques, characterization, stability and applications for circular bioeconomy," Dec. 01, 2020, *Elsevier B.V*, https://doi.org/10.1016/j.btre.2020.e00570.
36. S. Barbhuiya, B. Bhusan Das, and F. Kanavaris, "Biochar-concrete: A comprehensive review of properties, production and sustainability," *Case Studies in Construction Materials*, vol. 20, Jul. 2024, https://doi.org/10.1016/j.cscm.2024.e02859.
37. B. A. Akinyemi and A. Adesina, "Recent advancements in the use of biochar for cementitious applications: A review," Nov. 01, 2020, *Elsevier Ltd*, https://doi.org/10.1016/j.jobe.2020.101705.
38. J. W. C. Wong, J. B. W. Webber, and U. O. Ogbonnaya, "Characteristics of biochar porosity by NMR and study of ammonium ion adsorption," *J Anal Appl Pyrolysis*, vol. 143, Oct. 2019, https://doi.org/10.1016/j.jaap.2019.104687.
39. P. Conte *et al.*, "NMR techniques for the evaluation of biochar characteristics," in *The Environment in a Magnet*, Royal Society of Chemistry, 2024, pp. 353–380, https://doi.org/10.1039/bk9781837671250-00353.
40. C. H. Chia, B. Gong, S. D. Joseph, C. E. Marjo, P. Munroe, and A. M. Rich, "Imaging of mineral-enriched biochar by FTIR, Raman and SEM-EDX," *Vib Spectrosc*, vol. 62, pp. 248–257, Sep. 2012, https://doi.org/10.1016/j.vibspec.2012.06.006.
41. X. Ma *et al.*, "Study of biochar properties by scanning electron microscope – energy dispersive X-ray spectroscopy (SEM-EDX)," *Commun Soil Sci Plant Anal*, vol. 47, no. 5, pp. 593–601, Mar. 2016, https://doi.org/10.1080/00103624.2016.1146742.
42. R. K. Mishra, S. Chinnam, and A. Sharma, "Thermocatalytic pyrolysis of low-value waste biomass: Thermal decomposition, kinetics behaviour, and biochar characterization," *Results in Engineering*, vol. 25, Mar. 2025, https://doi.org/10.1016/j.rineng.2025.104210.
43. A. Y. Elnour *et al.*, "Effect of pyrolysis temperature on biochar microstructural evolution, physicochemical characteristics, and its influence on biochar/polypropylene composites," *Applied Sciences (Switzerland)*, vol. 9, no. 6, 2019, https://doi.org/10.3390/app9061149.
44. R. R. Nair, M. M. Mondal, and D. Weichgrebe, "Biochar from co-pyrolysis of urban organic wastes – investigation of carbon sink potential using ATR-FTIR and TGA," *Biomass Convers Biorefin*, vol. 12, no. 10, pp. 4729–4743, Oct. 2022, https://doi.org/10.1007/s13399-020-01000-9.
45. S. E. Ibitoye *et al.*, "An overview of biochar production techniques and application in iron and steel industries," Dec. 01, 2024, *Springer*, https://doi.org/10.1186/s40643-024-00779-z.
46. L. Chen, X. Zhu, Y. Zheng, L. Wang, C. S. Poon, and D. C. W. Tsang, "Development of high-strength lightweight concrete by utilizing food waste digestate based biochar aggregate," *Constr Build Mater*, vol. 411, Jan. 2024, https://doi.org/10.1016/j.conbuildmat.2023.134142.

47. Z. Zhao, A. El-Naggar, J. Kau, C. Olson, D. Tomlinson, and S. X. Chang, "Biochar affects compressive strength of Portland cement composites: A meta-analysis," *Biochar*, vol. 6, no. 1, pp. 1–18, Dec. 2024, https://doi.org/10.1007/S42773-024-00309-2/TABLES/2.
48. L. Chen *et al.*, "A review on the roles of biochar incorporated into cementitious materials: Mechanisms, application and perspectives," Dec. 15, 2023, *Elsevier Ltd*, https://doi.org/10.1016/j.conbuildmat.2023.134204.
49. W. Xu *et al.*, "Durability and micromechanical properties of biochar in biochar-cement composites under marine environment," *J Clean Prod*, vol. 450, p. 141842, Apr. 2024, https://doi.org/10.1016/J.JCLEPRO.2024.141842.
50. S. De Carvalho Gomes, J. L. Zhou, X. Zeng, and G. Long, "Water treatment sludge conversion to biochar as cementitious material in cement composite," *J Environ Manage*, vol. 306, Mar. 2022, https://doi.org/10.1016/j.jenvman.2022.114463.
51. A. Dixit, S. Gupta, S. D. Pang, and H. W. Kua, "Waste Valorisation using biochar for cement replacement and internal curing in ultra-high performance concrete," *J Clean Prod*, vol. 238, Nov. 2019, https://doi.org/10.1016/j.jclepro.2019.117876.
52. W. Liu, K. Li, and S. Xu, "Utilizing bamboo biochar in cement mortar as a bio-modifier to improve the compressive strength and crack-resistance fracture ability," *Constr Build Mater*, vol. 327, Apr. 2022, https://doi.org/10.1016/j.conbuildmat.2022.126917.
53. D. Ali *et al.*, "Thermo-physical properties and microstructural behaviour of biochar-incorporated cementitious material," *Journal of Building Engineering*, vol. 64, Apr. 2023, https://doi.org/10.1016/j.jobe.2022.105695.
54. S. Gupta, P. Krishnan, A. Kashani, and H. W. Kua, "Application of biochar from coconut and wood waste to reduce shrinkage and improve physical properties of silica fume-cement mortar," *Constr Build Mater*, vol. 262, Nov. 2020, https://doi.org/10.1016/j.conbuildmat.2020.120688.
55. M. Habibur Rahman Sobuz *et al.*, "Assessment of mechanical properties with machine learning modeling and durability, and microstructural characteristics of a biochar-cement mortar composite," *Constr Build Mater*, vol. 411, Jan. 2024, https://doi.org/10.1016/j.conbuildmat.2023.134281.
56. L. Chen, X. Zhu, Y. Zheng, L. Wang, C. S. Poon, and D. C. W. Tsang, "Development of high-strength lightweight concrete by utilizing food waste digestate based biochar aggregate," *Constr Build Mater*, vol. 411, Jan. 2024, https://doi.org/10.1016/j.conbuildmat.2023.134142.
57. L. Qing, H. Zhang, and Z. Zhang, "Effect of biochar on compressive strength and fracture performance of concrete," *Journal of Building Engineering*, vol. 78, Nov. 2023, https://doi.org/10.1016/j.jobe.2023.107587.
58. S. Praneeth, R. Guo, T. Wang, B. K. Dubey, and A. K. Sarmah, "Accelerated carbonation of biochar reinforced cement-fly ash composites: Enhancing and sequestering CO2 in building materials," *Constr Build Mater*, vol. 244, May 2020, https://doi.org/10.1016/j.conbuildmat.2020.118363.
59. S. Yang, S. Wi, J. Lee, H. Lee, and S. Kim, "Biochar-red clay composites for energy efficiency as eco-friendly building materials: Thermal and mechanical performance," *J Hazard Mater*, vol. 373, pp. 844–855, Jul. 2019, https://doi.org/10.1016/j.jhazmat.2019.03.079.
60. M. I. Haque, R. I. Khan, W. Ashraf, and H. Pendse, "Production of sustainable, low-permeable and self-sensing cementitious composites using biochar," *Sustainable Materials and Technologies*, vol. 28, p. e00279, Jul. 2021, https://doi.org/10.1016/J.SUSMAT.2021.E00279.
61. S. Gupta, P. Krishnan, A. Kashani, and H. W. Kua, "Application of biochar from coconut and wood waste to reduce shrinkage and improve physical properties of silica fume-cement mortar," *Constr Build Mater*, vol. 262, Nov. 2020, https://doi.org/10.1016/j.conbuildmat.2020.120688.

62. M. Habibur Rahman Sobuz *et al.*, "Assessment of mechanical properties with machine learning modeling and durability, and microstructural characteristics of a biochar-cement mortar composite," *Constr Build Mater*, vol. 411, Jan. 2024, https://doi.org/10.1016/j.conbuildmat.2023.134281.
63. S. Muthukrishnan, S. Gupta, and H. W. Kua, "Application of rice husk biochar and thermally treated low silica rice husk ash to improve physical properties of cement mortar," *Theoretical and Applied Fracture Mechanics*, vol. 104, Dec. 2019, https://doi.org/10.1016/j.tafmec.2019.102376.
64. S. Gupta, A. Kashani, A. H. Mahmood, and T. Han, "Carbon sequestration in cementitious composites using biochar and fly ash – effect on mechanical and durability properties," *Constr Build Mater*, vol. 291, Jul. 2021, https://doi.org/10.1016/j.conbuildmat.2021.123363.
65. H. Maljaee, H. Paiva, R. Madadi, L. A. C. Tarelho, M. Morais, and V. M. Ferreira, "Effect of cement partial substitution by waste-based biochar in mortars properties," *Constr Build Mater*, vol. 301, Sep. 2021, https://doi.org/10.1016/j.conbuildmat.2021.124074.
66. L. Mo, J. Fang, B. Huang, A. Wang, and M. Deng, "Combined effects of biochar and MgO expansive additive on the autogenous shrinkage, internal relative humidity and compressive strength of cement pastes," *Constr Build Mater*, vol. 229, Dec. 2019, https://doi.org/10.1016/j.conbuildmat.2019.116877.
67. A. Sirico *et al.*, "Mechanical characterization of cement-based materials containing biochar from gasification," *Constr Build Mater*, vol. 246, Jun. 2020, https://doi.org/10.1016/j.conbuildmat.2020.118490.
68. M. I. Haque, R. I. Khan, W. Ashraf, and H. Pendse, "Production of sustainable, low-permeable and self-sensing cementitious composites using biochar," *Sustainable Materials and Technologies*, vol. 28, Jul. 2021, https://doi.org/10.1016/j.susmat.2021.e00279.
69. Y. Qin, X. Pang, K. Tan, and T. Bao, "Evaluation of pervious concrete performance with pulverized biochar as cement replacement," *Cem Concr Compos*, vol. 119, May 2021, https://doi.org/10.1016/j.cemconcomp.2021.104022.
70. Y. Qin, X. Pang, K. Tan, and T. Bao, "Evaluation of pervious concrete performance with pulverized biochar as cement replacement," *Cem Concr Compos*, vol. 119, p. 104022, May 2021, https://doi.org/10.1016/J.CEMCONCOMP.2021.104022.
71. K. Tan, Y. Qin, and J. Wang, "Evaluation of the properties and carbon sequestration potential of biochar-modified pervious concrete," *Constr Build Mater*, vol. 314, p. 125648, Jan. 2022, https://doi.org/10.1016/J.CONBUILDMAT.2021.125648.
72. S. Song, Z. Liu, G. Liu, X. Cui, and J. Sun, "Application of biochar cement-based materials for carbon sequestration," Nov. 17, 2023, *Elsevier Ltd*, https://doi.org/10.1016/j.conbuildmat.2023.133373.
73. J. Liu, J. Liu, L. Cheng, H. Jin, and F. Xing, "Sustainable upcycling of artificial lightweight cold-bonded aggregates (ALCBAs) designed by biochar and concrete slurry waste (CSW) into porous carbons materials for CO_2 sequestration," *Constr Build Mater*, vol. 412, Jan. 2024, https://doi.org/10.1016/j.conbuildmat.2023.134736.
74. K. Tan, Y. Qin, and J. Wang, "Evaluation of the properties and carbon sequestration potential of biochar-modified pervious concrete," *Constr Build Mater*, vol. 314, Jan. 2022, https://doi.org/10.1016/j.conbuildmat.2021.125648.
75. S. S. Senadheera *et al.*, "Application of biochar in concrete – a review," *Cem Concr Compos*, vol. 143, Oct. 2023, https://doi.org/10.1016/j.cemconcomp.2023.105204.
76. J. Liu, G. Liu, W. Zhang, Z. Li, F. Xing, and L. Tang, "Application potential analysis of biochar as a carbon capture material in cementitious composites: A review," Oct. 03, 2022, *Elsevier Ltd*, https://doi.org/10.1016/j.conbuildmat.2022.128715.
77. Y. Zhang, Y. Maierdan, T. Guo, B. Chen, S. Fang, and L. Zhao, "Biochar as carbon sequestration material combines with sewage sludge incineration ash to prepare lightweight concrete," *Constr Build Mater*, vol. 343, Aug. 2022, https://doi.org/10.1016/j.conbuildmat.2022.128116.

78. A. Sirico *et al.*, "Effects of biochar addition on long-term behavior of concrete," *Theoretical and Applied Fracture Mechanics*, vol. 122, Dec. 2022, https://doi.org/10.1016/j.tafmec.2022.103626.
79. S. Gupta, P. Krishnan, A. Kashani, and H. W. Kua, "Application of biochar from coconut and wood waste to reduce shrinkage and improve physical properties of silica fume-cement mortar," *Constr Build Mater*, vol. 262, Nov. 2020, https://doi.org/10.1016/j.conbuildmat.2020.120688.
80. S. E. Ibitoye *et al.*, "An overview of biochar production techniques and application in iron and steel industries," Dec. 01, 2024, *Springer*, https://doi.org/10.1186/s40643-024-00779-z.
81. J. E. Amonette *et al.*, "Key Challenges and Opportunities," 2016. [Online], https://www.sciencedirect.com/topics/engineering/bio-oil.
82. EBC, "European Biochar Certificate–Guidelines for a Sustainable Production of Biochar," European Biochar Foundation (EBC), 2012. http://european-biochar.org.
83. J. E. Amonette *et al.*, "Biomass to Biochar | Chapter 2: Key Challenges and Opportunities," 2016, Accessed: Mar. 05, 2025. [Online], https://www.sciencedirect.com/topics/engineering/bio-oil.
84. L. Mo, J. Fang, B. Huang, A. Wang, and M. Deng, "Combined effects of biochar and MgO expansive additive on the autogenous shrinkage, internal relative humidity and compressive strength of cement pastes," *Constr Build Mater*, vol. 229, p. 116877, Dec. 2019, https://doi.org/10.1016/J.CONBUILDMAT.2019.116877.

3 Stabilization of Problematic Soils Using Industrial-Sludge-Derived Pozzolanic Biochar

Muneeb Farooq, Ubaid Rehman, Jasim Nisar and Mohammad Yousuf Shah

3.1 INTRODUCTION

Problematic soils encompass various types of soils, each posing unique challenges to construction and infrastructure. Among them, Karewa and Dredged soils (DS) are found in the North-west Indian Himalayan Region. The Karewa soils (KS) are composed of mixed glacial, fluvial, and lacustrine sediments that appear as raised tablelands or low, flat topped mounds (Bhatt, 1982; Chandra et al., 2016). This soil has a predominance of silt in its composition and is challenging to compact due to its negligible permeability (Jan and Raj, 2018). DS are residual soils formed as a result of the deposition of solid waste resulting from the dredging of water bodies (Wani and Mir, 2021). These soils are characterized by low compressive strength and pose difficult ground conditions for infrastructure development (Mirza and Mir, 2025; Wani and Mir, 2020). To address this issue, calcium-based stabilizers, viz., cement and lime, are employed to enhance the strength of these problematic soils (Jalal et al., 2020). However, manufacturing these traditional stabilizers requires substantial energy, and this results in significant carbon emissions. Cement and lime production contributes about 9% of global anthropogenic CO_2 emission, with cement production contributing approximately 8% of global CO_2 emission (Dinga and Wen, 2022) and lime production contributing about 1% of global CO_2 emission (Campo et al., 2021). Moreover, producing 1 tonne of cement and lime each results in about 1 ton and 1.2 tonnes CO_2 emissions, respectively (Simoni et al., 2022). Hence, research should focus on developing new environment-friendly, economically feasible soil stabilizers which reduce carbon footprints compared to conventional calcium-based binders in the construction industry.

Biochar is a carbon-negative material produced through the pyrolysis of feedstocks, including waste wood, manure, sewage sludge, peanut shells, and straw (Vuppaladadiyam et al., 2023). It is a fine-grained, porous material with a

DOI: 10.1201/9781003664284-3

charcoal-like appearance (Wani et al., 2022). Most studies focus on the role of biochar in carbon sequestration, reducing greenhouse gases, improving soil fertility, and increasing crop yields. However, not many studies look at how biochar affects soil properties from a geotechnical engineering perspective. Sadasivam and Reddy (2015) investigated how adding different amounts and sizes of biochar to soil affects its physical, chemical, and engineering properties. Their findings showed that soil mixed with biochar had better properties, making it suitable as a cover material in landfills. Recent studies have demonstrated the potential of biochar as a material for infrastructure development (Farooq et al., 2024a). Biochar alters the pH, cation exchange capacity, and point of zero charge of the soil. This study investigated the influence of varying dosages of industrial-sludge-derived biochar, added in 2.5% increments ranging from 5% to 15%, on the unconfined compressive strength (UCS) of two problematic soils, KS and DS, over curing periods of 1, 3, 7, 14, and 28 days.

3.2 MATERIALS

3.2.1 Soil

Two distinct locations were selected for soil sampling, viz., Pampore (33.98°N, 74.94°E) for KS and Shalimar (34.11°N, 74.87°E) for DS. Both disturbed and undisturbed (50 mm diameter × 100 length cylindrical cores) soil samples were collected from 1 metre below the ground surface to avoid any external contamination. Subsequently, the samples were wrapped in a bubble wrap to preserve their structural integrity and were transported to the laboratory for further analysis. All tests on the virgin soil sample were conducted in accordance with IS: 2720 standards. Figure 3.1 shows the particle distribution curves, compaction curves, and flow curves of both the soils. Based on the Indian Standard Soil Classification System (ISSCS), the soil is classified as silt of intermediate compressibility (MI), and the geotechnical properties are summarized in Table 3.1.

3.2.2 Industrial Sludge

The industrial sludge (IS) was collected from the common effluent treatment plant of industrial growth centre (IGC), Lassipora (33.80°N, 74.92°E). The IS samples were put in the airtight bags and were kept in shade during the transportation process. The physico-chemical and biological properties of the IS are summarized in Table 3.2.

3.2.3 Biochar

Biochar was synthesized via slow pyrolysis of IS using a muffle furnace. The temperature was steadily increased to 450°C at a controlled rate of 5°C per minute, and the material was maintained at this temperature for about 120 minutes. The resulting biochar exhibited a moderately alkaline character, with pH values ranging between 8.84 and 9.32. The proportions of fixed and labile organic matter, determined through volatile matter and ash content in proximate analysis, together with the elemental composition obtained from ultimate analysis, are summarized in Table 3.2.

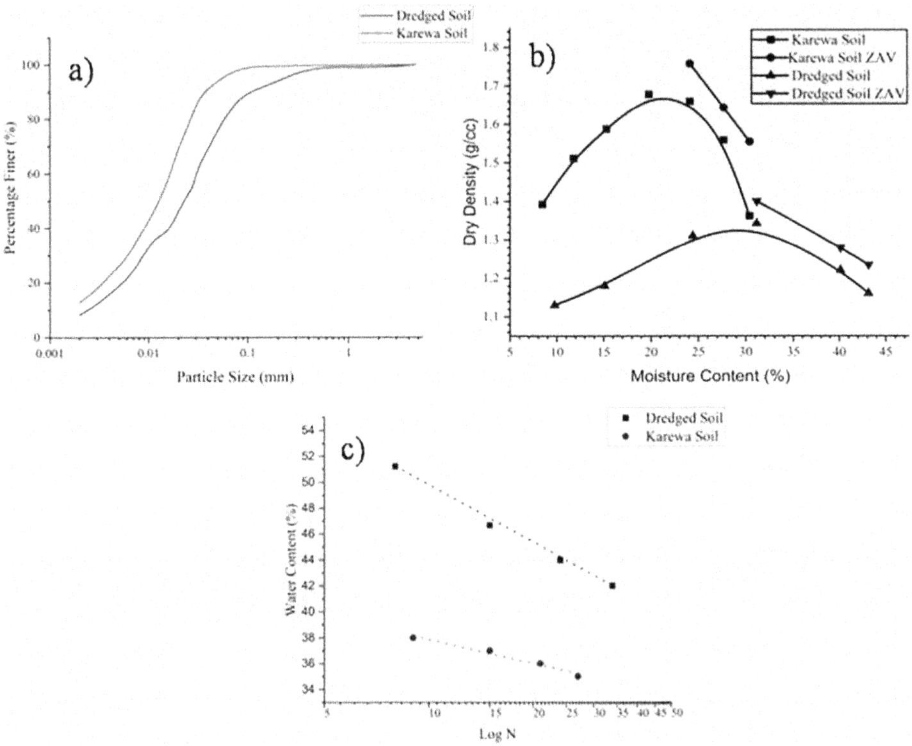

FIGURE 3.1 (a) Particle size distribution curves for KS and DS, (b) compaction plots of KS and DS, and (c) flow curves of KS and DS for the determination of liquid limit.

TABLE 3.1
Geotechnical Properties of the Soils Used in this Study

Parameters	Karewa soil	Dredged soil
Specific gravity (G)	2.56	2.40
Sand (%)	5.20	1.20
Silt (%)	80.10	90.60
Clay (%)	14.70	8.20
Liquid limit (%)	35.4	43
Plastic limit (%)	26.53	23.24
Shrinkage limit (%)	23	14
Plasticity index (%)	8.87	19.76
P.I, A-line	11.24	16.79
P.I, U-line	24.66	31.5
Soil classification (ISCS)	MI	MI
Clay mineral	Kaolinite	Kaolinite
Optimum moisture content (%)	19.78	31.13
Maximum dry density (g/cc)	1.68	1.34

TABLE 3.2
Characteristics of Industrial Sludge and Biochar

Parameters	IS	Biochar
Specific gravity	1.98	1.26
pH	7.7–8.2	8.84–9.32
BOD_5 (mg/l)	129	-
COD (mg/l)	1334.57	-
Ash content[a]	22 ± 2	54 ± 2
Volatile matter[a]	75 ± 2	39 ± 5
Fixed carbon[a,b]	3 ± 0.6	7 ±1
Carbon[c]	18.51 ± 0.8	27.2 ± 0.6
Hydrogen[c]	3.32 ± 0.2	3.7 ± 0.1
Nitrogen[c]	2.34 ± 0.2	2.4 ± 0.1
Oxygen[c]	36.86 ± 3	-
Sulphur[c]	0.42 ± 0.01	0.28 ± 0.01
H/C	0.18	0.16
Pyrolysis temperature (°C)	-	450
Residence time (hour)	-	2
Pyrolysis rate, β (°C/min)	-	5

[a]*Proximate analysis results;* [b]*Fixed Carbon = 100 − (volatile matter + ash);* [c]*Ultimate analysis results.*

3.3 METHODOLOGY

The methodology adopted in this study is shown in Figure 3.2 and is structured into three phases, viz., material procurement, sample preparation, and experimentation.

3.3.1 Material Procurement

In this initial stage, weak soils (KS and DS) and IS required for the study were procured. All material procurement and preparation were conducted in accordance with the relevant Indian Standards (IS: 2720). Simultaneously, all essential equipment were calibrated and set up to ensure reliable and repeatable results throughout the experimental program.

3.3.2 Sample Preparation for UCS

Upon collection, soil samples were air-dried, pulverized, and passed through a 2 mm IS sieve for consistency and ease of testing. For UCS testing, remoulded samples were prepared by mixing oven-dried soil with distilled water to achieve a moisture content which corresponds to 95% of the maximum dry unit weight. Homogeneous

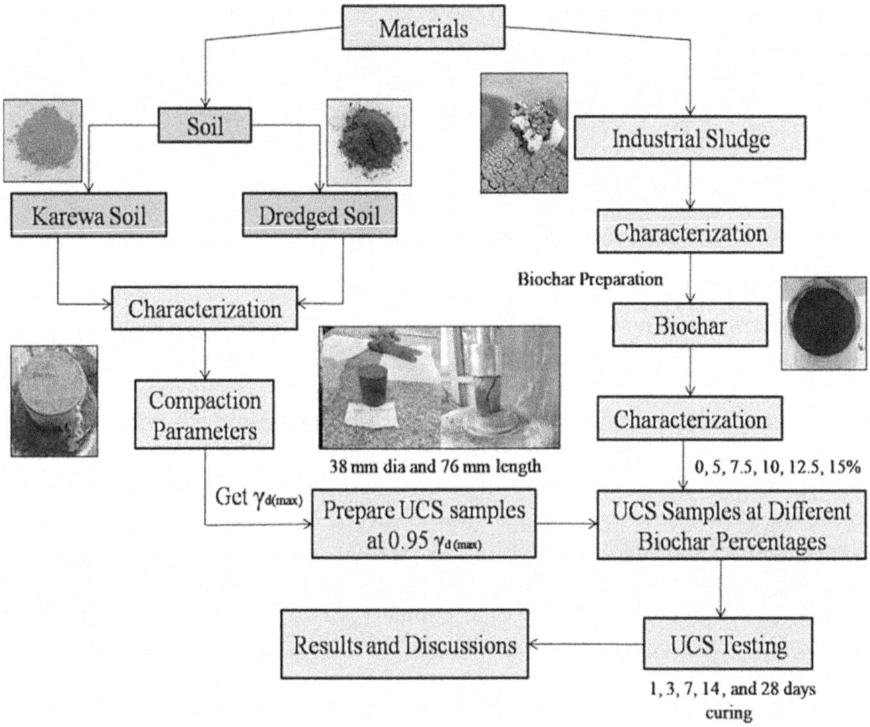

FIGURE 3.2 Methodology adopted in this study.

mixtures of soil and biochar were prepared by thoroughly blending the biochar with dry soil before water addition. These samples were then compacted in cylindrical UCS moulds (76 mm height × 38 mm diameter) in three equal layers to ensure uniformity. For KS, the samples were compacted at a dry unit weight of 15.96 kN/m^3 with a water content of 15.5%. For the Dredged soil, a dry unit weight of 12.73kN/m^3 and a water content of 25% were used. After compaction, all samples were sealed and placed in desiccators with a shallow layer of water at the base under controlled humidity conditions for the curing periods of 24 hours, 3 days, 7 days, 14 days, and 28 days. This equilibration ensured moisture uniformity and prevented drying prior to testing (Farooq et al., 2024b).

3.3.3 Experimentation

The experimental phase aimed to evaluate the influence of biochar dosage and curing period on the UCS and microstructural properties of the KS and DS. IS-derived biochar was incorporated into the KS and DS at dosages of 0%, 5%, 7.5%, 10%, 12.5%, and 15% by dry weight. Samples were subjected to five different curing durations: 1, 3, 7, 14, and 28 days. UCS tests were conducted in accordance with IS: 2720 (Part X) using a HEICO loading frame at a strain rate of 0.6 mm/min. Tests were continued until visible failure occurred or until the axial strain exceeded 12% without further increase in stress. Each test was repeated to ensure reproducibility and reliability. This methodology provided a structured approach to evaluating the effects of biochar with the focus on UCS at different curing periods.

3.4 RESULTS AND DISCUSSION

3.4.1 Effect of Biochar in Stress-Strain Behaviour of Soil

3.4.1.1 Karewa Soil

The stress-strain behaviour of KS amended with different biochar dosages under varying curing periods is shown in Figure 3.3(a–e). All samples exhibited a typical strain-softening response characterized by an initial linear rise in stress, followed by a distinct peak and gradual post-peak reduction. The untreated KS ($KS - B_0$) showed the lowest peak stress value and the most ductile behaviour with 7% strain at failure, reflecting its high deformability. With the addition of biochar, a noticeable improvement in the peak stress and stiffness was observed across all curing periods, indicating enhanced inter-particle bonding by biochar. Furthermore, the failure strain values were reduced up to 4% across all KS-biochar matrices. The strength gain was most pronounced for 7.5% biochar content ($KS - B_{7.5}$), which developed higher peak stresses and steeper initial stress-strain slopes, suggesting improved strength (Gonzalez et al., 2021). Beyond this dosage (7.5%), a slight decrease in peak stress was recorded for ($KS - B_{10}$) and ($KS - B_{12.5}$), likely due to excess biochar creating discontinuities and increased porosity within the soil matrix (Farooq et al., 2024a). Furthermore, the peak stress values increased progressively with curing time, reaching the maximum at 28 days, with no significant changes in the failure strain values. Overall, the inclusion of biochar transformed the stress-strain response of KS from ductile to semi-brittle, with 7.5% biochar identified as the optimum dosage yielding highest strength.

3.4.1.2 Dredged Soil

The stress-strain curves of DS amended with different biochar dosages under varying curing periods are illustrated in Figure 3.4(a–e). Each curve exhibits a strain-softening pattern, characterized by an initial linear rise in stress, a well-defined peak, and a subsequent gradual decline, as was the case with KS. The untreated DS ($DS\text{-}B_0$) demonstrated the lowest compressive strength and the most ductile response with 8% strain at failure, indicating weak particle cohesion and high deformability. With the inclusion of biochar, the stress-strain response of DS improved considerably, as evidenced by higher peak stresses, steeper initial slopes, and the strain values reducing up to 3% (Chen et al., 2020). Among all mixes, the specimen containing 12.5% biochar ($DS - B_{12.5}$) exhibited the highest peak stress at failure across all curing periods, indicating that this dosage offers the most effective soil-biochar bonding and strength enhancement. Beyond this optimum level, such as at 15% biochar content ($DS - B_{15}$), a slight reduction in peak stress was observed, likely due to the formation of discontinuities and increased porosity from excessive biochar, which impedes effective load transfer through the soil skeleton (Farooq et al., 2024a).

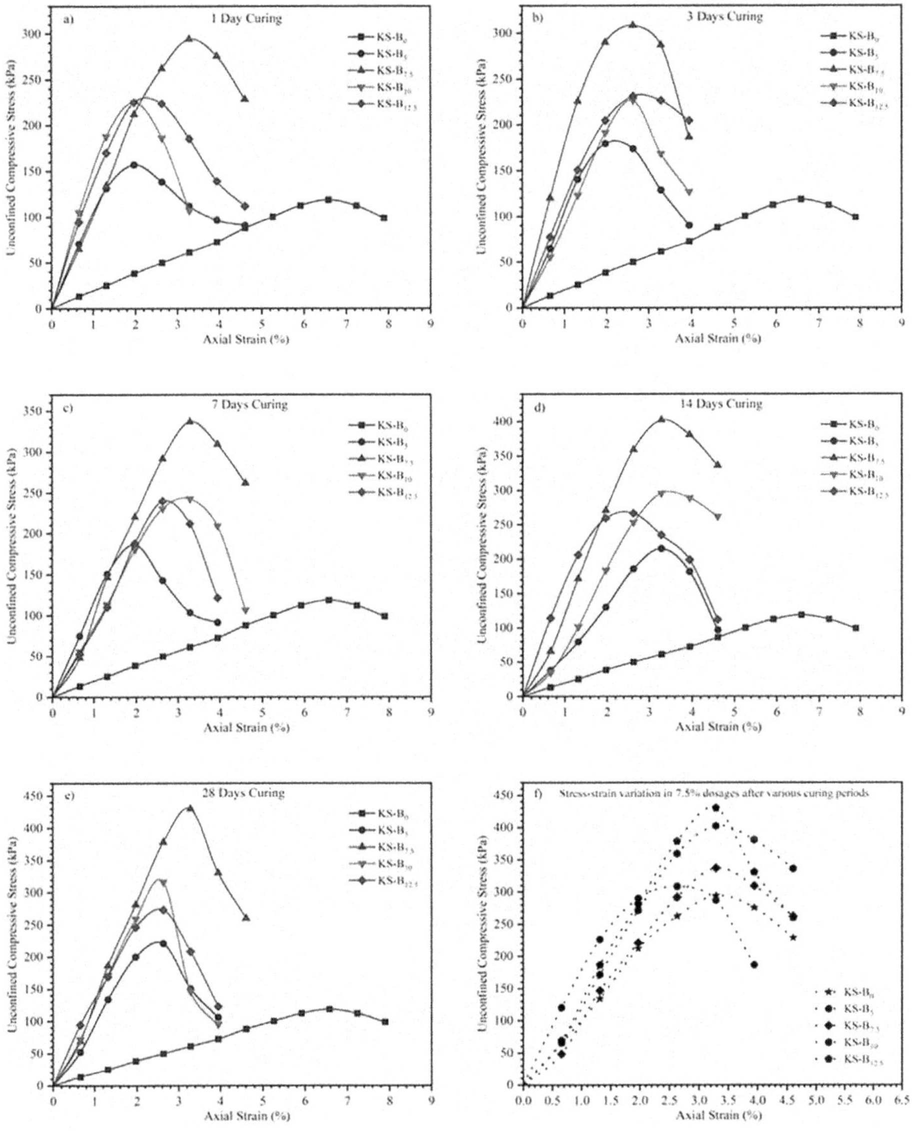

FIGURE 3.3 Stress-strain curves of biochar-amended KS over different curing periods and different biochar percentages. KS - B_0: KS with no biochar, KS - B_5: KS with 5% biochar, KS - $B_{7.5}$: KS with 7.5% biochar, KS - B_{10}: KS with 10% biochar, KS - $B_{12.5}$: KS with 12.5% biochar.

3.4.2 Effect of Biochar on UCS of Soil

The results of the UCS test revealed a significant improvement in the strength of both KS and DS upon biochar incorporation, as illustrated in Figure 3.5(a–d). The

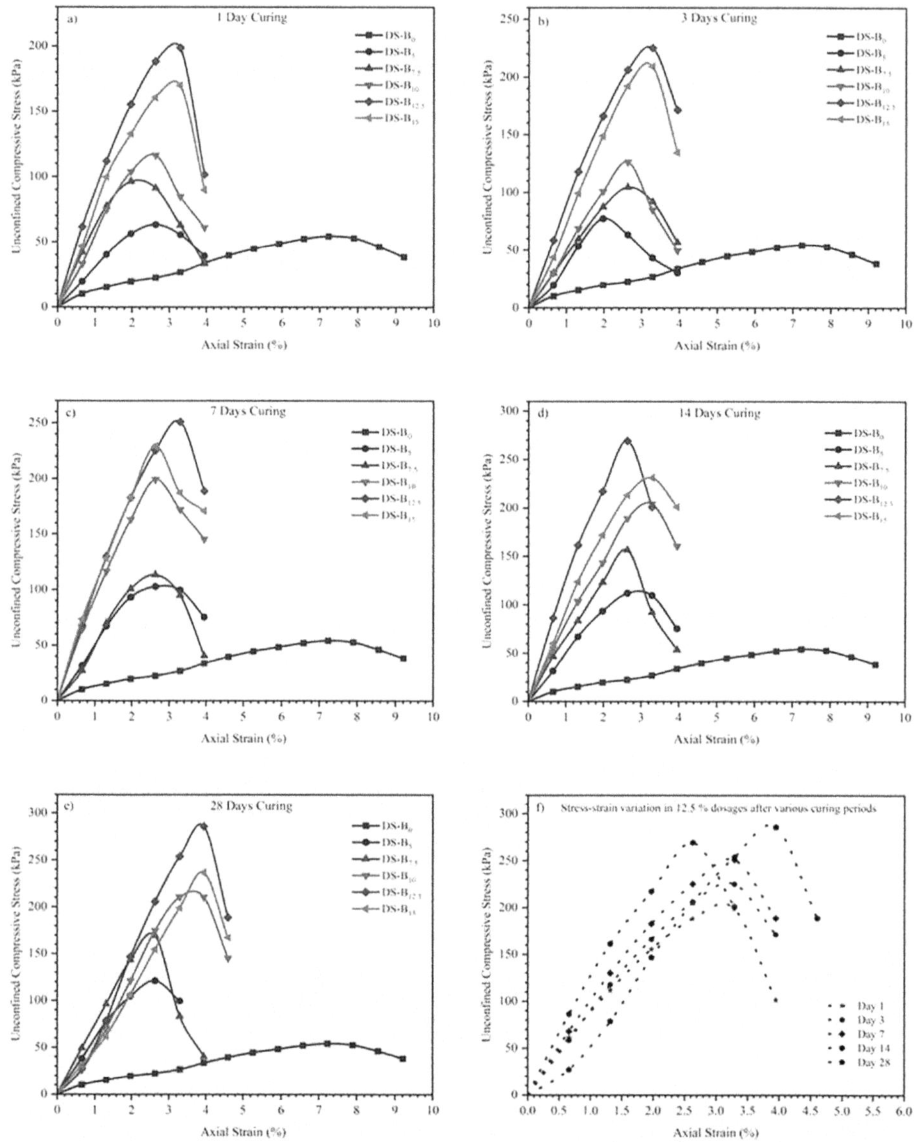

FIGURE 3.4 Stress-strain curves of biochar-amended DS over different curing periods and different biochar percentages. DS - B_0: DS with no biochar, DS - B_5: DS with 5% biochar, DS - $B_{7.5}$: DS with 7.5% biochar, DS - B_{10}: DS with 10% biochar, DS - $B_{12.5}$: DS with 12.5% biochar, and DS - B_{15}: DS with 15% biochar.

untreated KS and DS recorded UCS values of 54 kPa and 118 kPa, respectively. Upon biochar amendment, the maximum UCS achieved for KS was 429 kPa at a 7.5% dosage after 28 days of curing nearly a four times increase, while DS attained

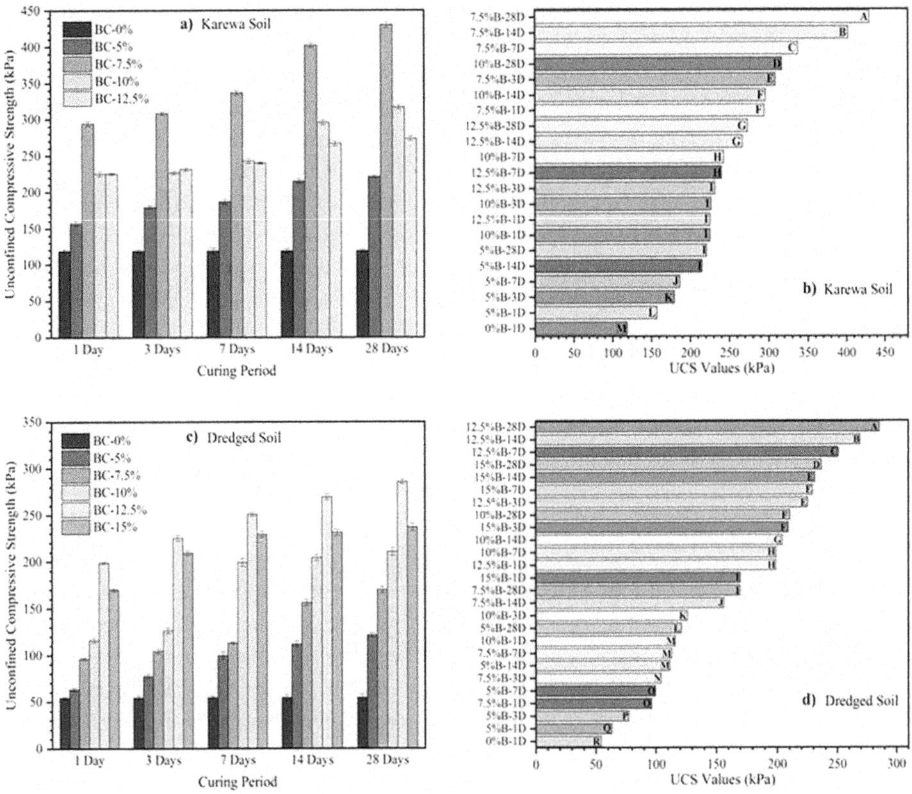

FIGURE 3.5 (a) UCS values of KS, (b) relationship between UCS values of KS samples, (c) UCS values of DS, and (d) relationship between UCS values of DS samples across all biochar percentages and curing periods.

a peak value of 285 kPa at 12.5% biochar content over the same curing period which was nearly a five times increase in the UCS. A progressive increase in strength was observed with extended curing durations, indicating that curing time plays a critical role in the development of strength in biochar-treated soils. The biochar dosage was initially varied between 5% and 12.5% for KS and up to 15% for DS. The optimum dosage for KS was determined to be 7.5%, beyond which a decline in strength was observed. Conversely, the optimum dosage for DS was found to be 12.5%, with strength decreasing beyond this level. This reduction in UCS may be attributed to the non-uniformity in the soil-biochar matrices due to an overdose of stabilizing agent and increase in pore water pressure as biochar retains the water as explained in the previous studies (Farooq et al., 2024a; Jyoti Bora et al., 2021; Reed et al. 2017; Sharma, 2017).

Despite these differences, both soils exhibited a consistent trend of strength enhancement with biochar addition. It was also learnt that the strength improvement was most pronounced during the early curing phase (one to seven days), followed by a gradual increase up to 14 days and a plateau thereafter. This trend suggests that major bond formation and structural rearrangement within the soil-biochar matrix primarily occur during the initial curing stage. By 28 days, the soils attained their maximum compressive strength, particularly for the optimal biochar mix, confirming that a 28-day curing period is adequate for achieving peak performance. Similar trend in the strength value was documented in the earlier studies.

The UCS data was further evaluated for standard error of mean (SEM) and coefficient of variation to confirm low dispersion within the dataset. Analysis of variance (ANOVA) was then performed to determine the statistical significance of the dosage and curing period. Following ANOVA, Tukey's post hoc test was applied to identify pairwise differences. Letter groupings assigned in Figures 3.5 (b and d) represent statistically homogeneous subsets, where UCS values sharing the same uppercase letter do not differ significantly at $p < 0.05$. Overall, the results demonstrate that biochar effectively enhances the UCS of both DS and KS, with the degree of improvement influenced by both biochar dosage and curing duration.

3.5 CONCLUSIONS

The study demonstrated that biochar can serve as an efficient and sustainable stabilizing agent for problematic soils such as Karewa and Dredged soils. Incorporation of biochar significantly enhanced the UCS of both soils. Optimal strength improvements were achieved at 7.5% biochar for KS and 12.5% for DS, with further additions leading to marginal strength reductions due to non-uniformity of the soil-biochar matrices. Curing duration played a crucial role in strength development, with major bonding and structural rearrangement occurring within the first 7 days and peak strength achieved at 28 days. Overall, biochar not only improved the geotechnical performance of soils but also offered an environmentally sustainable alternative to conventional stabilizers, contributing to carbon reduction and promoting circular waste utilization in construction applications.

REFERENCES

Bhatt, D.K., 1982. A review of the stratigraphy of the Karewa Group (Pliocene/Quaternary), Kashmir. *Man and Environment*, *6*(4).

Campo, F.P., Tua, C., Biganzoli, L., Pantini, S. and Grosso, M., 2021. Natural and enhanced carbonation of lime in its different applications: A review. *Environmental Technology Reviews*, *10*(1), pp.224–237.

Chandra, R., Ahmad, I. and Qurashi, A., 2016. Pedological and geochemical characterization of loess-paleosol sediments of Karewa basin: Implications for paleoclimatic reconstruction of Kashmir valley. *J. Geol. Soc. India*, *4*, pp.38–54.

Chen, R., Zhu, Y., Lai, H.P. and Bao, W., 2020. Stabilization of soft soil using low-carbon alkali-activated binder. *Environmental Earth Sciences*, *79*, pp.1–13.

Dinga, C.D. and Wen, Z., 2022. China's green deal: Can China's cement industry achieve carbon neutral emissions by 2060? *Renewable and Sustainable Energy Reviews*, *155*, p.111931.

Farooq, M., Shah, M.Y. and Gani, K.M., 2024a. Soft soil stabilization using co-pyrolytic biochars from sewage sludge and Nymphaeale, an invasive plant biomass. *Journal of Cleaner Production*, *480*, p.143981.

Farooq, M., Shah, M.Y. and Gani, K.M., 2024b. Utilizing industrial wastewater sludge-derived biochar for enhancing strength and microstructure of soft soil-An infrastructure application of wastewater sludge. *Journal of Environmental Management*, *370*, p.122577.

Gonzalez, J., Sargent, P. and Ennis, C., 2021. Sewage treatment sludge biochar activated blast furnace slag as a low carbon binder for soft soil stabilisation. *Journal of Cleaner Production*, *311*(December 2020), p.127553. https://doi.org/10.1016/j.jclepro.2021.127553.

IS: 2720, 1983. Methods of test for soils. *Bureau of Indian Standards*, Reaffirmed (2006).

IS: 2720, Part X:1991. Methods of test for soils, determination of unconfined compressive strength. *Bureau of Indian Standards*, (Reaffirmed 2006), pp.1–6.

Jalal, F.E., Xu, Y., Jamhiri, B. and Memon, S.A., 2020. On the recent trends in expansive soil stabilization using calcium-based stabilizer materials (CSMs): A comprehensive review. *Advances in Materials Science and Engineering*, *2020*(1), p.1510969.

Jan, O.Q. and Raj, S., 2018, October. Strength behavior of cement stabilized karewa soil. In *International Conference on Sustainable Waste Management through Design* (pp.273–284). Cham: Springer International Publishing.

Jyoti Bora, M., Bordoloi, S., Kumar, H., Gogoi, N., Zhu, H.H., Sarmah, A.K., Sreeja, P., Sreedeep, S. and Mei, G., 2021. Influence of biochar from animal and plant origin on the compressive strength characteristics of degraded landfill surface soils. *International Journal of Damage Mechanics*, *30*(4), pp.484–501.

Mirza, M.I. and Mir, B.A., 2025. Synergistic effect of cement stabilization and glass fiber reinforcement on the mechanical and durability characteristics of Karewa soil: MI Mirza, BA Mir. *Indian Geotechnical Journal*, pp.1–22.

Reed, E.Y., Chadwick, D.R., Hill, P.W. and Jones, D.L., 2017. Critical comparison of the impact of biochar and wood ash on soil organic matter cycling and grassland productivity. *Soil Biology and Biochemistry*, *110*, pp.134–142.

Sadasivam, B.Y. and Reddy, K.R., 2015. Adsorption and transport of methane in landfill cover soil amended with waste-wood biochars. *Journal of Environmental Management*, *158*, pp.11–23.

Sharma, R.K., 2017. Laboratory study on stabilization of clayey soil with cement kiln dust and fiber. *Geotechnical and Geological Engineering*, *35*, pp.2291–2302.

Simoni, M., Wilkes, M.D., Brown, S., Provis, J.L., Kinoshita, H. and Hanein, T., 2022. Decarbonising the lime industry: State-of-the-art. *Renewable and Sustainable Energy Reviews*, *168*, p.112765.

Vuppaladadiyam, A.K., Vuppaladadiyam, S.S.V., Sahoo, A., Murugavelh, S., Anthony, E., Bhaskar, T., Zheng, Y., Zhao, M., Duan, H., Zhao, Y. and Antunes, E., 2023. Bio-oil and biochar from the pyrolytic conversion of biomass: A current and future perspective on the trade-off between economic, environmental, and technical indicators. *Science of the Total Environment*, *857*, p.159155.

Wani, I., Kushvaha, V., Garg, A., Kumar, R., Naik, S. and Sharma, P., 2022. Review on effect of biochar on soil strength: Towards exploring usage of biochar in geo-engineering infrastructure. *Biomass Conversion and Biorefinery*, pp.1–32.

Wani, K.S. and Mir, B.A., 2020. Stabilization of weak dredged soils by employing waste boulder crusher dust: A laboratory study. *Geotechnical and Geological Engineering*, *38*(6), pp.6827–6842.

Wani, K.S. and Mir, B.A., 2021. Influence of microbial geo-technology in the stabilization of dredged soils. *International Journal of Geotechnical Engineering*, *15*(2), pp.235–244.

4 Self-Compacting Concrete for Sustainable Infrastructure

A Smart Strategy for Durable and Eco-Efficient Construction

Sivamalar Shijumon and V. Shijumon

4.1 INTRODUCTION

The 21st century has been characterized by rapid infrastructure construction, largely at the peripheries of emerging market economies, driving development and, simultaneously, producing environmental issues like the release of CO_2 and the degradation of natural resources. By itself, the construction sector accounts for near about 40% of global emissions (IEA, 2022), largely due to the high volume usage of classical concrete and energy-intensive applications. Self-Compacting Concrete (SCC), launched in Japan in the 1980s to tackle workman shortage and workmanship variation (Hajime & Masahiro, 2003), revolutionized concrete technology with self-compactibility through self-weight flow. With fine mixes, Polycarboxylate Ether (PCE)-based superplasticizers, and viscosity-modifying admixtures, SCC achieves quicker constructions, reduces labour and noise, and achieves superior surface finish, particularly in complex reinforcement structures. SCC also promotes sustainability through Supplementary Cementitious Materials (SCM) applications such as fly ash, GGBS, and silica fume in curbing the usage of cement and the consequent carbon imprint (Gambhir, 2013). Beyond sustainability, SCC boosts resilience through better compaction, lower voids, and better durability, and it also uses fibres and fast-setting admixtures to suit advanced applications such as prestressed elements. Uniting with global benchmarking in sustainability and resilience (e.g. ISO 21929, ISO 14040), SCC takes centre stage in low-carbon, durable, and efficient infrastructure.

4.2 MATERIAL INNOVATION IN SCC

SCC stands out with not only its unique rheology but also its ability to accommodate sustainable material innovation. SCC mix design innovations enable reduced

 DOI: 10.1201/9781003664284-4

environmental footprints, increased durability, and performance-based applications in construction.

4.2.1 Cementitious Materials and SCMs

The self-compacting character, strength, and durability of SCC are governed by the cementitious matrix. Although the Ordinary Portland Cement (OPC) binder is still predominant, its high production energy input and high CO_2 emissions have motivated SCM usage. The pozzolanic/latent hydraulic SCM materials generate supplementary C-S-H, fine-tuning microstructure towards enhanced behaviour. SCMs have differential influences upon SCC properties: fly ash offers high workability and ultimate strength but slow early strength; GGBS offers durability by reducing permeability but extending set; silica fume provides compact microstructure and very high early strength but is very costly and inconvenient; metakaolin offers high early strength and flexural strength but poor workability; rice husk ash's renewable by-product status amplifies impermeability, but its reactivity is conditioned by burning regime (Mehta & Monteiro, 2006; Siddique & Singh, 2011). In application, SCMs can replace 30–70% of OPC with serviceability retention. Binary mixes of 50–60% GGBS provide approximately 40% chloride permeability reduction whilst maintaining compressive strength above 50 MPa; ternaries outperform binaries by frequently providing 20–25% higher durability indexes (Danish & Mohan, 2020a). However, very high SCM replacement ratios may slow hydration and slow early strength such that SCM usage is limited to precast/prefabrication/make-fast construction works.

Properties and benefits of some common SCMs of SCC like fly ash, GGBS, and rice husk ash are explained in Table 4.1. They all contribute to strength, durability, and impermeability and establish sustainability through the optimization of the usage of cement, recycling of wastes, and reduction in the release of carbon.

TABLE 4.1
Properties and Benefits of Common Scms in Scc

SCM type	Source	Pozzolanic activity	Effect on strength	Sustainability benefit
Fly Ash (Class F)	Coal-fired power plants	Moderate	Improves long-term strength	Utilizes industrial waste, reduces cement content
GGBS	Blast furnace slag	High	Enhances durability	Lowers CO_2 emissions, improves workability
Silica Fume	Silicon metal/alloy production	Very high	Boosts early strength	Dense microstructure, improves impermeability
Metakaolin	Calcined kaolin clay	High	Enhances early/flexural strength	Enhances durability, boosts particle packing
Rice husk ash	Agricultural by-product	Moderate to high	Improves pozzolanic reaction	Renewable, low-carbon alternative

4.2.2 Aggregates and Fines

Aggregates that make up between 60% and 75% of the volume of SCC significantly influence workability, stability, and strength. Well-graded fine aggregates of size 10–20 mm improve deformability and inhibit segregation (EFNARC, 2002). Optimum proportion of spherical and angular particles optimize optimal strength and optimal flowability with aggregate-to-paste ratios perfectly optimized for cohesive maintenance (Khayat, 1999). Fine admixtures such as limestone powder and quarry dust improve packing density and stability but at the disadvantage of greater water demand, while nano-silica significantly improves strength and durability but at higher costs. Performance comparisons have shown that optimization of gradation with fine addition reduces segregation minimally but optimizes passing capabilities, but higher fines generate higher shrinkage with poor long-duration workability. Aggregate and fine judicious selection and proportioning are thus necessary for stability, sustainability, and flowability of SCC balance.

Figure 4.1 presents the flowchart of the aggregate and fines optimization strategy adopted in SCC to achieve improved grading, enhanced workability, and increased durability.

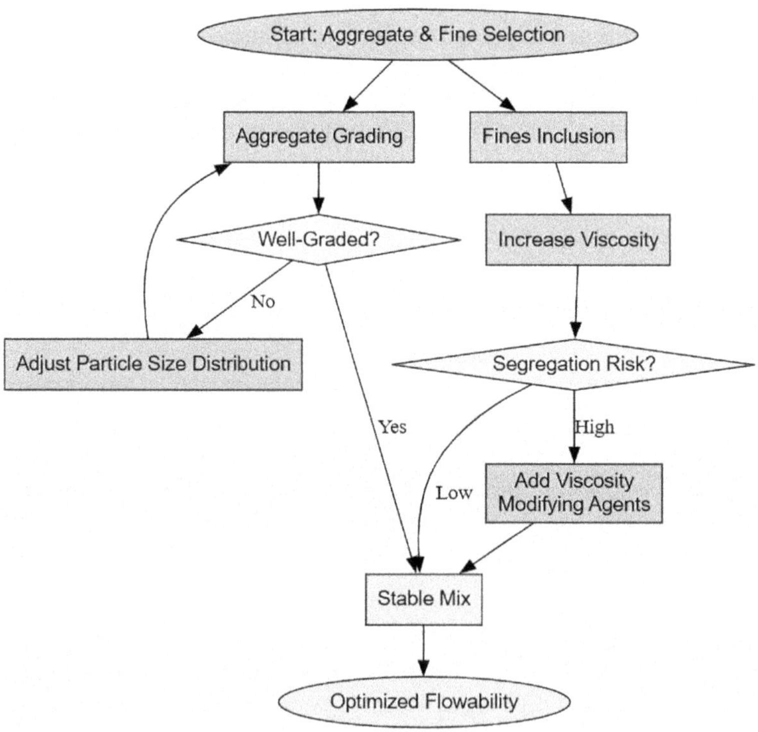

FIGURE 4.1 Flowchart of aggregate and fines optimization strategy in SCC.

4.2.3 Chemical Admixtures

Workability of SCC greatly depends on chemical admixtures that must also balance stability with flowability while not losing strength or durability. High-Range Water-Reducing Admixtures (HRWRAs), that is, PCE-based superplasticizers, present high fluidity at low dosages of water, low slump loss and bleeding, and robust mixes at low water-to-binder ratios (Khayat, 1999). Their effectiveness is dosage sensitive, however, with over-dosing causing segregation or retardation of set. Viscosity-Modifying Agents (VMAs), such as hydroxypropyl methylcellulosic (HPMC) and xanthan gum, stall high-fluid mixes of low fines of high stability (Assaad et al., 2003). Although good at preventing segregation, they may reduce flowability and increase mix cost if over-applied. PCE–VMA admixture blends are particularly useful for complex geometries and vertical placements but require strict quality control. Recent nano- and multifunctional admixtures accelerate hydration and early strength, widening the range of applicability of SCC at harsh exposures, although high cost and lack of standardized guidelines are concerns. In short, chemical admixtures are essential for SCC, but their application demands judicious proportioning and cost-performance analysis.

Figure 4.2 illustrates the chemical admixtures commonly used in SCC, including PCE-based superplasticizers, VMAs, and nano-additives in both powder and liquid

FIGURE 4.2 Chemical admixtures for self-compacting concrete.

forms. These admixtures enhance flowability, stability, and strength while maintaining a low water content, thereby improving the workability and performance of SCC in various applications.

4.2.4 Incorporation of Fibres and Hybrid Additives

Fibre addition strengthens SCC by improving ductility, toughness after crack formation, and resistance against seismic and dynamic load cracking. Fibre-reinforced SCC maintains flowability while providing toughness for application in industrial floors, bridge decks, and long-duration construction. Steel fibres provide high energy dissipation but lower tensile strength with lower flowability requiring precise adjustment of the superplasticizer dosage. Polypropylene (PP), PVA, and basalt fibres provide improved crack control and resistance against corrosives, but synthetic fibres tend to provide lower structural load contribution. Recent studies confirm that basalt fibre addition improves flexural strength and durability indices while providing an environmentally safe steel alternative for selected applications (Kiran Prabha et al., 2025). Broader reviews also document that fibre-admixture systems that include hybrid systems hold promise of significantly improving toughness with coupled crack resistance while balancing workability demands (Athiyamaan & Ganesh, 2017). Hybrid fibre systems of micro- and macro-fibres provide a balance of coupled crack resistance with ductility, while natural fibres such as coir, sisal, and jute provide potential application for low-load environmentally safe uses despite varying durability. Typically, fibre addition improves service life extension and optimizes SCC performance in adverse circumstances but successfully does so by the selection of type, content, and optimum mix design.

4.2.5 Integration of Waste-Derived Materials

The search for sustainability has driven the use of waste from agriculture and industry in SCC, both to decrease carbon footprints and because of compliance with principles of circular economies. Waste glass has been shown capable of improving particle packing and aesthetic appearance, but high content could add towards lower durability over the longer term due to alkali–silica reactions (Shayan & Xu, 2006). Industry wastes such as bottom ash, foundry sand, ceramic wastes, and copper slag could partially replace natural sand, improving strength while reducing reliance on river sand use but at times with unacceptable variation of particle size that prohibits large-scale application. Pozzolanic ashes such as palm oil fuel ash, bagasse ash, and rice husk ash are extremely popular due to their reactivity and potential towards increasing durability (Safiuddin & Jumaat, 2011). However, their effectiveness is greatly reliant on combustion protocol and end-product fineness that could vary from source to source. Recent technologies such as microbially induced calcite precipitation (MICP) and CO_2-reactive binds reduce embodied carbon very considerably and permit certification by IGBC, BREEAM, and LEED. These technologies are still at pilot-scale, and problems of higher costs, scale of application, and longer term effectiveness need resolution before widespread application.

4.2.6 Environmental and Performance Assessment

Sustainability and sustainability performance are central pillars of SCC's sustainable infrastructure contribution. Life Cycle Assessment (LCA) has proved that high SCM content mixes reduce Global Warming Potential (GWP) by 30–50% compared with conventional concrete (Witte & Garg, 2024). The elimination of mechanical vibration also reduces energy requirement and noise, giving credits in LEED and BREEAM assessment processes. However, the environmental gain will be compromised if SCMs must be transported over extremely long distances or if mix optimization is of inferior quality, consequently mandating exposure-specific evaluations. Performance-wise, SCC's compact matrix and low permeability present effective resistance against chloride ingress, carbonation, and freeze–thaw damage. Recycling and nano-materials hold promise of enhancing such properties, but their end-of-life characterization is unknown and requires verification by means of non-destructive testing and accelerated durability testing. Available data verify that SCC holds promise of being part of resilient low-carbon infrastructure, but standardized assessment processes continue to be needed for the assurance of consistent effects across a range of exposure situations.

4.3 DESIGN STRATEGIES FOR SCC MIX

SCC proportioning is more complex than traditional concrete with a requirement to balance flowability, passing ability, anti-segregating ability, and strength carefully through the use of low water-to-binder proportions (0.30–0.45), suitable paste volume, and efficient admixtures (EFNARC, 2002). Performance is defined with regular tests such as slump flow, T50, V-funnel, L-box, and J-ring according to EFNARC and ASTM C1611 specifications.

4.3.1 Empirical and Rational Design Approaches

Design approaches of SCC mixtures are broadly categorized as empirical and rational. Empirical methods such as the Okamura method rely on trial mixes and fixed aggregate ratios; while greatly practised, they become increasingly rigid with newer materials and are not very accurate. Rational methods involve particle packing theories and optimization procedures sometimes integrated with artificial intelligence that make SCC by design possible for sustainability, economy, and high performance. While such approaches require voluminous data inputs and computing effort, their practical application is limited. A balanced perspective is that empirical methods will still be useful for routine, but that bigger potential for sustainable and high-performance SCC is with the application of rational approaches.

As shown in Figure 4.3, the empirical approach to SCC mix design is primarily based on experimental trials and practical observations, while the rational approach incorporates modelling and simulation techniques to achieve greater flexibility, accuracy, and performance optimization.

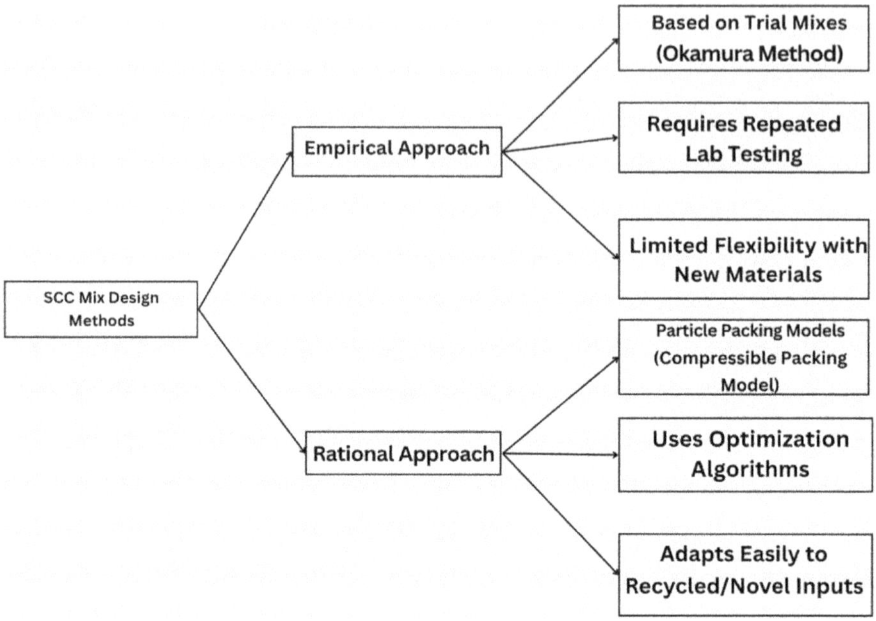

FIGURE 4.3 Comparison of empirical and rational approaches in SCC mix design.

4.3.2 Adaptability to Precast and On-Site Applications

SCC is appropriate for use both in precast work and in-situ construction but demands varying mix design strategies. Precast production demands early strength development and surface finishing which are achieved by low water-to-binder ratios, optimized gradings, and dosing with retarders or slow-release PCEs for minimum slump loss and early demoulding. Such mixes are dependable if tightly controlled factory environment exists but are more costly and very dose-ratio sensitive. Work on site stresses longer workability, pumpability, and accommodation of varying quality of formwork and varying aggregate quality by tight dose-ratio control of dosing of admixture as well as moisture corrections (Khayat, 1999). Table 4.2 shows that precast work stresses early strength development, quality assurance, and stable production but which uses on-site centres on the provision of longer workability accommodation of diverse site circumstances. Precast usually offers high quality in strictly controlled environments but suffers from high variabilities of environment as well as quality assurance at use.

4.4 FRESH PROPERTIES AND TESTING METHODS

The fresh performance of SCC is governed by flowability, passing capability, and resistance to segregation such that they could be placed in heavily congested reinforcements without vibration. Flowability, conventionally assessed by slump flow (650–750 mm) and T50 time (2–5 s), provides a quick measure of spread and viscosity

TABLE 4.2
Scc Mix Characteristics – Precast Versus on-Site

Criteria	Precast	On-site
Workability retention	Short (due to quick placement)	Long (to accommodate delays in placement)
Strength development	Accelerated using additives	Moderate, adjusted for ambient conditions
Admixture use	High-range water-reducers and accelerators	Retarders and extended-release admixtures
Aggregate grading	Highly optimized, uniform supply	Variable; needs batch adjustments
Formwork requirements	Tight tolerance moulds for dimensional accuracy	Varies; formwork may be irregular or congested
Quality control	Factory-controlled, consistent	Field-dependent, requires active monitoring

but is not very effective for segregations. Companion tests such as the V-funnel (6–12 s) and T5-minute tests assess the viscosity and time-dependent stability of SCC, although their results are mix-dependent. Passing ability is evaluated using the L-box (blocking ratio ≥ 0.80), J-ring (≤ 50 mm variation), and U-box tests, which simulate reinforcement congestion and are sensitive to aggregate grading and mix composition. Segregation resistance is normally carried out by the Visual Stability Index (VSI), whereby scores of 0–1 indicate stable mixes, but the procedure is subjective. Additional verification of fresh density, air content, and temperature offers broad quality control. Arguing that no test by itself is sufficient and holistic, SCC testing entails an ensemble of tests that compensate for practicality over that of accuracy and reliability. Independent tests on deformability (slump flow, T50), passing capability (L-box, J-ring), viscosity (V-funnel), and stability (VSI) are tabulated in Table 4.3 for the purpose of veritable quality manipulation and placement.

4.5 HARDENED PROPERTIES AND DURABILITY

Hardened properties of SCC are responsible for its structural adequacy and durability. Compressive strength is also likely to be equivalent to or greater than that of conventionally vibrated concrete due to superior particle packing, reduced voids, and greater homogeneity. Structural-quality SCC is likely to develop compressive strengths of around 30–80 MPa depending on mix proportioning as well as curing regimes (Domone, 2006). Apart from strength, compact microstructure of SCC reduces honeycombing, segregation, and permeability improving resistance towards chloride penetration as well as freeze–thaw effects. Behaviour of SCC is, however, very sensitive towards curing and is likely to be more sensitive towards shrinkage or cracking if not proportioned adequately. A comparative study indicates that while SCC has certain strengths pertaining towards homogeneity as well as durability, consistent hardened properties are only achievable with judicious mix proportioning as well as quality control.

TABLE 4.3
Fresh State Tests and Acceptable Ranges for Scc

Test	Parameter measured	Typical value range	Significance
Slump flow	Flowability	550–850 mm	Indicates SCC's ability to spread under its own weight without obstruction
T50 time	Viscosity (flow speed)	2–5 seconds	Lower values mean faster flow; higher indicates more viscous mix
V-funnel	Viscosity and flow time	6–12 seconds	Measures resistance to flow and segregation risk
L-box	Passing ability	≥0.80 (H_2/H_1 ratio)	Assesses ability to pass through tight reinforcement without blockage
J-ring	Passing ability with obstructions	<50 mm flow difference	Detects aggregate blocking during flow
V-funnel (T5min)	Segregation resistance over time	≤ +3 seconds change	Checks stability after 5-minute delay; higher time = possible segregation
VSI (Visual Stability Index)	Stability grade	0–1 preferred	Evaluates visual signs of segregation (0 = very stable, 3 = high segregation)

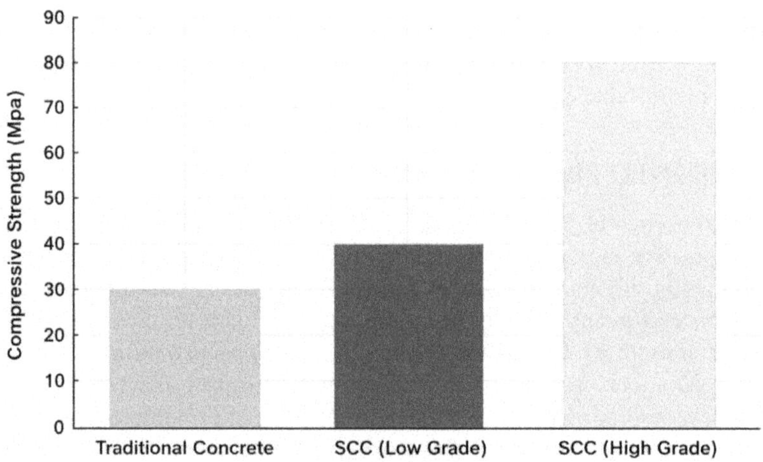

FIGURE 4.4 Compressive strength range of SCC compared to traditional concrete.

Figure 4.4 compares the compressive strength range of self-compacting concrete (SCC) with that of traditional concrete. The figure illustrates that even low-grade SCC can achieve higher compressive strength than conventional concrete, while high-grade SCC can attain strengths of up to 80 MPa for specialized applications. SCC also demonstrates improved flexural strength; however, its modulus of elasticity typically ranges from 25 to 35 GPa and may decrease with increasing paste content unless fibres or hard aggregates are incorporated to enhance stiffness.

4.5.1 Durability Aspects of SCC

SCC proves to be more durable than ordinary concrete, basically because of its dense microstructure, low permeability, and narrow interfacial transition zone. These properties endow greater resistance towards chloride penetration, carbonation, freeze–thaw effects, as well as sulphate attack, enhancing service life as well as durability of the embedded steel. Applications of SCMs such as fly ash, GGBS, and silica fume also lower chloride permeability (ASTM C1202) and enhance resistance against sulphate attack, while suitable curing and air entrapment enhance resistance against creep as well as shrinkage. Recent publications related to ternary mineral powder blends reaffirm that SCC is capable of reaching up to around 35–40% lower chloride permeability as well as much higher resistance against sulphate attack than binary mixes (Danish & Mohan, 2020b). However, durability benefits of SCC still remain extremely mix-design- as well as curing-practice-dependent, whereby sub-standard quality control is able to nullify its potential. By comparison, SCC is capable of outperforming conventional concrete even at harsh exposures but requires judicious selection of materials as well as site control.

4.6 SUSTAINABILITY ASPECTS

SCC offers principal sustainability advantages over conventional concrete by damping mechanical vibration, lowering energy requirement, CO_2 emissions, noise, and dust hazards (Khayat, 1999). Its superior flow and appearance also reduce material waste and rework, particularly in vertical and precast applications. Replacement of SCMs such as fly ash, GGBS, silica fume, rice husk ash, and metakaolin for 50–70% of OPC minimizes embodied carbon at a time when cement production is responsible for about 8% of the world's CO_2 emissions (Andrew, 2019). LCAs confirm that high SCM content SCC delivers a 30–50% reduction of GWP over conventional concrete. Those conclusions assume local SCM availability, transport energy, and optimization of cost against performance trade-offs. In general terms, SCC has good potential for low-carbon, resilient infrastructure, but its sustainability advantages are best attained only with optimum mix design and place-specifying assessment.

4.7 APPLICATIONS IN INFRASTRUCTURE PROJECTS

SCC has become a key solution to modern-day infrastructure, with enhanced constructability, durability, and sustainability in complicated, high-performance applications.

4.7.1 SCC in Cast-In-Place Infrastructure

SCC has also been successful in cast-in-place uses like bridges, tunnels, retaining walls, and high-rise buildings, wherein the congestion of the reinforcement prohibits the use of vibration. An interesting case is that of the Akashi Kaikyo Bridge in Japan, wherein SCC allowed consistent compaction of densely reinforced anchorages (Hajime & Masahiro, 2003). Because of its flowing character without vibration, coupled with thixotropic stability, SCC is of use for vertical structural members like bridge piers and tower columns, particularly in seismically active areas. Usage in metro systems and airports from Asia and the Middle East also shows its benefit of minimum noise, efficiency, and smooth finishes. Though its use is successful for such projects depending upon mix design care and rigorous quality control, higher material costs and dosage sensitivity will act as limitations towards its use on a very large scale.

4.7.2 SCC in Precast and Modular Systems

SCC increases precast production tremendously by enhancing dimensional consistency, surface quality, and mechanization. It is widely adopted for bridge sections, tunnel linings, facade panels, and modular components where fast mould changeover and low vibration are imperative. For example Delhi Metro Rail Corporation adopted SCC for precast tunnel linings for faster construction and higher quality at tight reinforcement areas (Pierson et al., 2022). Its self-filling abilities also enable automation and BIM-based modular construction integration with robot placement at lower labour requirement. These are however subject to correct mix design, intense curing, and quality control because higher admixture cost and mix variation sensitiveness tend to inhibit large-scale usage.

4.8 RECENT ADVANCES AND INNOVATIONS

SCC surpassed optimizing flowability to fulfilling intelligent, durable, and sustainable infrastructure requirements. New trends with nanomaterials, fibres, self-healing admixtures, and digital technologies streamlined SCC's functionality and environmental responsibility. With building headed in the direction of automation and low-carbon building, SCC is a multi-facet platform for joining multifunctional material and speeded-up fabricating methods with latest system structures.

4.8.1 Nano-Engineered SCC and Functional Additives

Recent advancements of SCC aim at utilizing nanomaterials and functional additives for improving both sustainability and workability. Nanomaterials such as nano-silica, nano-alumina, and carbon nanotubes improve packing density, strength, and microstructural refinement. Nano-silica is specifically advantageous at improving pozzolanic activity and early-force strength while maintaining flowability suitable for precast and high-rise construction. Self-healing materials such as micro-encapsulated polymers and bacteria-based agents contribute service life by sealing

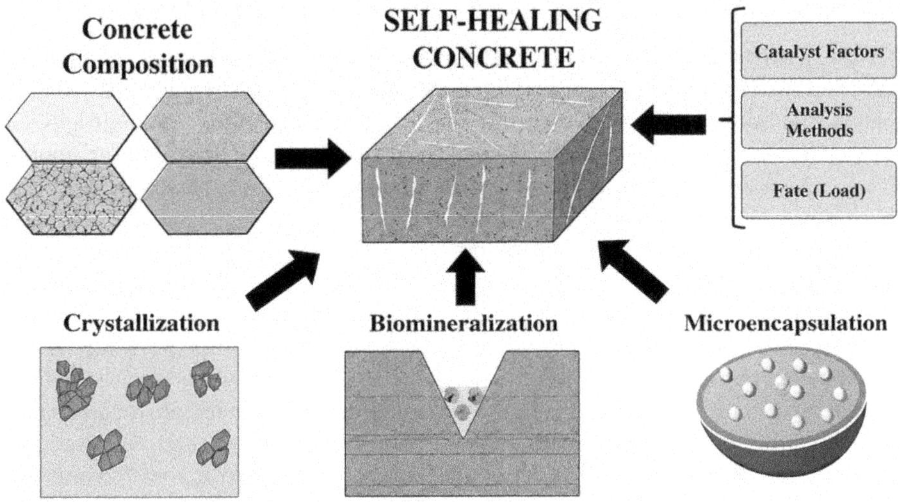

FIGURE 4.5 Mechanisms and components of self-healing concrete.

cracks while reducing chloride and sulphate penetration (Wiktor & Jonkers, 2011). Phase-change materials are also studied for thermal stress control of mass concrete. Although those technologies extend the vision of SCC as a multifunctional material with sustainability and durability benefits, their large-scale use is constrained by high costs, lack of standard guidelines, and lack of evidence of sustained field performance.

Figure 4.5 illustrates the key mechanisms and components of self-healing concrete. The figure highlights the self-repair processes of crystallization, biomineralization, and microencapsulation, which enable the automatic sealing of cracks and contribute to enhanced durability, reduced maintenance requirements, and improved long-term structural performance.

4.8.2 Digitalization, 3D Printing, and AI-Based Optimization

Integration of SCC with digital technologies is revolutionizing the field of construction, particularly with 3D concrete printing (3DCP) and mix optimization by AI. Pumpability and flowability of SCC suit its applicability with robot placement, wherein printable concretes are designed for thixotropy, buildability, and open time for elimination of formwork as well as minimization of wastes (Bos et al., 2016). Development continues towards overcoming the limitations of interlayer bonding, shrinkage, and dimensional stability that deter large-scale structural use. AI-based technologies enhance mix optimization by predicting SCC ratios with lower trial-and-error and, when integrated with BIM, enable real-time quality and sustainability monitoring. Though such technologies hold promise of automation as well as low-carbon benefit, practical deployment gets constrained by high costs, lack of standardized processes, as well as the need of verification by site conditions.

4.9 CHALLENGES, LIMITATIONS, AND FUTURE OUTLOOK

SCC, with all its positives, still faces practical limitations to wider usage. High chemical admixtures such as polycarboxylate ether-based superplasticizers and agents to control viscosity elevate up-front costs appreciably, particularly in small applications. SCC's intrinsic susceptibility to variability in aggregate and moisture content dictates exacting quality control and skilled labour, which are typically in short supply in situ. Lack of standardized mix design codes means reliance on empirical rules of thumb and trial mixes, making the process long winding. SCC's greater volume of paste generates more shrinkage and reduced workability in the work, particularly with high temperatures or with the incorporation of (recycled content and fibres that require precise proportioning). SCC, however, holds bright prospects. Material science breakthroughs such as nano-enriched admixtures, bio-based additives, and carbon-negative cements are improving its functionality and green image. Incorporation with technologies such as 3DCP, automated quality monitoring, and AI-based mix design is making SCC more aligned with future trends in construction. Towards net-zero carbons and resilient infrastructure, SCC increasingly presents itself as a base material for future-proof and digitized strategies in construction.

4.10 CONCLUSION

SCC is a technical breakthrough that embodies a distinctive blend of workability, inherent potential for structure, and environmental responsibility. Through the elimination of the need for vibration through the self-compacting feature, overcoming of surface finish, and maximizing durability, SCC stands uniquely equipped to face the challenge of the modern era of infrastructure. Through SCMs, use of wastes, nanotechnology, and information technology, SCC employs low-carbon construction, optimization of lifecycle costs, and uniformity in performance. Even though cost, material non-differentiation, and standardization are deterrents, innovation and greater awareness are increasingly succeeding in overcoming barriers. SCC will thus be a constructor of long-lasting, affordable, and environmentally safe infrastructure of the coming decades.

REFERENCES

Andrew, R. M. (2019). Global CO_2 emissions from cement production, 1928–2018. *Earth System Science Data, 11*(4), 1675–1710.

Assaad, J., Khayat, K. H., & Mesbah, H. (2003). Assessment of thixotropy of flowable and self-consolidating concrete. *Materials Journal, 100*(2), 99–107.

ASTM C1202–22. (2022). *Standard test method for electrical indication of concrete's ability to resist chloride ion penetration.* West Conshohocken, PA: ASTM International. https://doi.org/10.1520/C1202-22

ASTM International. (2021). *ASTM C1611/C1611M–21: Standard test method for slump flow of self-consolidating concrete.* West Conshohocken, PA: ASTM International. https://doi.org/10.1520/C1611_C1611M-21

Athiyamaan, V., & Ganesh, G. M. (2017, November). Statistical and detailed analysis on fiber reinforced self-compacting concrete containing admixtures-A state of art of review. In

IOP conference series: Materials science and engineering (Vol. 263, No. 3, p. 032037). IOP Publishing.

Bos, F., Wolfs, R., Ahmed, Z., & Salet, T. (2016). Additive manufacturing of concrete in construction: Potentials and challenges of 3D concrete printing. *Virtual and Physical Prototyping*, *11*(3), 209–225.

Danish, P., & Ganesh, M. G. (2020a). Behaviour of self-compacting concrete using different mineral powders additions in ternary blends. *Revista Romana de Materiale*, *50*(2), 232–239.

Danish, P., & Ganesh, G. M. (2020b). Durability properties of self-compacting concrete using different mineral powders additions in ternary blends. *Revista Romana de Materiale*, *50*(2), 369–378.

Domone, P. L. (2006). Self-compacting concrete: An analysis of 11 years of case studies. *Cement and Concrete Composites*, *28*(2), 197–208.

EFNARC, S. (2002). *Guidelines for self-compacting concrete*. London, UK: Association House, 32, 34.

Gambhir, M. L. (2013). *Concrete technology: Theory and practice*. Tata McGraw-Hill Education.

Hajime, O., & Masahiro, O. (2003). Self-compacting concrete. *Journal of Advanced Concrete Technology*, *1*(1), 5–15.

International Energy Agency. (2022). *Global CO emissions rebounded to their highest level in history in 2021*. https://www.iea.org/news/global--emissions-rebounded-to-their-highest-level-in-history-in-2021co2

International Organization for Standardization. (2006). *ISO 14040:2006 – Environmental management – Life cycle assessment – Principles and framework*. Geneva, Switzerland: ISO.

International Organization for Standardization. (2011). *ISO 21929–1:2011 – Sustainability in building construction – Sustainability indicators – Part 1: Framework for the development of indicators and a core set of indicators for buildings*. Geneva, Switzerland: ISO.

Khayat, K. H. (1999). Workability, testing, and performance of self-consolidating concrete. *Materials Journal*, *96*(3), 346–353.

Kiran Prabha, M., Vishnu Vardhan, K., Santhi, A. S., & Mohan Ganesh, G. (2025). Advancements in self-compacting concrete reinforced with basalt fiber: A comprehensive review. *Engineering Reports*, *7*(5), e70147.

Mehta, P. K., & Monteiro, P. J. M. (2006). *Concrete: Microstructure, properties, and materials* (3rd ed.). McGraw-Hill Education.

Pierson, R., Dehn, F., Dahlhaus, F., & Sanft, S. (2022). Research and testing of self-compacting concrete (SCC) for tunnelling under realistic conditions. In *Reclaiming the Underground Space* (Vol. 2, pp. 681–685). Routledge.

Safiuddin, M., & Jumaat, M. (2011). Fresh properties of self-consolidating concrete incorporating palm oil fuel ash as a supplementary cementing material. *Chiang Mai J. Sci*, *38*(3), 389–404.

Shayan, A., & Xu, A. (2006). Performance of glass powder as a pozzolanic material in concrete: A field trial on concrete slabs. *Cement and Concrete Research*, *36*(3), 457–468.

Siddique, R., & Singh, G. (2011). Utilization of waste foundry sand (WFS) in concrete manufacturing. *Resources, Conservation and Recycling*, *55*(11), 885–892.

Wiktor, V., & Jonkers, H. M. (2011). Quantification of crack-healing in novel bacteria-based self-healing concrete. *Cement and Concrete Composites*, *33*(7), 763–770.

Witte, A., & Garg, N. (2024). Quantifying the global warming potential of low carbon concrete mixes: Comparison of existing life cycle analysis tools. *Case Studies in Construction Materials*, *20*, e02832.

5 Performance and Sustainability Assessment of Self-Compacting Concrete With Mineral Powder Additions for Resilient Infrastructure

Prithiviraj Chidambaram

5.1 INTRODUCTION

Infrastructure is a critical enabler of socio-economic development, yet it also represents one of the largest contributors to global sustainability challenges. It accounts for nearly 70% of global greenhouse gas emissions and over 30% of natural resource consumption, making it central to the achievement or failure of global climate goals and the United Nations Sustainable Development Goals (SDGs). Despite growing investments in green infrastructure, inconsistencies in environmental, social, and governance (ESG) frameworks and standards continue to hinder measurable progress. In many developing countries, implementation is further constrained by limited technical capacity, inadequate stakeholder engagement, and high initial costs.[1]

Infrastructure decisions made today have long-term implications due to the "lock-in effect" of extended asset lifespans. Therefore, ensuring infrastructure sustainability requires integrated, system-level approaches that balance environmental, social, and economic objectives from the earliest stages of planning.[2]

Among the materials that dominate infrastructure development, concrete stands out as both indispensable and problematic. As the second most consumed substance globally after water, it is integral to construction yet imposes a substantial environmental burden.[3] Cement, the primary binder in concrete, accounts for approximately 8–10% of global CO_2 emissions[4,5] along with significant impacts on water resources, air quality, and material depletion.[6] Life Cycle Assessment (LCA) studies demonstrate that cement production adversely affects multiple environmental indicators, including climate change, toxicity, acidification, and fossil fuel depletion.[4,7]

DOI: 10.1201/9781003664284-5

In developing regions, additional issues arise from dust emissions, water pollution, and solid waste linked to weak regulatory enforcement.[8] While high-strength concrete may reduce overall material volume, it often increases carbon intensity per unit volume because of its higher cement demand.[9] Recent innovations such as Supplementary Cementitious Materials (SCMs), Recycled Concrete Powder (RCP), and carbonation curing offer promising routes to reduce emissions and enhance material circularity. Studies show that replacing 15% of cement with RCP can reduce CO_2 emissions by up to 12%, while recycling end-of-life concrete can lower lifecycle impacts by up to 30%.[4,7]

Durability also plays a vital role in sustainability. Structures with extended service lives require fewer repairs, reducing embodied carbon over time.[3] These findings highlight the urgent need for innovative concrete technologies that deliver both environmental and performance benefits. Among these, Self-Compacting Concrete (SCC), particularly when enhanced with SCMs and recycled materials, offers a practical pathway towards sustainable infrastructure. Table 5.1 summarizes the environmental impacts and corresponding mitigation strategies, while Figure 5.1 compares the CO_2 emissions associated with different concrete mix types.[4,7,9]

SCC emerged in Japan during the late 1980s to address labour shortages and quality inconsistencies in construction. It is a highly flowable, non-segregating concrete that compacts under its own weight without mechanical vibration.[10] This characteristic improves build quality, reduces noise, and enhances safety on construction sites.[11] SCC differs from conventional concrete primarily in its mix design, which employs High-Range Water Reducers (HRWRs), low water-to-powder ratios, and sometimes Viscosity-Modifying Admixtures (VMAs) to achieve optimal filling and passing ability.[12,13]

TABLE 5.1
Environmental Impacts of Concrete and Mitigation Strategies

Environmental impact	Primary source	Mitigation strategy	Estimated reduction	Source
CO_2 emissions	Cement production	Use of SCMs (fly ash, slag, silica fume)	↓ up to 30%	4,5
Fossil fuel depletion	Clinker burning, material transport	Use of RCP, energy-efficient processes	↓ 10–15%	7
Water consumption	Mixing and curing processes	Rainwater harvesting, SCC with lower w/c	↓ 10–20%	6,8
Acidification and toxicity	Cement and heavy metal emissions	SCMs, low-emission alternatives	↓ 15–25%	4
Solid waste generation	Construction & demolition debris	Recycled aggregates, concrete reuse	↓ 25–35%	9
Lifecycle carbon footprint	All phases (production to EOL)	Durability-focused mix designs	↓ over full service life	3

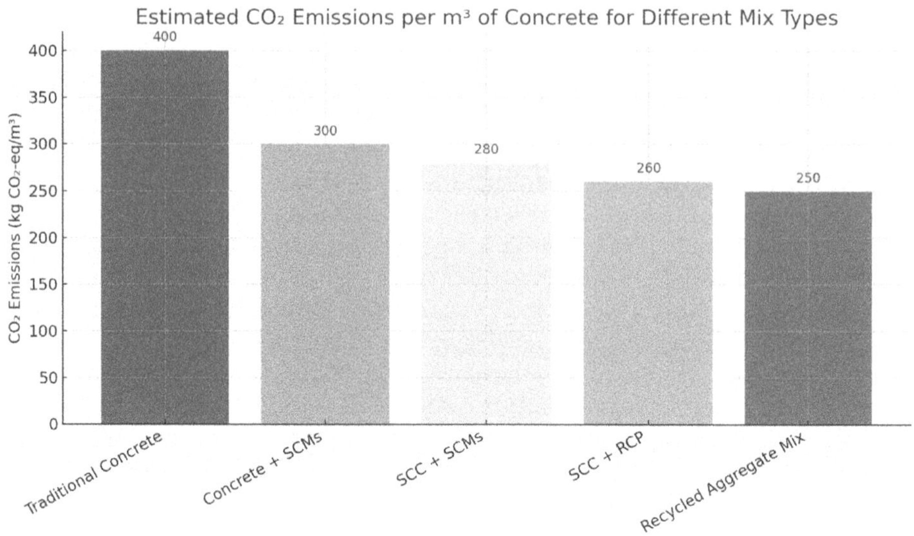

FIGURE 5.1 Estimated CO_2 emissions per m³ of concrete for different mix types.

Recent advancements in SCC mix design have focused on improving performance consistency and sustainability through rational and data-driven optimization methods. Danish and Ganesh (2021) proposed a modified rational mix design approach that enhances SCC workability and strength balance by systematically adjusting binder content, powder proportion, and rheological parameters.[14] Their model supports sustainable concrete production by reducing cement consumption without compromising mechanical or durability performance.

Beyond its mechanical efficiency, SCC supports sustainability objectives when combined with SCMs or recycled components such as fly ash, silica fume, alccofine, and recycled aggregates.[12,15,16] Empirical studies confirm that SCC incorporating industrial by-products can reduce CO_2 emissions, enhance resource efficiency, and align with circular economy principles.[17,18] Therefore, SCC represents a viable solution for achieving structural performance alongside environmental stewardship.

This chapter examines the performance and sustainability advantages of SCC in modern infrastructure. It explores SCC's role in addressing global challenges such as environmental degradation, labour shortages, and construction inefficiencies. The discussion outlines how SCC's self-compacting behaviour eliminates the need for vibration, thereby reducing energy consumption, noise, and labour dependence. It also analyses the incorporation of SCMs, recycled aggregates, and advanced additives such as nano-silica, fibres, and geopolymers, which enhance strength and sustainability. The chapter concludes with a discussion of challenges in mix design, durability, and cost and highlights emerging innovations such as self-healing concrete and AI-driven mix optimization.

5.2 COMPOSITION AND KEY PROPERTIES OF SCC

SCC is a major advancement in concrete technology, offering superior workability, strength, and sustainability. It flows under its own weight and fills formworks without vibration, eliminating the need for mechanical compaction.

SCC typically consists of cement, water, fine and coarse aggregates, HRWRs, and occasionally VMAs. Incorporating SCMs such as fly ash, silica fume, GGBS, alccofine, and metakaolin improves both performance and environmental efficiency. Recycled materials like rice husk ash and glass powder further enhance sustainability by reducing cement consumption.[19–21]

The water-to-binder ratio generally ranges from 0.25 to 0.45, adjusted with superplasticizers to maintain flowability without loss of strength. Mix design approaches include empirical, rheology-based, particle-packing, and eco-efficient optimization methods, such as the modified rational model proposed by Danish and Ganesh (2021), which balances workability and strength through systematic binder adjustments.[14,22–24,25]

Key fresh properties such as filling ability, passing ability, and segregation resistance are evaluated through slump flow (650–800 mm), T_{500} (2–5 s), L-box (≥0.8), and V-funnel (6–12 s) tests. Fibres (steel or recycled) improve tensile and flexural strength but may reduce workability at higher dosages.[20,26]

Compressive strength of SCC typically ranges from 30 to over 100 MPa, influenced by curing conditions and SCM composition. Silica fume enhances early strength, while fly ash supports long-term gain. Due to reduced coarse aggregates, SCC usually exhibits a slightly lower elastic modulus than conventional concrete.[27,28]

SCC shows excellent durability, with lower permeability, greater chloride resistance, and improved freeze and thaw performance.[29] The use of self-curing agents such as PEG 4000 enhances hydration under limited-curing conditions.[30] Microstructural analyses using Scanning Electron Microscopy (SEM), X-Ray Diffraction (XRD), and TGA reveal pore refinement and denser ITZ, while smart additives like glass powder, bottom ash, and carbon nanotubes promote intelligent, climate-resilient concrete.[20,31]

A summary of global SCC mix compositions and performance parameters is presented in Table 5.2, illustrating its adaptability to diverse construction requirements.

5.3 ENVIRONMENTAL BENEFITS AND SUSTAINABILITY ASPECTS

SCC contributes significantly to sustainable construction by optimizing both material use and environmental performance. A primary benefit lies in the incorporation of SCMs such as fly ash, silica fume, GGBS, limestone powder, and coal bottom ash, many of which are industrial by-products. These not only replace large portions of Portland cement, thereby reducing CO_2 emissions, but also divert waste from landfills and promote a circular economy.[21,34] SCC eliminates the need for mechanical vibration during placement, leading to substantial energy savings on-site and reducing associated noise pollution. This energy efficiency is particularly advantageous in precast and high-rise construction projects.[23] Moreover, SCC's ease of placement

TABLE 5.2
SCC Composition and Key Properties

Source	Binder Type	Admixtures	Water/Binder Ratio	Compressive Strength (Mpa)	Fresh Properties
10	Cement + SCMs (FA, SF, slag)	HRWR, optional VMA	0.25–0.50	40–100	Slump flow 600–750 mm, V-funnel 8–15 s
28	Cement + FA, SF	HRWR, VMA	0.30–0.50	30–100	Slump flow 600–750 mm
24	Cement + SF	HRWR	< 0.48	72	Slump 600 mm, V-funnel ~10 s
32	Pozzolanic cement + SF	PCE-based SP	0.32–0.46	43–65.5	Slump 589–793 mm, L-box 0.79–1.00
27	Cement + Class C FA	SP	0.30	101.6 (90 days)	Slump 750 mm, T500 4.2 s
33	Cement + limestone powder	SP	0.35–0.70	20–70	Slump 680–730 mm, L-box 0.83–0.92
34	Cement + FA, GGBS, M, LP	PCE-SP, VMA	0.30–0.50	30–60	Slump 560–720 mm, VSI 0–1
35	Cement + variable SCMs	Varies by method	Varied	Variable	Depends on method
36	Cement	SP	0.35–0.63	30, 60, 80	Slump 680 mm, V-funnel 7–10 s
37	Cement + silica fume	SP, VMA	0.30–0.38	65–70	Slump flow 711–755 mm, V-funnel 8–13 s
22	Cement + SCMs (FA, SF, GGBS)	SP, rheology aids	0.22–0.50	15–160	Slump 650–800 mm, L-box ≥0.8
26	Various including waste SCMs	SP, VMA	0.30–0.45	30–92	Slump 650–800 mm, SF1–SF3
30	Cement + fly ash + silica fume	SP, self-curing agents	0.345	45–70	Slump 700–755 mm, L-box ≥0.85
23	Cement + fly ash, SF	SP, VMA	0.32–0.45	40–92	Slump 550–850 mm, V-funnel 6–12 s
38	Cement + fly ash	SP	0.345	45.87	Slump 700 mm, V-funnel 13 s
20	Cement + SCMs + RPF	SP	0.30–0.45	30–60	Slump ↓ with fibres; V-funnel ↑
21	Cement + SF, FA, GGBS, CBA	SP	0.30–0.50	30–70	Slump 550–800 mm
31	Various SCM blends	None reported	0.30–0.45	Varies	Microstructural focus

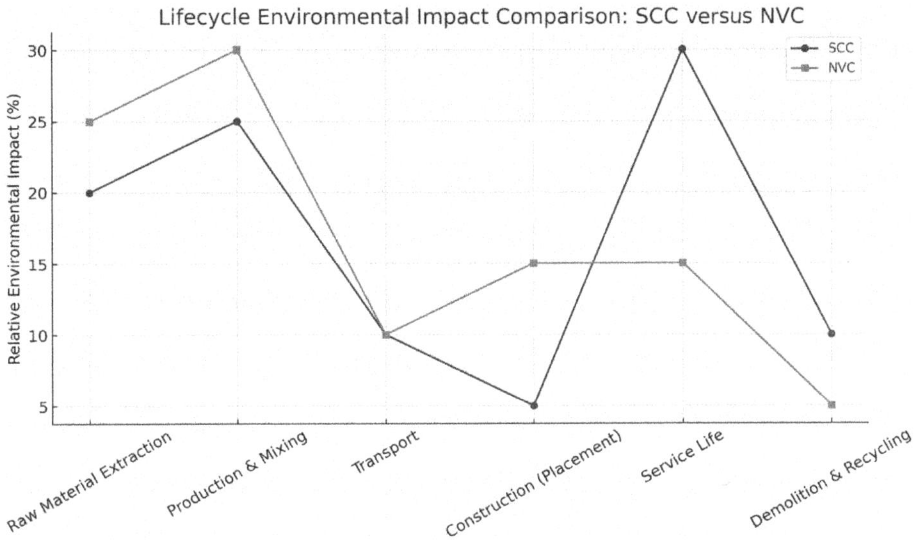

FIGURE 5.2 Lifecycle environmental impact comparison: SCC versus NVC.

reduces labour dependency and mitigates occupational hazards, which is especially valuable in congested or complex reinforcement areas.[38]

Durability benefits further support long-term sustainability. SCC exhibits lower permeability and enhanced resistance to chloride ingress and carbonation, particularly when mineral admixtures are used.[29] These properties extend service life and reduce the maintenance needs. They connect SCC with global goals for resilient and low-carbon infrastructure.[20,31] Figure 5.2 demonstrates the environmental impact at six lifecycle stages of SCC and normal vibrated concrete (NVC). SCC has a lower impact while placement and offers a longer service life due to its enhanced durability. Utilization of industrial by-products in SCC also significantly cuts the environmental impact during raw material extraction and production. Studies from[21,23,26,29,30,34] support these findings.

5.4 INNOVATIONS IN SCC FOR RESILIENT INFRASTRUCTURE

SCC continues to evolve to meet sustainability and performance demands. Nanomaterials like nano-silica, nano-alumina, and nanoclay improve compressive strength, reduce porosity, and densify the microstructure. Studies show that 2–4% nano-silica optimizes the ITZ and enhances long-term durability.[39–41] When combined with steel or basalt fibres, SCC gains improved crack resistance and toughness.

Geopolymer and alkali-activated SCC (SCGC) have emerged as greener alternatives, using fly ash, slag, and metakaolin to significantly cut CO_2 emissions. While these mixes offer good mechanical and durability traits, especially with fibre or recycled aggregate inclusion, achieving optimal workability and permeability remains a challenge.[42–44] To overcome heat-curing limitations, ambient-curable SCGC with

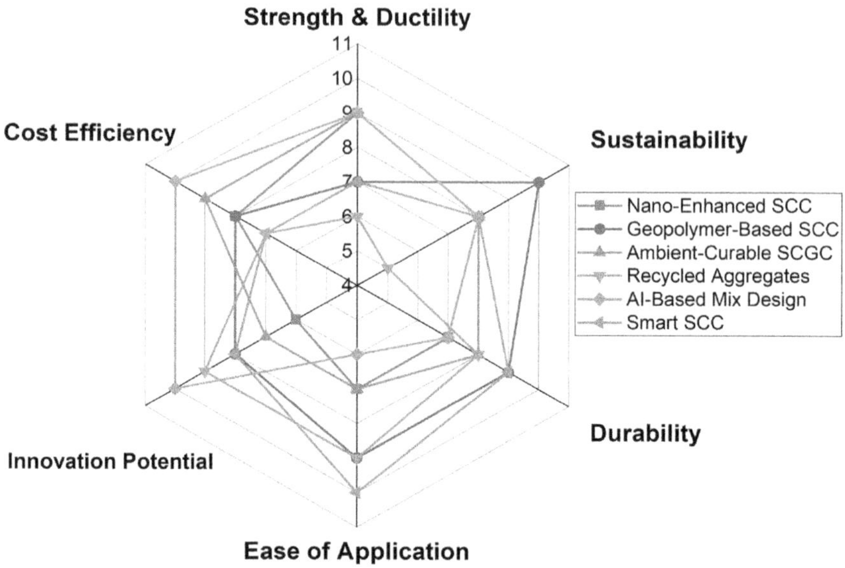

FIGURE 5.3 Performance comparison of SCC innovations.

partial Ordinary Portland Cement (OPC) replacement (5–20%) has been proposed, offering better acid resistance and sorptivity in aggressive environments.[45]

Further innovation involves using recycled materials such as M-sand, steel slag, and lime waste, which improve microstructure and life cycle performance.[46,47] Recent reviews also highlight the potential of plastic waste as a partial aggregate substitute in sustainable concrete, contributing to waste reduction and enhanced ductility without compromising strength.[48]

Machine learning models like K-star and M5Rules now help predict SCC strengths, reducing the need for manual trials.[49] Additionally, smart SCC with nano-silica, self-healing agents, and embedded sensors enables real-time monitoring and adaptive performance. These advancements position SCC as a future-ready material for intelligent, low-carbon infrastructure. Figure 5.3 compares these innovations across key performance metrics.

5.5 GLOBAL APPLICATIONS OF SCC

SCC has become a critical material in modern construction, offering vibration-free placement, improved durability, and labour efficiency. Since its development in Japan during the 1980s, SCC has been widely used in earthquake-resistant infrastructure, including the Akashi Kaikyo Bridge and Roppongi Hills Mori Tower, where its self-flowing nature ensured compaction in dense reinforcement zones while reducing noise.[10,28] In India, SCC has been adopted in major projects like the Delhi Metro, Mumbai Coastal Road, and Bandra-Worli Sea Link, where it improved speed and environmental performance in complex urban and marine conditions. The Sitabuldi Interchange of Nagpur Metro further showcased its strength in densely reinforced cast-in-place work.[50,51] The Netherlands applied SCC in flood-control structures under the Room for the River Project, meeting long-term durability needs.[52] Scandinavian countries, including Norway and Sweden, used SCC in cold-climate projects such as the Hardanger Bridge to reduce labour and enhance finish quality.[53] Across Europe and the United States, SCC is used in precast elements, retrofitting, and infrastructure like viaducts and bridge decks. Its proven seismic resilience, offering enhanced ductility, energy dissipation, and reduced cracking, makes it ideal for high-risk zones globally.[54]

5.6 CHALLENGES AND OPPORTUNITIES OF SCC

The practical use of SCC still faces certain technical, economic, and institutional issues. One of the main technical issue is mix design optimization. The properties of SCC in its fresh state are balanced with mechanical and long-term durability performance. The absence of a universal design method leads engineers to rely on empirical trials, AI tools, or particle packing models, particularly when incorporating multiple SCMs or recycled materials, each with distinct rheological and hydration behaviours.[55–57] The availability of material added complicates performance. SCC is particularly sensitive to variations in raw materials such as recycled aggregates, and industrial by-products affect the SCC behaviour.[58,59] Inconsistent dispersion of additives like silica fume or alumina can result in uneven microstructures.[60] Likewise, SCC is sensitive in field applications. The fresh and hardened state of SCC is influenced by temperature, humidity, and aggregate moisture content.[57,58] These issues become critical in conditions like mass pours, tall buildings, or hot weather where pressure buildup and early-age hydration are critical.[44,59] From an economic view, the use of advanced admixtures, fibres, nano-modifiers, and controlled aggregates may increase material production cost. Even though SCC offers long-term savings through faster construction and reduced labour, upfront costs still discourage widespread use.[61,62] Institutional gaps compound the issue. As mentioned earlier, the SCC-specific codes are lacking, and limited awareness among site engineers often leads to poor handling and inconsistent results in many regions. Even in developed countries, traditional practices and limited training hinder SCC adoption. These barriers, along with potential solutions, are summarized in Table 5.3.

SCC offers strong potential for meeting modern construction demands. AI and ML are increasingly used to optimize mix designs and predict performance, minimizing the need for trial mixes.[49,57] Eco-friendly SCC, incorporating recycled aggregates and SCMs like fly ash and GGBS, supports low-carbon goals without compromising strength.[55] Innovations such as self-healing agents, embedded sensors, and adaptive mixes enable real-time monitoring and improved durability. With growing emphasis on green certifications and digital construction methods like 3D printing, SCC is positioned to play a central role in sustainable infrastructure.

5.7 FUTURE TRENDS AND STRATEGIC RECOMMENDATIONS

SCC is evolving beyond placement efficiency to address the priorities of digital construction, sustainability, and the circular economy. Future SCC development will

TABLE 5.3
Challenges and Opportunities in Implementing Self-Compacting Concrete

Challenge Area	Key Challenge Description	Opportunities	Source
Mix design complexity	No universal mix design approach; balancing flow, segregation, and strength is labour-intensive.	Use of AI, ML, and rheology-based design tools	49,56,57
Material variability and compatibility	Inconsistent properties of SCMs, recycled aggregates, and industrial waste affect performance.	Tailored local mix adjustments and nanotechnology to stabilize performance	53,55
Climate sensitivity	SCC's rheology is highly affected by ambient temperature and humidity during field applications.	Use of curing agents, mix stabilizers, and smart sensors for consistency	43,45,58,59
Formwork pressure	Higher lateral pressure on formwork; requires stricter planning and structural integrity	Precast use, staged casting, and digital pressure prediction tools	53,59,61
Cost considerations	Upfront cost is higher due to premium materials and admixtures, despite long-term lifecycle savings.	Cost offset through reduced labour, speed, and enhanced durability	55,61,62
Regulatory and training gaps	Lack of national standards and low awareness in many regions hinder implementation.	Development of codes, training modules, and field adoption programmes	60,62

depend on integrating smart technologies, material innovations, and policy frameworks that promote sustainable infrastructure.

Artificial Intelligence (AI) tools, including Artificial Neural Networks (ANNs), have shown strong potential for predicting SCC properties and optimizing mix designs based on local material characteristics. Such models can reduce the number of trial mixes by up to 50%, conserving both materials and emissions.[57,63]

Next-generation SCC will increasingly adopt self-healing systems which use bacterial agents and calcium-based compounds to restore tensile strength and extend service life.[64] Nanomaterials, such as nano-silica, enhance microstructural densification, while hybrid fibres improve ductility with minimal impact on workability.[57]

High-volume SCM utilization represents another key sustainability strategy. Replacing up to 60% of cement with materials such as fly ash, GGBS, rice husk ash (RHA), and palm oil fuel ash (POFA) can reduce CO_2 emissions by approximately 35%.[65] Likewise, the use of locally sourced materials within a 50 km radius supports circular value chains and reduces transport-related impacts (Table 5.4). When properly combined with SCMs, recycled aggregates can successfully replace up to 50% of natural aggregates without compromising performance.[63]

TABLE 5.4
Regional Availability Of Dominant Waste Streams and Their Supplementary Cementitious Material (Scm) Potential In Scc[57,63–65]

Region	Dominant waste stream	SCM type	Cement replacement potential	Performance benefits
South-East Asia	Rice husk ash (RHA)	Pozzolanic SCM	20–30%	↑ Strength, ↓ permeability
Europe and North America	Fly ash, Ground Granulated Blast Furnace Slag (GGBS)	Latent hydraulic and pozzolanic	40–50%	35% CO_2 reduction, improved durability
Middle East	Treated Palm Oil Fuel Ash (POFA)	Pozzolanic SCM	20–25%	↑ flowability, sulphate resistance
India	Sugarcane bagasse ash, RHA, GGBS	Mixed SCMs	30–45%	Lower heat of hydration, ↑ workability
China	Fly ash (Class F), slag	Latent hydraulic SCMs	30–50%	Enhanced strength, ↓ water demand
Africa	Volcanic ash, agricultural wastes	Natural pozzolans	20–35%	↑ Resistance to Alkali-Silica Reaction (ASR)
Latin America	Sugarcane bagasse ash, metakaolin	Reactive pozzolans	25–40%	↑ Early strength, good workability

Emerging digital construction technologies also play a transformative role. IoT-enabled sensors embedded in SCC allow real-time monitoring of strength, crack development, and hydration. When integrated with digital twin systems, these tools enable predictive maintenance and could lower lifecycle costs by up to 80%.[57]

To accelerate adoption, policy interventions are essential. Incentives such as tax benefits, green certifications, and public procurement mandates for SCC with high SCM or smart additive content can enhance market confidence. Updating material standards to accommodate recycled and nanomaterial-based components will further support scalability and performance assurance.

A phased implementation strategy is recommended to align technical, economic, and regulatory progress (Table 5.5). By 2030, SCC is projected to evolve into an intelligent, self-sustaining material system which underpins autonomous, resilient, and low-carbon infrastructure. In this transformation, SCC will serve not merely as a construction material but as a core component of sustainable construction ecosystems.

5.7.1 Future Scope

Future research should focus on developing standardized global guidelines for SCC production using regional materials, incorporating machine learning and LCA tools

TABLE 5.5
Strategic Roadmap for Scc Deployment Across Time Horizons[57,63–65]

Timeline	Key actions	Stakeholders	Expected impact
Short-term (2025)	• Deploy AI-powered SCC mix design tools in precast plants • Establish standards for recycled aggregates & nano-SCMs	Industry, standards bodies, R&D institutes	• ↓ 10% construction cost • ↓ 20% CO_2 emissions
Mid-term (2027)	• Codify performance standards for self-healing SCC • Scale regional SCM processing hubs	Policymakers, academia, material suppliers	• ↓ 40% maintenance costs • ↑ 25% mechanical performance
Long-term (2030)	• Launch global SCC materials database • Enforce policy mandates for ≥40% SCMs in public works	Governments, international agencies	• ↑ 70% SCC market share • Support net-zero concrete ecosystems

for predictive performance modelling. Experimental work on carbon-negative binders, biogenic self-healing agents, and sensor-integrated SCC will further support the transition towards intelligent, low-carbon infrastructure systems. At the policy level, integrating SCC into green rating frameworks and public procurement standards can accelerate its adoption in mainstream construction.

As global infrastructure advances towards carbon neutrality, circularity, and climate resilience, SCC is poised to become a cornerstone material in next-generation sustainable construction practices.

5.8 CONCLUSION

SCC is a transformative material for achieving sustainable and resilient infrastructure. Its unique ability to flow and compact under its own weight enhances constructability and reduces labour requirements. Moreover, by incorporating SCMs, industrial by-products, and recycled aggregates, SCC contributes significantly to environmental sustainability. This study has deliberated on the challenges of modern infrastructure, the environmental impact of normal concrete, and the growing significance of SCC in tackling these problems.

The mix and properties of SCC can transform recognitions to new additives, nanomaterials, and geopolymer chemistry. Recent ideas like AI-driven mix design, smart SCC with built-in sensors, and self-repairing technologies exhibit an optimistic future for this material. Simultaneously, issues like material differences, climate effects, high costs, and a lack of standards need solutions through joint research, policy help, and industry training. Considering future possibilities, using SCC in regular construction will depend on data-based methods, regional material choices, and supportive regulations. As global infrastructure develops towards carbon neutrality, circularity, and climate resilience, SCC could become a key material in the construction practices of the future.

REFERENCES

1. Meko, B., Ighalo, J. O. & Ofuyatan, O. M. Enhancement of self-compactability of fresh self-compacting concrete: A review. *Clean. Mater.* 1, 100019 (2021).
2. Yanamandra, S. Sustainable infrastructure: An overview placing infrastructure in the context of sustainable development. *Univ. Cambridge Inst. Sustain. Leadersh.* 57 (2020).
3. Adesina, A. & Zhang, J. Impact of concrete structures durability on its sustainability and climate resiliency. *Next Sustain.* 3, 100025 (2024).
4. Tran, C. N. N., Illankoon, I. M. C. S. & Tam, V. W. Y. Decoding concrete's environmental impact: A path toward sustainable construction. *Buildings* 15, 1–23 (2025).
5. Belaïd, F. How does concrete and cement industry transformation contribute to mitigating climate change challenges? *Resour. Conserv. Recycl. Adv.* 15, 200084 (2022).
6. Eštoková, A., Wolfová Fabiánová, M. & Ondová, M. Concrete structures and their impacts on climate change and water and raw material resource depletion. *International Journal of Civil Engineering* 20, 735–747. https://doi.org/10.1007/s40999-022-00701-8 (2022).
7. Manan, A. *et al.* Environmental and human health impact of recycle concrete powder: An emergy-based LCA approach. *Front. Environ. Sci.* 12, 1–20 (2024).
8. Solaimanian, S. Environmental risk assessment of concrete construction projects in developing countries based on Analytical Hierarchy Process method. *Green Technol. Sustain.* 3, 100178 (2025).
9. Mostafaei, H., Chamasemani, N. F., Mashayekhi, M., Hamzehkolaei, N. S. & Santos, P. Sustainability enhancement and evaluation of a concrete dam using recycling. *Appl. Sci.* 15 (2025).
10. Ouchi, M., Nakamura, S., Osterberg, T., Hallberg, S.-E. & Lwin, M. Applications of self-compacting concrete in Japan, Europe and the United States. In *Proceedings of International Symposium on High Performance Concrete-ISHPC*, 147–173 (2003).
11. EFNARC. The European guidelines for self-compacting concrete. *Eur. Guidel. Self Compact. Concr.* 63 (2005).
12. Prithiviraj, C. *et al.* Assessment of strength and durability properties of self-compacting concrete comprising alccofine. *Sustain.* 14, 5895 (2022).
13. Prithiviraj, C. & Saravanan, J. Influence of W/B ratio and chemical admixture on fresh and hardened properties of self compacting concrete using alccofine. *J. Xidian Univ.* 14, 4906–4915 (2020).
14. Danish, P. & Ganesh, G. M. Self-compacting concrete – optimization of mix design procedure by the modifications of rational method. *RILEM Bookseries* 29, 369–396 (2021).
15. Adesina, A. & Awoyera, P. Overview of trends in the application of waste materials in self-compacting concrete production. *SN Appl. Sci.* 1, 1–18 (2019).
16. Okokpujie, I. P., Kelechi, S. E. & Tartibu, L. K. A review of self-compacting concrete incorporating waste materials. *Ann. Chim. Sci. des Mater.* 47, 305–323 (2023).
17. Prithiviraj, C., Swaminathan, P., Kumar, D. R., Murali, G. & Vatin, N. I. Fresh and hardened properties of self-compacting concrete comprising a copper slag. *Buildings* 12 (2022).
18. Santos, S., da Silva, P. R. & de Brito, J. Self-compacting concrete with recycled aggregates – A literature review. *J. Build. Eng.* 22, 349–371 (2019).
19. Chidambaram, P. & Jagadeesan, S. Analysis of GFRP-reinforced beams in enhanced SCC. *Gradjevinar* 76, 515–530 (2024).
20. Frhaan, W. K. M., Abu Bakar, B. H., Hilal, N. & Al-Hadithi, A. I. Relation between rheological and mechanical properties on behaviour of self-compacting concrete (SCC) containing recycled plastic fibres: A review. *Eur. J. Environ. Civ. Eng.* 26, 4761–4793 (2022).
21. I. Albiajawi, M., Embong, R. & Muthusamy, K. Influence of mineral admixtures on the properties of self-compacting concrete: An overview. *Construction* 1, 62–75 (2021).

22. Ashish, D. K. & Verma, S. K. An overview on mixture design of self-compacting concrete. *Struct. Concr.* 20, 371–395 (2019).
23. Geiker, M. & Jacobsen, S. *Self-Compacting Concrete (SCC). Developments in the Formulation and Reinforcement of Concrete* (Elsevier LTD, 2019). https://doi.org/10.1016/B978-0-08-102616-8.00010-1.
24. Sebaibi, N., Benzerzour, M., Sebaibi, Y. & Abriak, N. E. Composition of self compacting concrete (SCC) using the compressible packing model, the Chinese method and the European standard. *Constr. Build. Mater.* 43, 382–388 (2013).
25. Dinakar, P. & Manu, S. N. Concrete mix design for high strength self-compacting concrete using metakaolin. *Mater. Des.* 60, 661–668 (2014).
26. Han, B. *et al.* Self-compacting concrete. In *Smart and Multifunctional Concrete Toward Sustainable Infrastructures*, vol. 1, 11–36 (Japan Concrete Institute, 2017).
27. Soleymani Ashtiani, M., Scott, A. N. & Dhakal, R. P. Mechanical and fresh properties of high-strength self-compacting concrete containing class C fly ash. *Constr. Build. Mater.* 47, 1217–1224 (2013).
28. Domone, P. L. A review of the hardened mechanical properties of self-compacting concrete. *Cem. Concr. Compos.* 29, 1–12 (2007).
29. Rodríguez Viacava, I., Aguado De Cea, A. & Rodríguez De Sensale, G. Self-compacting concrete of medium characteristic strength. *Constr. Build. Mater.* 30, 776–782 (2012).
30. Chand, M. S. R., Kumar, P. R., Giri, P. S. N. R. & Kumar, G. R. Performance and microstructure characteristics of self-curing self-compacting concrete. *Adv. Cem. Res.* 30, 451–468 (2018).
31. Badgujar, S. & Kumar Dwivedi, A. Microstructural analysis of self-compacting concrete – a review. *Mater. Today Proc.* 65, 1250–1259 (2022).
32. Sfikas, I. P. & Trezos, K. G. Effect of composition variations on bond properties of self-compacting concrete specimens. *Constr. Build. Mater.* 41, 252–262 (2013).
33. Nikbin, I. M. *et al.* A comprehensive investigation into the effect of water to cement ratio and powder content on mechanical properties of self-compacting concrete. *Constr. Build. Mater.* 57, 69–80 (2014).
34. Long, G., Gao, Y. & Xie, Y. designing more sustainable and greener self-compacting concrete. *Constr. Build. Mater.* 84, 301–306 (2015).
35. Shi, C., Wu, Z., Lv, K. & Wu, L. A review on mixture design methods for self-compacting concrete. *Constr. Build. Mater.* 84, 387–398 (2015).
36. Alyhya, W. S., Abo Dhaheer, M. S., Al-Rubaye, M. M. & Karihaloo, B. L. Influence of mix composition and strength on the fracture properties of self-compacting concrete. *Constr. Build. Mater.* 110, 312–322 (2016).
37. Kostrzanowska-Siedlarz, A. & Gołaszewski, J. Rheological properties of high performance self-compacting concrete: Effects of composition and time. *Constr. Build. Mater.* 115, 705–715 (2016).
38. Kefelegn, A. & Gebre, A. Performance of self-compacting concrete used in congested reinforcement structural element. *Eng. Struct.* 214, 110665 (2020).
39. Ansari Rad, T., Tanzadeh, J. & Pourdada, A. Laboratory evaluation of self-compacting fiber-reinforced concrete modified with hybrid of nanomaterials. *Constr. Build. Mater.* 232, 117211 (2020).
40. Konsta-Gdoutos, M. S. Nanomaterials in self-consolidating concrete: A state-of-the-art review. *J. Sustain. Cem. Mater.* 3, 167–180 (2014).
41. Lone, S. S., Gupta, M. & Mehta, G. Comparative study of nano-silica induced mortar, concrete and self-compacting concrete – a review. *Indian J. Sci. Technol.* 9 (2016).
42. Ayub, T., Mahmood, W. & Khan, A. U. R. Durability performance of scc and scgc containing recycled concrete aggregates: A comparative study. *Sustain.* 13 (2021).

43. Elbialy, S., El-Latief, A. A., Al-Jabali, H. M., Elsayed, H. A. & Shawky, S. M. M. Enhancing the properties of steel fiber self-compacting NaOH-based geopolymer concrete with the addition of metakaolin. *Civ. Eng. J.* 10, 2244–2260 (2024).
44. Gülşan, M. E., Alzeebaree, R., Rasheed, A. A., Niş, A. & Kurtoğlu, A. E. Development of fly ash/slag based self-compacting geopolymer concrete using nano-silica and steel fiber. *Constr. Build. Mater.* 211, 271–283 (2019).
45. Jeyaseela, J. & Vishnuram, B. G. Study on workability and durability characteristics of self-compacting geopolymer concrete composites. *Int. J. Adv. Technol. Eng. Sci.* 03, 1246–1256 (2015).
46. Kumar, B. N., Reddy, A. M. & Vineela, P. S. development of sustainable high performance self compacting concrete using construction & demolition waste and steel slag. *Discov. Civ. Eng.* 2 (2025).
47. Bahmani, H. & Mostofinejad, D. Innovative fiber-reinforced self-compacting concrete using activated slag from sugar factory lime waste. *Case Stud. Constr. Mater.* 22, e04752 (2025).
48. Almohana, A. I., Abdulwahid, M. Y., Galobardes, I., Mushtaq, J. & Almojil, S. F. Producing sustainable concrete with plastic waste: A review. *Environ. Challenges* 9, 100626 (2022).
49. Onyelowe, K. C. *et al.* Mechanical properties of self compacting concrete reinforced with hybrid fibers and industrial wastes under elevated heat treatment. *Sci. Rep.* 15, 1–40 (2025).
50. Maharashtra Engineering Research Institute. Self compacting concrete. In *WRD Handbook Chapter No. 3 Self Compacting Concrete* (2019). Water Resources Department, Government of Maharashtra, Nashik, India.
51. Sonule, S., Tayade, K. C., Vidyasagar, N. V. P. & Gupta, A. B. A case study on performance of self-compacting concrete in highly congested reinforcement of cast in situ structure. *Asian J. Civ. Eng.* 22, 1423–1432 (2021).
52. Goodier, C. I. Development of self-compacting concrete. *Proc. Inst. Civ. Eng. Struct. Build.* 156, 405–414 (2003).
53. Moravvej, M. & Rashidi, M. *Structural Performance of Self-Compacting Concrete. Self-Compacting Concrete: Materials, Properties and Applications* (Elsevier Inc., 2019). https://doi.org/10.1016/B978-0-12-817369-5.00013-1.
54. Ghaboussi, D. & Lee, K. Y. Performance analysis of self-compacting concrete in earthquake-resistant buildings. *J. Mech. Energy, Ind. Technol.* 16, 19–28 (2024).
55. Kumar, K. R., Shyamala, G., Awoyera, P. O., Vedhasakthi, K. & Olalusi, O. B. Cleaner production of self-compacting concrete with selected industrial rejects-an overview. *Silicon* 13, 2809–2820 (2021).
56. Velichko, E. G. Problems and prospects of self-compacting concrete mixes for high-strength concrete. *IOP Conf. Ser. Mater. Sci. Eng.* 896 (2020).
57. Sonebi, M. & Abdalqader, A. Recent development and future trends in self-compacting concrete. In *The Institute of Concrete Technology Annual Technical Symposium*, 1–23 (2022).
58. Emborg, M. & Simonsson, P. Industrial concrete construction for a better economy and working environment – possibilities and obstacles with self compacting concrete. *Nord. Concr. Res.* 38, 47–60 (2008).
59. Nehdi, M. L. Only tall things cast shadows: Opportunities, challenges and research needs of self-consolidating concrete in super-tall buildings. *Constr. Build. Mater.* 48, 80–90 (2013).
60. Bumajdad, A., Bouhamra, W., Alsayegh, O. A., Kamal, H. A. & Alhajraf, S. F. *Gulf Conference on Sustainable Built Environment* (2020). https://doi.org/10.1007/978-3-030-39734-0.

61. Peterson, M. *High-Performance and Self-Compacting Concrete in House Building: Field Tests and Theoretical Studies of Possibilities and Difficulties* (Lund University, 2008).
62. Rabehi, R., Fidjah, M., Belakhdar, A. R. & Amieur, M. Analysis of the use of self-compacting concrete in Algeria: Opportunities and challenges. In *4th International Conference on Engineering and Applied Natural Sciences* (2023). https://doi.org/10.59287/as-proceedings.313.
63. Sfikas, I. P. Self-compacting concrete: History and current trends. *Concr.* 1–264 (2017) https://doi.org/10.1002/9781118602164.
64. Bashaveni, B. & Pannem, R. M. R. Development of self-healing property in self compacting concrete. *Case Stud. Constr. Mater.* 20, e02942 (2024).
65. Nandhini, K. & Karthikeyan, J. Influence of industrial and agricultural by-products as cementitious blends in self-compacting concrete – a review. *Silicon* 14, 2431–2452 (2022).

6 The Rise of Zero-Cement Geopolymer Concrete
A Sustainable Revolution in Construction

Shreekanth Birgonda and Senthilkumar

6.1 INTRODUCTION

The construction industry is a major contributor to the world's carbon dioxide (CO_2) emissions due to the extensive usage of Ordinary Portland Cement (OPC), where the production of OPC is energy-intensive and causes large-scale pollution (Dasnish and Mohan, 2020; Kiran Prabha et al., 2025). With the need to curb climate change, reduce the impact on the ecosystem, and shift to greener construction practices, there is a need for new, sustainable practices. In the face of these challenges, geopolymer concrete has come up as an innovative alternative that can change construction practices.

In recent years, the construction industry has embraced several innovative approaches to minimize carbon dioxide emissions, with geopolymer concrete emerging as one of the most sustainable alternatives to traditional concrete production. Unlike conventional concrete that depends on OPC, geopolymer concrete doesn't use any OPC at all. Instead, it makes use of waste materials like fly ash (FA), Ground Granulated Blast Furnace Slag (GGBFS), and silica fume. These materials are rich in aluminosilicates, and, when mixed with alkaline liquids like sodium hydroxide or sodium silicate, they work like glue to hold the concrete together. This not only helps reduce waste going to landfills but also cuts down the carbon emissions linked to cement production.

The scientific base of geopolymer concrete is in geopolymerization a chemical process that turns aluminosilicate materials into a largescale polymeric structure. Out of this process comes a binder that presents great mechanical performance, improved chemical stability, better high temperature resistance, and long-term durability. These traits make geopolymer concrete a very attractive material for infrastructure development that in turn is very much what regions looking for sustainable and resilient construction solutions are after.

The last few years have added remarkable advancements with respect to geopolymer concrete's formulation, performance, and applications in the field. Confidence in adoption has been further aided by pilot projects and real-world structures proving the feasibility and effectiveness of geopolymer technology. Still, the lack of standardization, large-scale production, curing practices, and consistency in the availability

DOI: 10.1201/9781003664284-6

of precursor materials pose ongoing concerns. This chapter describes the development of zero-cement geopolymer concrete, focusing on the composite materials, their reactions, and performance attributes. It studies the mechanical behavior, durability in aggressive environments, thermal properties, and environmental impact relative to OPC-based concrete. The chapter also discusses the barriers to mainstream implementation and the strategies needed to overcome them. Ultimately, this chapter aims to position geopolymer concrete not simply as a substitute for traditional cementitious materials, but as a sustainable revolution with the potential to reshape the future of global construction.

6.2 EMERGENCE OF GEOPOLYMER TECHNOLOGY

Geopolymer technology has emerged as a sustainable and technically promising alternative to traditional OPC systems. First introduced by Joseph Davidovits in the late 1970s, the core concept of geopolymers lies in the alkali activation of aluminosilicate-rich materials such as FA, ground granulated blast furnace slag (GGBFS), and calcined clays. Under highly alkaline conditions, these materials undergo dissolution, reorganization, and polycondensation reactions that lead to the formation of three-dimensional amorphous to semi-crystalline polymeric networks. Unlike the hydration-based setting of OPC, geopolymerization forms binding gels like N–A–S–H and C–A–S–H, which provide high mechanical strength, excellent chemical resistance, and thermal stability. A key advantage of geopolymer concrete is its significantly lower carbon footprint, primarily due to the use of industrial by-products and the absence of energy-intensive clinker production. This makes geopolymer systems particularly attractive for eco-friendly and low-carbon infrastructure applications.

Currently, most geopolymer concretes are produced using two-part systems where liquid alkali activators (e.g., sodium hydroxide and sodium silicate) are mixed with solid precursors. These systems have shown strong performance in both precast and in-situ applications, especially where curing can be controlled and skilled labor is available. Although construction methods and equipment are generally compatible with geopolymer concrete, challenges such as limited workability, raw material variability, and lack of compatible admixtures still exist. Looking toward broader industrial adoption, significant research is now focused on developing one-part or "just-add-water" geopolymer systems. These systems incorporate solid activators within the dry mix, simplifying handling and enabling geopolymer concrete to be prepared in a manner similar to OPC-based mixes. As reported by Mahendra et al. (2024), one-part alkali-activated materials (AAMs) display high strength at ambient cure that also removes the need to work with corrosive liquid alkalis. In terms of the geopolymer field, we see that which goes into the future is a result of admixture development, low-alkali activator formulation, and long-term durability studies that in turn will improve the performance and reliability of the materials in many climate and structure types. Also with the global trend toward sustainability and the growth of performance-based design specs, geopolymer binders are at a point of great change that sees them take a large role in the push for green, more sustainable construction practices.

6.2.1 Chemistry of Geopolymerization: Role of Activators and Precursors

Geopolymerization is a chemical process where aluminosilicate-rich precursors – such as FA, ground granulated blast furnace slag (GGBFS), and metakaolin – react with alkaline activators to form a hardened binder. This reaction occurs through three main stages: dissolution, gelation (polycondensation), and hardening. In the dissolution phase, the alkaline environment – typically created using sodium hydroxide (NaOH), sodium silicate (Na_2SiO_3), or potassium hydroxide (KOH) – breaks the Si–O–Si and Al–O–Si bonds in the raw materials, releasing $Si(OH)_4$ and $Al(OH)_4$ –. These reactive species undergo polycondensation to form a three-dimensional network of aluminosilicate gels. In low-calcium systems, the dominant product is sodium aluminosilicate hydrate (N–A–S–H), as represented in Equation (6.1), where wH_2O denotes the bound water molecules.

$$nSi(OH)_4 + mAl(OH)_4 - +Na^+ \rightarrow Na^+.(Si\text{–}O\text{–}Al\text{–}O)n.wH_2O \quad (6.1)$$

In calcium-rich systems (e.g., GGBFS-based), calcium ions lead to the formation of calcium aluminosilicate hydrate (C–A–S–H) gel:

$$xCa_2^+ + ySi(OH)_4 + zAl(OH)_4^- \rightarrow C\text{–}A\text{–}S\text{–}H\,gel + H_2O \quad (6.2)$$

Hybrid gels such as $C\text{–}(N)\text{–}A\text{–}S\text{–}H$ may also form in mixed-precursor systems, contributing to improved mechanical and durability performance. The precursor type and composition significantly affect gel chemistry and setting behaviors. Class F FA favors $N\text{–}A\text{–}S\text{–}H$ formation and provides good long-term durability, but requires heat curing for early strength. GGBFS promotes $N\text{–}A\text{–}S\text{–}H$ formation, enabling ambient curing and faster strength gain. Ternary blends with silica fume improve packing density, reduce porosity, and increase strength. Such systems have demonstrated compressive strengths exceeding 130 MPa under ambient curing (Jagad et al., 2023; Mahendra et al., 2024).

The alkaline activator system – comprising NaOH and Na_2SiO_3 – greatly influences reaction kinetics and final properties. NaOH assists in dissolving precursors, while Na_2SiO_3 supplies soluble silica to accelerate gelation. Optimal NaOH molarity typically ranges from 8 M to 14 M, with 12 M often producing the best strength. However, higher molarity can reduce workability and cause efflorescence, while lower molarity may result in incomplete reactions (Oleiwi et al., 2018; Oyejobi et al., 2023).

The sodium silicate to sodium hydroxide ratio (SS/SH), generally between 1.5 and 2.5, also affects workability and strength. Higher SS/SH ratios improve compressive strength but may reduce flowability and accelerate setting. In warm climates or large pours, retarders may be necessary (Hadi et al., 2019).

Sodium-based activators are typically favored for scalability and cost-efficiency. However, solid activators like powdered sodium metasilicate have enabled one-part geopolymer systems, which only require water addition and offer compressive strengths above 60 MPa under ambient curing (Mahendra et al., 2024). Similarly, low-silica-modulus sodium silicate without NaOH has produced strengths over 100 MPa in slag-based systems (Ukritnukun et al., 2020).

6.3 MIX DESIGN OF ZERO-CEMENT GEOPOLYMER CONCRETE

The mix design method for zero-cement geopolymers concrete is different compared to conventional concrete. The choice of activator, whether it's sodium hydroxide, potassium hydroxide, or sodium silicate, also impacts strength gain, setting time, and overall performance of the concrete. Activator-to-binder (A/B) ratios are usually between 0.4 and 0.6, and sodium silicate-to-NaOH (SS/SH) ratios are between 2.0 and 2.5. Low porosity and good workability are ensured by a water-to-binder (W/B) ratio of 0.30 to 0.40. Aggregates are proportioned similar to OPC mixes (approx. 1:1.5:2.5–3.0 for binder: fine: coarse), with a 30–35% paste volume recommended, especially in self-compacting variants. Curing plays a vital role, with heat curing benefiting low-calcium systems, while high-calcium mixes (e.g., GGBFS-rich) perform well under ambient curing. Molarities of 8M–16M NaOH are commonly used, with 12M offering an effective balance between reactivity and durability. Studies such as those by Mahendra et al. (2024), Jagad et al. (2023), and Hadi et al. (2019) highlight the benefits of blending multiple binders and optimizing mix ratios to achieve strengths exceeding 100 MPa. Overall, the performance of geopolymer concrete is governed by the interaction between precursor chemistry, activator composition, and mix design. Achieving the right balance between $N-A-S-H$ and $C-A-S-H$ gel formation, through a careful selection of raw materials and activators, is key to producing durable, strong, and sustainable alternatives to OPC-based concrete.

6.4 SETTING TIME OF GEOPOLYMER CONCRETE

The type of precursor material, alkaline activator composition, and curing conditions are the main factors that affect how long geopolymer concrete takes to set. Setting time in conventional concrete is controlled by hydration, and geopolymer concrete undergoes dissolution and polymerization reactions. Mixes rich in GGBFS tend to set faster due to their high calcium content and greater reactivity in alkaline conditions (Hadi et al., 2019; Ukritnukun et al., 2020). On the other hand, systems based mainly on low-calcium Class F fly ash often have delayed setting at room temperature, as the aluminosilicates dissolve more slowly. To overcome this, additives like alccofine, known for their high reactivity, have been used to boost early strength and speed up gel formation (Parveen et al., 2018). One practical way to strike a balance between workability and performance is by blending FA with GGBFS (Rafeet et al., 2017; Mahendra et al., 2024). Junaid et al. (2015) reported that lowering the water-to-geopolymer solids (W/GPS) ratio and increasing the alkaline liquid-to-fly ash (AL/FA) ratio helped accelerate setting under heat-curing conditions. A higher molarity of NaOH or a higher sodium silicate-to-NaOH (SS/SH) ratio can trigger rapid polymerization, leading to quicker setting, but this can also reduce working time and make the mix harder to handle. For example, while 12M NaOH improves strength, it can also lead to very early setting (Oyejobi et al., 2023). SS/SH ratios above 3.0 can be helpful in some cases but may also reduce setting time too much. Overall, setting time in geopolymer concrete can range from just a few minutes to several hours, depending on curing conditions, activator concentration, and the type of precursor used.

6.5 FRESH AND HARDENED PROPERTIES

6.5.1 Fresh Properties

There are no specified fresh property tests available for geopolymer concrete right now. All the tests are carried out similar to conventional concrete tests only. The behavior of conventional concrete and geopolymer concrete is not similar, so following similar fresh properties tests as conventional will not be suitable because geopolymer systems often have higher viscosity, mainly due to the thick nature of sodium silicate, which can hinder flow and make testing much difficult. Geopolymer concrete mixes with fly-ash-based show better workability, due to their spherical shape. Whereas mixes with high GGBFS content set rapidly, which makes the concrete more difficult to cast and finish (Hadi et al., 2019; Ukritnukun et al., 2020). SS/SH is the main factor that effects the fresh properties. Higher SS/SH ratio improves strength but reduces the flow properties similar to setting time behavior. According to Jagad et al. (2023), by the addition of microfine materials like silica fume, we can increase the workability by reducing the slump loss by 28%. Eltantawi et al. (2024) demonstrated that ternary blended systems comprising FA, GGBFS, and silica fume can exhibit excellent workability when the precursor and activator proportions are properly optimized. Using a water-to-binder (W/B) ratio between 0.30 and 0.40, along with alkali-resistant superplasticizers or specially formulated polycarboxylate ethers, the mix achieved a high slump flow of 740 mm, indicating exceptional flowability suitable for self-compacting applications. Several studies reported how chemical ratios can enhance flow. Higher SiO_2/Na_2O modulus enhanced both flow and strength, whereas higher Na_2O/binder ratios accelerated setting speed but decreased workability, according to research by Sun et al. (2022) and Pavithra et al. (2016). Excellent flow is achieved for self-compacting applications by keeping the liquid-to-binder (L/B) ratio between 0.4 and 0.5 and the SS/SH ratio between 3 and 4, particularly when using finely divided materials like alccofine or silica fume.

6.5.2 Density of Geopolymer Concrete

The density of geopolymer concrete usually falls between 2,200 and 2,500 kg/m³. GGBFS-rich mixes, along with other dense supplementary cementitious materials like silica fume and microfine particles, tend to produce heavier and more compact concretes because of their refined pore structures and improved gel formation (Jagad et al., 2023; Mahendra et al., 2024). Lightweight aggregates, on the other hand, like foamed glass, pumice, or expanded clay, can drastically lower density to as low as 1,500–1,800 kg/m³, which makes them perfect for thermal insulation or nonstructural uses. For example, Mermerdaş et al. (2017) achieved densities in the range of 1,800–2,100 kg/m³. Although this reduction in density helps lower the dead load and improves thermal performance, it may compromise mechanical strength if not properly optimized. Overall, geopolymer concrete allows for flexible density tailoring through appropriate selection of aggregates and binder systems. High-density systems are suitable for structural use, while low-density formulations meet the requirements of lightweight and energy-efficient construction. Furthermore, higher activator concentrations and optimized NaOH molarity have been shown to increase fresh unit weight, indirectly indicating a denser matrix and reduced porosity (Oleiwi et al., 2018; Oyejobi et al., 2023).

6.5.3 Compressive, Tensile, and Flexural Strength

Geopolymer concrete exhibits excellent compressive strength, often comparable to or exceeding that of traditional OPC-based systems when properly designed. Compressive strengths ranging from 30 MPa to over 100 MPa have been widely reported, depending on the type of precursor, activator composition, and curing conditions. This high strength results from the formation of dense N–A–S–H and C–A–S–H gel networks. Jagad et al. (2023) reported a well-optimized binder blend consisting of 25% FA, 10% SF, 55% GGBFS, and 10% MF to create high-strength concrete (104 MPa). Eltantawi et al. (2024) reported a mixture of GGBFS, FA, and silica fume to achieve 132.7 MPa at 28 days; high-calcium binders like GGBFS contribute to early strength gain even under ambient curing by promoting the rapid formation of C–A–S–H gel. According to Mermerdaş et al. (2017), high curing temperatures (60–90°C) are frequently required for low-calcium systems, such as Class F FA, in order to achieve comparable strength levels. Strength development is significantly influenced by factors like activator-to-binder ratios, NaOH molarity (10–12 M), and the Na_2SiO_3/NaOH ratio (2.0–2.5) (Hadi et al., 2019; Oyejobi et al., 2023). As is common with brittle cementitious materials, geopolymer concrete typically shows tensile strength values between 2 and 4 MPa. However, the addition of reactive additives or optimized binder chemistry can greatly improve this property. Parveen et al. (2018) found that FA-based concrete containing alccofine had a split tensile strength of 3.5 MPa, improving the bonding and microstructure. In one-part self-compacting geopolymer concrete, Mahendra et al. (2024) reported tensile strengths ranging from 3.5 to 4.6 MPa, which they attributed to the formation of compact C/N–A–S–H gels and an appropriate mix balance. Tensile performance is further enhanced in ultra-high-performance systems. Eltantawi et al. (2024) reported the synergy between FA, slag, and silica fume by documenting a 4.9 MPa tensile strength in a ternary-blended geopolymer mix. Mahendra et al. (2024) achieved flexural strength values between 5.5 and 7.2 MPa, while Eltantawi et al. (2024) reported 9.6 MPa in self-compacting ultra-high-performance geopolymer concrete. Flexural strength in geopolymer concrete typically ranges from 3 to 6 MPa, but values as high as 9.6 MPa have been recorded in specialized high-performance mixes. Flexural strength is inherently influenced by the same factors impacting tensile performance, including precursor type, curing, and activator composition.

In summary, geopolymer concrete can achieve high compressive strength along with competitive tensile and flexural performance, making it suitable for structural applications when paired with proper mix design and reinforcement strategies. To unlock its full mechanical potential, a careful optimization of curing conditions, activator concentration, and precursor combinations is essential.

6.5.4 Modulus of Elasticity of Geopolymer Concrete

Modulus of elasticity (MOE) is influenced by several factors, including the type of precursor, curing method, stiffness of the aggregates, and the reactivity of the binder. Fly-ash-based systems often show lower MOE values at early ages due to slower polymerization. In contrast, mixes containing GGBFS or calcined clays tend to achieve higher stiffness, and it is due to the formation of denser C–A–S–H gel networks (Hadi et al., 2019; Ukritnukun et al., 2020). The MOE in geopolymer concrete typically ranges from 20 to 35 GPa, which is comparable to or slightly lower than that of OPC concrete with similar strength levels. Eltantawi et al. (2024) reported MOE values above 35–40 GPa, for self-compacting ultra-high-performance geopolymer concrete, and this is due to the fact that dense microstructures and well-developed gel networks contribute to increased stiffness. Furthermore, better matrix compactness and indirectly higher MOE have been linked to higher activator molarity and improved packing density. Slag-rich, well-optimized geopolymer mixes can provide MOE values appropriate for structural applications, but more standardized testing is required.

6.6 DURABILITY CHARACTERISTICS

Geopolymer concrete has shown outstanding durability in a variety of chemically aggressive environments, often outperforming traditional conventional concrete. Geopolymer concrete exhibits superior resistance to sulfate and acid attacks compared to conventional concrete. In traditional Portland cement systems, the presence of calcium hydroxide leads to the formation of expansive products like ettringite when exposed to sulfate ions, resulting in cracking and deterioration. In contrast, geopolymer concrete – particularly low-calcium systems based on FA – lacks free calcium hydroxide. Instead, it forms stable N–A–S–H or C–A–S–H gels, which enhance chemical durability and reduce vulnerability to aggressive environments. Multiple studies have confirmed that geopolymer concrete retains its strength and integrity even in sulfate-rich environments. For example, Sun et al. (2022) found that sealed curing combined with optimized SiO_2 /Na_2O ratios resulted in dense microstructures that resist sulfate penetration. Similarly, Eltantawi et al. (2024) reported excellent sulfate resistance in ternary blends of FA, GGBFS, and silica fume, attributing it to a refined pore structure and increased silica content. In acidic environment, the performance of geopolymer concrete largely depends on the type of precursor and acid. GGBFS, having high calcium content, can be somewhat vulnerable due to Ca^{2+} leaching and partial degradation of C–A–S–H gels. In contrast, Fly ash which have low calcium generally offer better acid resistance. Parveen et al. (2018) observed that FA-based geopolymer concrete cured at ambient temperatures had minimal strength and mass loss when exposed to acid. This resistance was further improved by adding alccofine, which densifies the matrix and reduces porosity. Dense microstructures reported by Mahendra et al. (2024) and Jagad et al. (2023) also contributed to better chemical resistance.

Geopolymer concrete also demonstrates superior resistance to chloride ion penetration compared to conventional concrete. In traditional systems, calcium hydroxide reacts with chloride ions to form Friedel's salt, which can destabilize over time and contribute to the corrosion of embedded steel reinforcement. In contrast, geopolymer concrete forms a dense microstructure with tightly bound aluminosilicate gel networks such as N–A–S–H and C–A–S–H that significantly reduce chloride ingress. Additionally, the binding capacity of GGBFS further restricts chloride mobility, while the high-pH environment helps maintain the passive layer on steel reinforcement, enhancing long-term durability in chloride-rich environments. Eltantawi et al. (2024) demonstrated that a ternary blend of GGBFS, FA, and silica fume formed a dense matrix with minimal chloride ingress. Blends based on FA and GGBFS, especially those enhanced with silica fume or alccofine, have demonstrated strong resistance to chloride penetration, further proving their potential in aggressive environments. The year 2022 was critical for minimizing capillary pores and improving gel formation. Although direct chloride diffusion coefficients were not always reported, improvements were inferred from indicators like low sorptivity, reduced permeability, and compact microstructures.

Carbonation is the major problem when high-volume SCMs are used in conventional concrete. Geopolymer concrete demonstrates superior performance over conventional concrete systems when it comes to carbonation resistance. In OPC concrete, carbonation typically occurs as CO_2 reacts with calcium hydroxide, leading to a drop in pH and an increased risk of steel corrosion. In contrast, geopolymer concretes made with low-calcium precursors contain little to no free $Ca(OH)_2$, which alters the carbonation process. In high-calcium mixes, carbonation can lead to the decalcification of C–A–S–H gels, but ternary blends and the addition of silica fume have been shown to reduce CO_2 ingress and improve durability (Sun et al., 2022; Eltantawi et al., 2024).

Geopolymer concrete shows better performance when concrete is subjected to elevated temperatures. Jagad et al. (2023) reported that geopolymer mixes made with FA, glass powder, and steel slag showed strength gains at 400°C and retained over 60–70% of their strength at 800°C. This resilience is linked to continued geopolymerization at moderate heat levels and the thermal stability of N–A–S–H gels. Thermal analysis tools like TG-DTA and X-Ray Diffraction (XRD) confirm that while calcium-based gels may degrade under heat, aluminosilicate networks remain mostly intact (Eltantawi et al., 2024; Jagad et al., 2023).

Geopolymer concrete offers exceptional durability, resisting sulfate and acid attack, chloride ingress, carbonation, and thermal damage far better than conventional OPC concrete. Its dense microstructure, chemically stable gels, and low permeability make it well suited for demanding environments such as marine structures, sewage systems, wastewater treatment facilities, and high-temperature zones. With careful selection of materials, activator chemistry, and curing practices, geopolymer concrete stands out as a sustainable and high-performance solution for critical infrastructure exposed to harsh conditions.

6.7 CURING METHODS AND TEMPERATURE EFFECTS

Curing has a major impact on how well geopolymer concrete works. Strength development, setting, and durability are all impacted by the choice between heat and ambient curing. As demonstrated by Junaid et al. (2015) and Mermerdaş et al. (2017), heat curing, usually at 60 to 90°C, speeds up geopolymerization and permits quick early strength, particularly in low-calcium FA systems. However, excessive heat can cause moisture loss, shrinkage, and microcracking, making it more suitable for controlled precast applications. Ambient curing, though slower in strength gain, is more practical for in-situ and sustainable construction when reactive binders like GGBFS, metakaolin, and silica fume are incorporated. Studies by Jagad et al. (2023), Mahendra et al. (2024), and Kong et al. (2024) demonstrate high compressive strengths (>60 MPa) under ambient conditions using optimized slag-rich or machine-learning-designed mixes. Sealed curing further improves moisture retention and gel formation. Temperature strongly influences gel development (N–A–S–H, C–A–S–H), and while heat aids early strength, ambient curing often results in better long-term performance when high-reactivity binders and proper curing practices are used, making it a viable, energy-efficient alternative for structural applications.

6.8 APPLICATIONS AND CASE STUDIES

Geopolymer concrete has proven effective in diverse real-world applications, including pavements, precast elements, retaining walls, marine infrastructure, and sewer systems, particularly in countries like Australia, India, China, and the Netherlands. Early large-scale implementations, such as VicRoads' ambient-cured FA pavements, demonstrated excellent strength and durability. In India, GGBFS-rich geopolymer mixes are used in precast bridge components and railway sleepers due to their rapid setting and early strength. Its acid and sulfate resistance has led to its use in wastewater and chemical environments across Europe and China, while 3D printing with geopolymer pastes is emerging for sustainable, complex construction. The material's rapid setting and high early strength make it ideal for precast components like blocks, sleepers, and panels, especially when heat cured. Its chemical durability also suits marine and industrial environments. Additionally, geopolymer concrete is gaining ground in 3D concrete printing for its extrudability, shape retention, and low carbon footprint. In harsh conditions – such as coastal zones, waste treatment plants, and fire-prone areas – it offers superior durability due to its dense microstructure, low permeability, chemical resistance, and thermal stability up to 1,000°C. Its absence of $Ca(OH)_2$ prevents damaging expansive reactions, making it highly suitable for long-term performance in aggressive environments. These case studies confirm geopolymer concrete as a sustainable and durable alternative to OPC systems across a range of demanding applications.

6.9 ENVIRONMENTAL AND ECONOMIC BENEFITS

Geopolymer concrete offers substantial environmental and economic benefits over traditional OPC concrete, primarily through reduced carbon emissions, resource efficiency, and long-term performance. Life Cycle Assessment (LCA) studies show a 40–80% reduction in CO_2 emissions due to the elimination of clinker production and the use of industrial by-products like FA and GGBS, supporting circular economy goals. However, the production and transport of alkaline activators – especially sodium silicate – can raise environmental impacts, though this is being mitigated through low-carbon alternatives and local sourcing. Geopolymer concrete also exhibits extended service life and lower maintenance needs, further improving life cycle sustainability. In terms of carbon footprint, the avoidance of high-temperature processing and the use of waste-based binders make geopolymer systems a low-carbon construction option. Optimized activator ratios and ambient curing can maximize these benefits. Economically, while alkaline activators can increase initial costs – particularly in two-part systems – the use of local by-products, energy savings in curing, and reduced repair needs make geopolymer concrete cost-competitive, especially in precast and durability-critical applications. Advances in one-part mixes and low-cost activators are expected to further reduce overall costs, solidifying geopolymer concrete as a sustainable and economically viable alternative to OPC.

6.10 CHALLENGES AND FUTURE PERSPECTIVES

Widespread adoption of geopolymer concrete faces key challenges, including the lack of standardized codes and the difficulty of scaling up production. Unlike OPC systems with established guidelines, geopolymer concretes suffer from variability in raw materials and reaction mechanisms, complicating the development of universal standards. Although organizations like RILEM, ASTM, and some national bodies (e.g., IRC SP:115–2018 in India) are working on performance-based guidelines, further efforts are needed in material classification, durability benchmarks, and training. Scaling up is hindered by inconsistent precursor quality, safety concerns in handling alkaline activators, and incompatibility with conventional batching systems. Solutions such as one-part mixes, modular production units, and automation can improve scalability. Coordinated efforts in code development, supply chain optimization, and infrastructure upgrades – supported by government incentives – are vital for mainstreaming geopolymer concrete.

6.11 CONCLUSION

Zero-cement geopolymer concrete offers a transformative solution to the environmental and technical limitations of conventional concrete. By leveraging industrial by-products and embracing a novel chemical binding mechanism, geopolymer technology presents a credible, high-performance, and low-carbon alternative for the future of sustainable construction. While challenges remain in terms of standardization and large-scale implementation, the growing body of scientific evidence and successful applications indicate a strong potential for its mainstream adoption in global infrastructure development.

Geopolymer concrete has emerged as a promising sustainable alternative to OPC concrete, offering significant advantages in terms of reduced carbon emissions, superior chemical durability, and thermal stability. Through the alkali activation of aluminosilicate-rich industrial by-products such as FA, GGBS, and rice husk ash, geopolymer binders eliminate the need for energy-intensive clinker production and support circular economy practices. Extensive research has demonstrated that, with appropriate mix design and curing conditions, geopolymer concrete can achieve comparable or superior mechanical properties, along with excellent resistance to acid, sulfate, and chloride attack, making it highly suitable for infrastructure exposed to harsh environments.

However, widespread adoption is still hindered by challenges such as high activator costs, lack of standardization, limited supply chain infrastructure, and variability in raw materials. Innovations in low-carbon activators, one-part dry mix systems, admixture development, and data-driven design tools offer pathways for overcoming these barriers. Standardization efforts, industry training, and performance-based codes will be essential to support market acceptance and scalability.

In conclusion, geopolymer concrete represents a next-generation green construction material, aligning with global goals for carbon neutrality, resource efficiency, and infrastructure resilience. Continued interdisciplinary research, supported by policy and industry collaboration, will be key to realizing its full potential in mainstream construction.

REFERENCES

Dasnish, P., & Mohan, G. G. (2020). Behaviour of self-compacting concrete using different mineral powders additions in ternary blends. *Romanian Journal of Materials*, *50*(2), 232–239.

Eltantawi, I., Sheikh, M. N., & Hadi, M. N. S. (2024). Design of a novel ternary blended Self-compacting ultra-high-performance geopolymer concrete. *Construction and Building Materials*, *451*. https://doi.org/10.1016/j.conbuildmat.2024.138819

Hadi, M. N. S., Zhang, H., & Parkinson, S. (2019). Optimum mix design of geopolymer pastes and concretes cured in ambient condition based on compressive strength, setting time and workability. *Journal of Building Engineering*, *23*, 301–313. https://doi.org/10.1016/j.jobe.2019.02.006

Jagad, G., Modhera, C., Patel, D., & Patel, V. (2023). Mechanical and microstructural behavior of high strength geopolymer concrete inclusion of various industrial wastes. *Innovative Infrastructure Solutions*, *8*(6). https://doi.org/10.1007/s41062-023-01149-y

Junaid, M. T., Kayali, O., Khennane, A., & Black, J. (2015). A mix design procedure for low calcium alkali activated fly ash-based concretes. *Construction and Building Materials*, *79*, 301–310. https://doi.org/10.1016/j.conbuildmat.2015.01.048

Kiran Prabha, M., Vishnu Vardhan, K., Santhi, A. S., & Mohan Ganesh, G. (2025). Advancements in self-compacting concrete reinforced with basalt fiber: A comprehensive review. *Engineering Reports*, *7*(5), e70147. https://doi.org/10.1002/eng2.70147

Kong, Y. K., Kurumisawa, K., Tokoro, C., Li, Z., & Chu, S. H. (2024). Data-driven inverse mix design for sustainable alkali-activated materials. *Journal of Sustainable Cement-Based Materials*, *13*(12), 1857–1878. https://doi.org/10.1080/21650373.2024.2416962

Mahendra, K., Narasimhan, M. C., Prakash, G. B., & Das, A. K. (2024). Experimental investigation and optimization of one-part alkali-activated self-compacting concrete mixes. *Case Studies in Construction Materials*, *21*. https://doi.org/10.1016/j.cscm.2024.e04062

Mermerdaş, K., Algın, Z., Oleiwi, S. M., & Nassani, D. E. (2017). Optimization of lightweight GGBFS and FA geopolymer mortars by response surface method. *Construction and Building Materials*, *139*, 159–171. https://doi.org/10.1016/j.conbuildmat.2017.02.050

Oleiwi, S. M., Algın, Z., Nassani, D. E., & Mermerdaş, K. (2018). Multi-objective optimization of alkali activator agents for FA- and GGBFS-based geopolymer lightweight mortars. *Arabian Journal for Science and Engineering*, *43*(10), 5333–5347. https://doi.org/10.1007/s13369-018-3170-x

Oyejobi, D., Jameel, M., Adewuyi, A., Aina, S., Avudaiappan, S., & Maureira-Carsalade, N. (2023). Analyzing influence of mix design constituents on compressive strength, setting times, and workability of geopolymer mortar and paste. *Advances in Civil Engineering*, *2023*. https://doi.org/10.1155/2023/5522056

Parveen, Singhal, D., Junaid, M. T., Jindal, B. B., & Mehta, A. (2018). Mechanical and microstructural properties of fly ash based geopolymer concrete incorporating alccofine at ambient curing. *Construction and Building Materials*, *180*, 298–307. https://doi.org/10.1016/j.conbuildmat.2018.05.286

Pavithra, P., Srinivasula Reddy, M., Dinakar, P., Hanumantha Rao, B., Satpathy, B. K., & Mohanty, A. N. (2016). A mix design procedure for geopolymer concrete with fly ash. *Journal of Cleaner Production*, *133*, 117–125. https://doi.org/10.1016/j.jclepro.2016.05.041

Rafeet, A., Vinai, R., Soutsos, M., & Sha, W. (2017). Guidelines for mix proportioning of fly ash/GGBS based alkali activated concretes. *Construction and Building Materials*, *147*, 130–142. https://doi.org/10.1016/j.conbuildmat.2017.04.036

Sun, B., Sun, Y., Ye, G., & De Schutter, G. (2022). A mix design methodology of slag and fly ash-based alkali-activated paste. *Cement and Concrete Composites*, *126*. https://doi.org/10.1016/j.cemconcomp.2021.104368

Ukritnukun, S., Koshy, P., Rawal, A., Castel, A., & Sorrell, C. C. (2020). Predictive model of setting times and compressive strengths for low-alkali, ambient-cured, fly ash/slag-based geopolymers. *Minerals*, *10*(10), 1–21. https://doi.org/10.3390/min10100920

7 Alkali-Activated Permeable Geopolymer Concrete Technologies

Carbon Neutral Solutions Towards Construction of Climate Resilient Cities

Sheh Ching, Khong, Jia Jun Yee, Kong Fah Tee, Jolius Gimbun and Siew Choo Chin

7.1 INTRODUCTION

Permeable concrete, also known as pervious concrete or porous concrete, is a specially engineered material that allows rainwater to flow directly through its open void structure into the soil below. Designed primarily for stormwater infiltration, it reduces surface runoff by channelling water into the ground rather than letting it accumulate on hard surfaces. Over the past decade, climate change has intensified rainfall patterns, bringing heavier and more frequent storms. At the same time, urban expansion has led to extensive paving of land, which increases impervious surface area and exacerbates flooding risks. As a result, urban environments face more frequent issues with water ponding and flash floods. Consequently, these concretes help alleviate water ponding on road surfaces and lower the risk of flash flooding. They are increasingly heralded as a forward-looking material for building climate-resilient infrastructure that supports the United Nations' Sustainable Development Goal 11: Sustainable Cities and Communities.

To produce these concretes, coarse aggregates were used with minimal or no fine aggregates and cement as the binding agent. Cement is a crucial component in concrete production; its manufacturing process consumes non-renewable raw minerals and emits substantial amounts of greenhouse gases. To mitigate the construction industry's reliance on cement, recent studies introduced an innovative binder called geopolymer or alkali-activated cement. This alternative reportedly eliminates the need for traditional cement while maintaining comparable binding and hardening characteristics. The adoption of this novel binder aims to decrease carbon emissions associated with cement production and strives for carbon neutrality by utilizing industrial waste materials (Tee and Mostofizadeh, 2021; Gharehbaghi et al., 2023).

DOI: 10.1201/9781003664284-7

The emergence of this innovative geopolymer binder spurred a surge of research focused on integrating geopolymer technology into permeable concrete, paving the way for eco-friendly, sustainable, and resilient infrastructure development. This chapter aims to thoroughly review the underlying chemistry and engineering principles of alkali-activated and geopolymer binders, focusing in particular on advances in their application within permeable concrete systems. It synthesizes current research on how these sustainable binding solutions, especially those derived from fly ash (FA), slag, metakaolin (MK), and other industrial by-products, can be formulated to deliver both permeability and strength in pervious concrete applications. The chapter concludes by introducing an ongoing experimental study led by the authors and other relevant studies and presents preliminary studies conducted on the utilization of spent coffee grounds (SCG) in cementitious materials and its potential towards the realization of SCG in permeable geopolymer concrete (PGC) mixes.

7.2 HISTORICAL CONTEXT OF ALKALI-ACTIVATED MATERIALS AND GEOPOLYMER TECHNOLOGY

Alkali-activated materials (AAMs) are synthetic mineral binders produced through a chemical process known as alkali activation. They have gained significant attention in recent literature as a sustainable alternative to traditional Ordinary Portland Cement (OPC), primarily due to their potential to reduce CO_2 emissions and utilize industrial by-products or waste materials with little to no commercial value. These materials can harden and exhibit mechanical and durability properties comparable to conventional cementitious systems (Nehdi and Yassine, 2020; Chuang et al., 2023; Tee and Mostofizadeh, 2024; Mostofizadeh and Tee, 2024). The earliest recorded work on alkali activation dates back to a patent by Whiting in 1895 (Whiting, 1895). In the 1950s, Victor Dmytrovych Glukhovsky and his collaborators expanded on this concept by investigating the alkali activation of natural soils and industrial wastes, culminating in the publication of the book *Ground Silicates* in 1952 (Glukhovsky, 1952). Their research primarily explored the production of construction materials through the reaction of sodium silicate with locally sourced solid precursors (Krivenko, 2017). In his work on soil silicates, Glukhovsky emphasized the importance of incorporating additives into soil–silicate mixtures to accelerate setting times without requiring elevated temperature curing, as well as to enhance water resistance and long-term stability.

The conceptual groundwork for geopolymer technology was established in the early 1970s by Professor Joseph Davidovits (Davidovits, 2002). He introduced the term "geopolymer" around 1978 to describe a class of inorganic polymeric materials derived through the alkali-driven activation of aluminosilicate sources such as MK, FA, or slag, which transform into three-dimensional, cement-like binders. Davidovits initially sought to develop non-flammable ceramic materials, but as the binder's cement-like properties and notably low carbon footprint became evident, it quickly gained attention as a sustainable alternative within the construction sector. His pioneering research ultimately inspired numerous studies on geopolymer applications in concrete as a promising low-CO_2 substitute for traditional OPC.

There has been a marked increase in studies investigating the properties of geopolymers and AAMs (Gökçe, 2024). However, much of the literature blurs the terminology, often using "geopolymer" as a catch-all for AAMs despite important differences between the two. Both classes arise from the reaction of aluminosilicate precursors with alkaline activators, but they differ significantly in composition and chemical mechanism. Geopolymers typically derive from low-calcium precursors and form a covalently bonded, three-dimensional aluminosilicate polymer network. In contrast, many AAMs, including high-calcium formulations, solidify through hydration reactions rather than true polymerization. Provis (2014) clarifies that geopolymers are a specific subset of AAMs, distinguished by their low-calcium origin and true inorganic polymer structure, not merely hydration-based products. In summary, while both geopolymers and AAMs are synthetically activated mineral binders, geopolymers are the low-calcium, polymer-type subset, whereas AAMs encompass a wider range, including high-calcium systems formed through hydration processes.

Ground granulated blast furnace slag (GGBFS) is frequently preferred over FA or MK as a precursor for alkali-activated binders due to its relatively higher reactivity in moderately alkaline environments. Slag's high calcium content facilitates faster reaction kinetics and broadens the choice of alkaline activators – including carbonate and sulphate-based solutions, in addition to hydroxides and silicates commonly used in low-calcium geopolymer systems. Because slag contains substantial amounts of CaO, alkali activation of slag typically yields calcium-containing C–A–S–H gels, often accompanied by C–(N)–A–S–H gels in hybrid systems incorporating both slag and FA. In contrast, geopolymer systems – particularly low-calcium formulations – are usually produced from FA and metakaolin (often metakaolin grade MK 750), activated with strong alkali solutions like NaOH, KOH, or LiOH and sodium or potassium silicates. According to Davidovits and widely adopted in geopolymer chemistry, the process begins with an "alkalination" (alkaline dissolution) of aluminosilicate precursors, followed by depolymerization of silicate structures. This leads to the formation of oligomeric sialates that undergo polycondensation into a three-dimensional polymer network, ultimately resulting in geopolymer solidification, distinct from hydrate formation in conventional binders.

As this chapter focuses on reviewing alkali-activated binders, geopolymers, and soil-silicate technologies specifically in the context of permeable concrete, it is unnecessary to explore in depth the distinctions among these three fields. Instead, this section provides essential background and context on the evolution of low- or no-cement binder systems, summarized in Table 7.1. Subsequent sections will discuss applications of these materials in permeable concrete, employing the terminology and conceptual frameworks preferred by each respective research group.

7.3 ALKALI-ACTIVATED MATERIALS AND GEOPOLYMER TECHNOLOGY IN PERMEABLE CONCRETE

Recent studies, such as Xu et al. (2023), have explored the integration of AAM and geopolymer technologies into permeable concrete applications. This approach

TABLE 7.1
Summary of the Different Aam and Geopolymer Concepts by Different Researchers

	V.D. Glukhovsky/ Ponomar et al. (2023)	J. Davidovits/ Davidovits (1994, 2002)	J. L. Provis/Provis (2014)
Main concept	Soil silicates	Geopolymerization	Alkali-activation
Chemical reaction	Adhesive properties of sodium silicate	(Ca, K)-poly(sialate-siloxo) polycondensation	C-A-S-H gel C–(N)–A–S–H gel N-A-S-H gel
Precursors	Fly ash, granulated blast furnace slag, bottom ash	Fly ash and metakaolin	Blast furnace slag
Alkaline medium	Alkali silicates	Alkali hydroxides and alkali silicates	Alkali carbonates, alkali sulphates, alkali hydroxides, and alkali silicates
Additives	Metal oxides, salt solutions, sodium hydroxides, surfactants, and solid powder additives (such as GBFS)	-	-

involves aluminosilicate precursors sourced from single or multiple origins and combining them with alkaline activators such as alkali metal carbonates, sulphates, hydroxides, or silicates. Natural or recycled coarse aggregates are incorporated, while fine aggregates are deliberately omitted to create internal pores within the concrete matrix. These pores facilitate water infiltration, enhancing the permeability of the concrete. Such characteristics make AAM and geopolymer-based permeable concretes effective for constructing resilient urban infrastructures that efficiently manage water seepage.

7.4 ALTERATION TO PRECURSOR MATERIALS OF AAM AND GEOPOLYMER TECHNOLOGY

Based on the required raw materials for AAM and geopolymer-permeable concrete, current researchers are investigating alternative solid precursors capable of replacing the traditional FA and MK. In a study by Huang and Wang (2021), various precursor materials such as MK, FA, and granulated blast furnace slag (GBFS) were evaluated for their impact on geopolymer permeable concrete, using sodium hydroxide (NaOH) and sodium silicate $\left(Na_2SiO_3\right)$ as the alkaline activators. MK and FA were not blended; rather, MK was used in the range of 70–100% with GBFS up to 30%, and FA was similarly used between 70 and 100% with GBFS comprising up to 30%. The optimal results were achieved with a mix of 90% MK and 10% GBFS, resulting in a strength of around 22 MPa. In contrast, the best performance for FA was achieved with a mix of 70% FA and 30% GBFS, yielding approximately 15 MPa. The findings confirm that permeable concrete can be produced using either MK or FA separately, each combined with GBFS. Moreover, using 100% MK under ambient-temperature curing resulted in a strength of around 15 MPa; using 100% FA achieved about 12 MPa after 28 days of curing.

In a study by Marathe et al. (2025), ground granulated blast furnace slag (GGBFS) and sugarcane bagasse ash (SBA) were used as precursors and NaOH plus Na_2SiO_3 solution as alkaline activators to produce PGC. When GBFS made up 90% of the precursor with 10% SBA, the compressive strength reached approximately 37.5 MPa, split tensile strength around 2.47 MPa, and flexural strength of 4.1 MPa. In contrast, a mix containing 80% GBFS and 20% SBA yielded compressive strength near 27.8 MPa, split tensile strength of 1.99 MPa, and flexural strength of 3.41 MPa. The higher strengths are attributed to the formation of a denser C–A–S–H matrix driven by an increased Ca/Si ratio and silica-rich hydration products from both precursors. The study demonstrated that SBA-free mixes could yield workable permeable concrete with 100% GBFS, with compressive strength of about 32.2 MPa, split tensile strength around 2.14 MPa, and flexural strength approximately 3.92 MPa. This highlights the potential for using both organic and agricultural wastes, such as SBA, to enhance the mechanical performance of PGC.

In a study by Azad et al. (2022), FA and silica powder were combined to develop a PGC. Silica powder was added at 2–5% by weight of FA, and activators consisted of Na_2SiO_3 and NaOH. The resulting mixes yielded compressive strengths ranging from 23.5 MPa to 39.9 MPa and flexural strengths between 1.5 MPa and 2.2 MPa. The highest compressive strength (39.9 MPa) was achieved with the silica powder additive, demonstrating that such additives can improve PGC performance. The mix with 39.9 MPa strength suggests potential for structural applications.

Across the reviewed studies, researchers have applied various strategies to enhance PGC performance, including substituting conventional precursors with locally abundant industrial or agricultural wastes, employing organic agricultural residues as aluminosilicate sources, and incorporating additives to improve mechanical and durability properties.

7.5 ALTERATION TO THE ALKALINE MEDIUM OF AAM AND GEOPOLYMER TECHNOLOGY

Across studies, including the work of Zaetang et al. (2015), researchers have explored the impact of NaOH concentration in the alkaline activator on PGC produced from FA and bottom ash precursors. These investigations consistently show that raising the NaOH molarity from about 10 M to 15 M generally enhances compressive strength, indicating improved geopolymerization and binder performance under higher alkaline concentrations. The strength gain was attributed to the geopolymer binder forming a denser, low-porosity matrix as NaOH concentration increased, as higher NaOH levels enhance the dissolution of silica and alumina ions and accelerate geopolymerization. However, researchers noted that incorporating OPC could be advantageous, as it promotes the formation of hydration products such as C–S–H and C–A–S–H gels, which complement geopolymer bonds and reinforce the matrix.

In a study by Chen et al. (2019), using red mud and GBFS as precursors, it was demonstrated that the molar ratio of the alkaline activator significantly influences the properties of PGC. Specifically, increasing the Na_2O/SiO_2 ratio – that is, using more NaOH relative to Na_2SiO_3 – initially improved compressive strength, but beyond a certain point, further increases in the ratio led to a reduction in strength. Although the study did not delve deeply into this mechanism, it likely relates to silica content in the geopolymer mix. When silica levels are high, unreacted free silica may persist, which can undermine strength (Wang et al., 2022). Reducing silica content to a certain threshold can improve mechanical performance by optimizing reactivity, but lowering it too much limits available silica ions, disrupting geopolymerization and preventing the formation of complete polysialate-siloxo networks (Ergeshov et al., 2025), thus weakening the matrix.

7.6 ALTERATION TO AGGREGATES OF PERMEABLE AAM AND GEOPOLYMER CONCRETE

Another aspect that has been studied is the suitable sizes of aggregates to be used for the application into the AAM binder and geopolymer permeable concrete. In a study by Luo et al. (2023), coarse aggregates were sieved and categorized into three size groups – large (15–30 mm), medium (10–15 mm), and small (6–10 mm) before being coated with cementitious material and combined with geopolymer gel from alkaline activator and precursor slurry. The optimum aggregate gradation comprised 40% large, 30% medium, and 30% small particles, which produced a PGC with compressive strength reaching approximately 11.7 MPa, which is acceptable for non-structural applications, all achieved without using any OPC.

Huang and Wang (2021) assessed the influence of aggregate size on geopolymer permeable concrete by dividing aggregates into four size ranges: A1 (0.15–4.75 mm), A2 (0.15–12.5 mm), A3 (0.30–12.5 mm), and A4 (A3 plus 10 % sand). Their results showed group A4 exhibited the highest compressive strength, while group A3 achieved the highest split tensile strength. This indicates that using smaller aggregate sizes increases compressive strength but reduces split tensile performance. Adding 10% sand raised compressive strength further, but decreased tensile strength. The likely explanation is that finer particles have a higher specific surface area, which requires more geopolymer paste for coating, reducing bond thickness and connectivity between coated aggregates, thereby lowering split tensile strength. At the same time, finer particles fill voids more effectively and increase contact points in the geopolymer matrix, reducing porosity and enhancing compressive strength.

Both studies collectively indicate that using smaller coarse aggregates generally improves strength properties, but this choice may negatively affect porosity in permeable concrete. Specifically, smaller aggregate sizes tend to increase compressive strength due to enhanced paste bonding and reduced void space, while larger aggregates decrease compressive strength but raise porosity and permeability.

7.7 INCLUSION OF FIBRES INTO PERMEABLE AAM AND GEOPOLYMER CONCRETE

Another strategy implemented in the PGC is the use of fibre reinforcement. Xu et al. (2023) explored fibre reinforcement in PGC by incorporating steel fibres into an MK-based binder activated with Na_2SiO_3 and NaOH, accompanied by naphthalene as a water-reducing agent. The results showed that adding steel fibres at 0.3% by volume delivered the optimal unconfined compressive strength, exceeding 20 MPa and significantly outperforming mixes without fibres. However, increasing the fibre content to 0.4 % led to a decline in strength. Additionally, porosity decreased as fibre content increased, likely because the fibres bridged the paste and aggregates, which improved strength but reduced permeability.

7.8 POTENTIAL OF FUTURE STUDIES ON PERMEABLE AAM AND GEOPOLYMER CONCRETE

In addition to the strategies discussed above, several other approaches have yet to be explored in the development of PGC. Current research on AAM and geopolymers has primarily concentrated on the use of alkali hydroxides as part of the alkaline activator. While the inclusion of hydroxides has been shown to significantly enhance geopolymer properties, it also introduces several challenges. For instance, Longhi et al. (2019) reported that the use of alkali hydroxides may lead to efflorescence formation on the surface of the geopolymer. Similarly, Aljuaydi et al. (2024) noted that unreacted or weakly bonded alkali hydroxides in AAM can result in the formation of soluble salts on the concrete surface. This occurs due to reactions between alkali hydroxides, carbon dioxide, and water, which can produce alkali carbonates that ultimately compromise the strength and durability of the material (Jeyasangar et al., 2024). Additionally, the reaction between water and alkali hydroxides is highly exothermic (Epstein et al., 2001), which can cause inconsistent geopolymer performance even when identical raw materials are used (Mohamed et al., 2024). These factors necessitate a pre-mixing process before casting to ensure a better control over the final concrete quality.

To avoid these issues, Neupane and Hadigheh (2021) developed a geopolymer binder using only Na_2SiO_3, completely omitting NaOH. Their sodium-hydroxide-free formulation achieved compressive strengths exceeding 60 MPa, outperforming typical OPC samples, demonstrating that high-strength structural-grade geopolymers can be produced without alkali hydroxides. While this promising approach shows that strength comparable to structural applications is attainable, its application in PGC remains underexplored, revealing an important gap in current research.

7.9 POTENTIAL OF SPENT COFFEE GROUNDS (SCGS) IN PERMEABLE AAM AND GEOPOLYMER CONCRETE

The reviewed studies demonstrate that geopolymer binders effectively bond aggregates in permeable concrete. Prior research further indicates that certain organic materials can serve as viable precursors in geopolymer formulations specifically designed for permeable AAM and geopolymer concrete. Building on this, the properties of SCG have been examined for their potential role as an additive in PGC, highlighting an emerging opportunity to integrate organic waste into sustainable binder systems.

SCG was considered for this study due to the availability of SCG wastes globally. Coffee is consumed all over the world, and there is evidence that the coffee culture is even expanding beyond regions that were traditionally known to have no coffee drinking culture (Samoggia and Riedel, 2019). This resulted in the presence of SCGs all over the world. Compared to other wastes, which are not typically shipped overseas, coffee is a globally traded commodity. When these coffees are extracted to make the coffee beverage, it leads to the production of SCG waste. It is estimated that yearly, eight million tons of coffee are produced (Tofalo et al., 2016). For every coffee bean, 65% of it ends up as SCG (Tian et al., 2021). This brings the estimated yearly production of SCG up to 5.2 million tons. When SCGs are left untreated and discarded into landfills, the acidic leachate could lead to soil pollution and even the production of dangerous greenhouse gases (Khan et al., 2023; Forcina et al., 2023). Therefore, a solution is needed to reduce the volume of spent coffee waste, and one potential direction would be its application as an additive in the permeable geopolymer concrete.

7.10 PROPERTIES OF SCGS AND THEIR POTENTIAL IN GEOPOLYMERS

Previous studies have demonstrated that water added during the mixing of geopolymer concrete plays no structural role in bond formation. Instead, it facilitates the dissolution of aluminosilicate precursors and enables the geopolymerization reaction, after which most of the mix water is expelled (Li et al., 2017). If substantial unreacted water remains, its evaporation during curing can create tiny air pockets within the matrix, resulting in micropores that weaken inter-particle bonding and reduce the overall strength of the geopolymer.

To address this issue, SCG can be added to the geopolymer matrix as an additive. The chemical composition of SCG is presented in Figure 7.1. From the chemical composition, SCG is known to have high carbon content (Yee et al., 2024a). According to a study by Soffian et al. (2022), organic materials high in carbon are usually highly absorbent materials and, hence, known for their high fibre content and excellent water absorption capacity, around 70% by weight (Koay et al., 2023). When SCG is incorporated into the geopolymer mix, it can absorb excess mixing water, allowing the geopolymer binder to occupy more space within the matrix. This results in a more compact, denser structure, improving overall strength.

SCGs are rich in phenolic compounds, typically between 4.6 and 9.9 mg per gram (or 16–173.3 mg gallic acid equivalents per gram), making raw SCG potentially toxic when disposed of untreated (Solomakou et al., 2022). Although phenolic compounds offer strong antioxidant and antimicrobial properties, their presence in untreated SCG can harm soil microorganisms, reduce soil fertility, and inhibit plant growth. Nonetheless, these same phenolic characteristics could be harnessed beneficially by using SCG as an additive in geopolymer concrete, turning a problematic waste into a valuable resource.

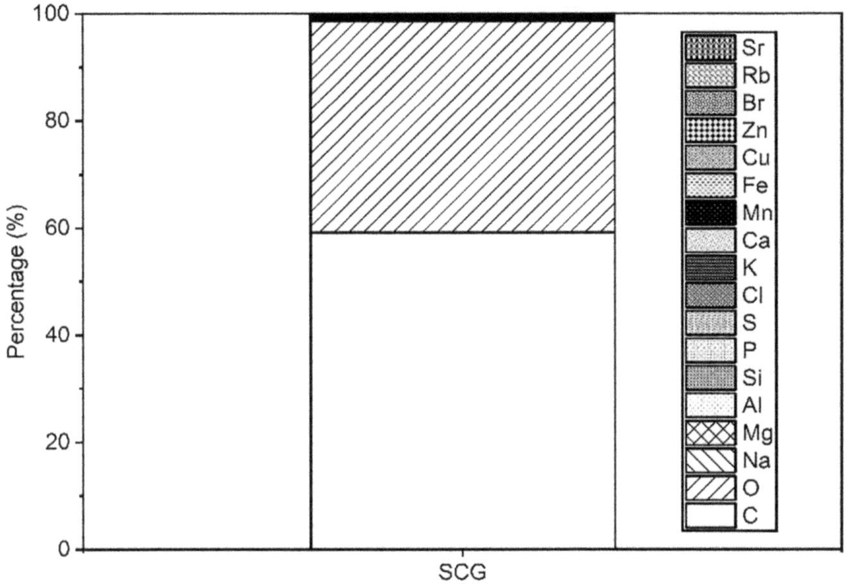

FIGURE 7.1 Chemical composition of SCG particles.

One study found that adding phenolic compounds can enhance the properties of FA-based geopolymers. Typically, FA geopolymer surfaces bear a negative charge due to oxygen-containing groups. Incorporating positively charged phenols (e.g. hydrogen-bearing aromatic rings) can interact electrostatically with the geopolymer surface (Hijazi et al., 2023). This interaction helps form more complex siloxane linkages within the geopolymer network, creating a stronger, better-connected matrix (Zhang et al., 2025). In addition to strengthening the matrix, phenol-enhanced FA geopolymers were also reported to exhibit improved thermal insulation, suggesting enhanced thermal durability (Ren et al., 2023).

Based on the reviewed studies, it is evident that incorporating SCG into the geopolymer matrix offers significant benefits. SCG is known for its high fibre content and phenolic compounds, which contribute positively to the properties of geopolymer binders. Research shows that when SCG is treated and added appropriately, its porous and fibrous structure can enhance mechanical performance and thermal behaviour in concrete applications. This also presents the potential of SCG in not impacting the permeability of permeable concretes while possibly providing improved mechanical strength when applied in permeable geopolymer concretes.

7.11 RESEARCH CONDUCTED ON SCGS AND CONCRETE RESEARCH

Recognizing the SCGs' inherent characteristics and their potential for use in concrete formulations, the following section examines prior research on cementitious materials that incorporated SCG. The studies across different applications of SCG

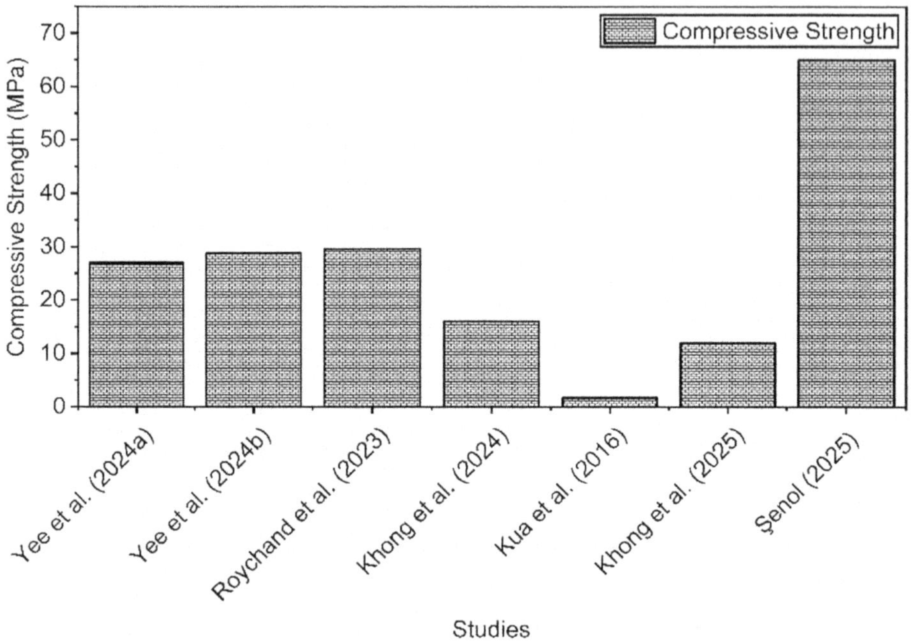

FIGURE 7.2 Best compressive strength of SCG cementitious mix from different studies.

in cementitious materials were reviewed, and the findings of the studies are summarized in Figure 7.2.

In a study by Yee et al. (2024a), SCG was added as an additive to cement mortar mixes. The study shows that SCG with 1% additives served as a filler material by increasing internal frictional surfaces within the mortar matrix, which improved strength. Additionally, SCG was found to absorb surplus mixing water, helping to optimize the water content in the mix and enhance overall performance. With the attempt to identify the best percentage of SCG additives to be added into the cement mortar, Yee et al. (2024b) further investigated its properties through RSM optimization. Upon optimization, it was reported that at 1.12% SCG additives, the strength of the cement mortar could achieve 32.40 MPa, which is higher compared to cement mortar without SCG, presenting a strength of 28.77 MPa.

Roychand et al. (2023) illustrated the application of SCG into conventional concrete mixes by replacing sand. In the research, SCG was turned into biochar through an incineration process at 350°C. Upon 28 days of curing, the concrete, having 15% of its sand replaced by volume, shows a strength of 29.6 MPa. The study also noted that raw SCG and SCG incinerated at temperatures of 500°C are not suitable to be used as sand replacement. Khong and Chin (2024) investigated the use of SCG in permeable concrete, finding that at a water–cement (w/c) ratio of 0.35 and 1% SCG, compressive strengths exceeded 15 MPa, outperforming mixes without SCG. At lower w/c ratios, SCG tended to absorb too much water in the mix, reducing workability and strength across all SCG levels. Conversely, at higher w/c ratios, excess water hindered proper setting, similarly resulting in reduced strength values. Beyond their use in standard cementitious mixes, SCGs have also been evaluated as precursors in geopolymer concretes. In the study by Kua et al. (2016), SCG was used as a precursor in geopolymer cement mixed with 70% SCG, 10% slag, and 20% FA. The alkaline activator comprised equal parts of Na_2SiO_3 and NaOH. After curing for 28 days at 50°C, the geopolymer achieved an unconfined compressive strength of approximately 1.69 MPa.

Another study by Khong et al. (2025) shows the potential application of SCG as additives in a low calcium FA geopolymer mix. In the study, geopolymer concretes made entirely of 100% Class F FA were used as the precursors, while alkali hydroxides were not used in their mix. SCG added as additives into the alkali-hydroxide-free (AHF) geopolymer concretes reported a strength of approximately 12 MPa. Upon curing in hydrochloric acid solution, the AHF SCG mix reported 43.45% of strength loss. The results were better compared to the AHF geopolymer concrete without SCG, with a 50.21% loss. Şenol (2025) investigated replacing sand with SCG at 5% by mass in a slag-based geopolymer mortar. Using equal proportions of Na_2SiO_3 and NaOH as the alkaline activator, the geopolymer achieved compressive strength exceeding 65 MPa after 28 days of curing, substantially higher than the 55 MPa observed in control mixes without SCG. However, samples containing SCG exhibited greater mass loss upon heating, although the residual compressive strength at elevated temperatures remained lowest in the 5% SCG specimens.

From the studies reviewed, it is evident that SCG has been incorporated into cement mortar, conventional concrete, permeable concrete, and geopolymer formulations. These studies also discussed that the application of SCG is feasible due to the SCG wastes having high accessibility and low reutilization costs and promoting environmental friendliness. However, research specifically focused on SCG in permeable geopolymer concrete (SCG-PGC) remains very limited, creating a notable gap that the authors of this chapter are currently investigating. Despite the growing body of literature on SCG applications in other contexts, little attention has been given to integrating SCG into permeable geopolymer structures, underscoring the novelty and importance of the current research effort.

7.12 RECOMMENDATIONS AND CONCLUSIONS

From the literature reviewed, the following recommendations and conclusions were made:

1. While research suggests that AAM and geopolymer technologies hold promise for permeable concrete applications, research on permeable geopolymer concrete (PGC) remains limited. More studies should explore applying AAM and geopolymer concepts specifically to permeable concrete, which could offer carbon-neutral construction solutions for resilient cities.
2. There is significant potential for replacing FA with alternative precursors or additive materials, particularly organic and agricultural wastes such as SCG. Therefore, studies should investigate the potential use of locally available

wastes, such as SCG, in PGC mixes. Waste abundance varies regionally; for example, palm oil fuel ash is often abundant in tropical areas, while SCG may be more accessible elsewhere.

3. Investigating alternative alkaline activators, such as using sodium silicate alone, could expand formulations for permeable geopolymer concretes and enhance performance options. Future studies can explore alternative alkaline activators beyond the traditional NaOH and sodium silicate combinations. Examining different activator systems could expand industrial options and promote real-world adoption of geopolymer technologies.
4. Conducting further experiments on permeable geopolymer concrete with various fibre types to assess their effects on mechanical and durability properties.
5. The influence of aggregate gradation on permeable geopolymer concrete mirrors that of conventional permeable concrete, whereby small aggregate sizes tend to lead to increased strength but decreased permeability properties.
6. SCG shows promise as an additive in geopolymer systems; however, its application in permeable geopolymer concrete (SCG-PGC) is largely unexplored. This represents a clear research gap, one that the authors of this chapter are actively investigating.

REFERENCES

Aljuaydi, F., Abousnina, R., Siddique, R., Alajarmeh, O., & Alajmi, A. (2024). Microstructural analysis of geopolymer pastes contaminated with crude oil: Impact on efflorescence, physical properties, and mechanical performance. *Construction and Building Materials*, 452, 138965. https://doi.org/10.1016/j.conbuildmat.2024.138965

Azad, A. M., Anshul, A., Azad, N., Samarakoon, S. M. S. M. K., Yadav, R., Bherwani, H., Gupta, A., & Gebremariam, K. F. (2022). Pervious geopolymer concrete as sustainable material for environmental application. *Materials Letters*, 318. https://doi.org/10.1016/j.matlet.2022.132176

Chen, X., Guo, Y., Ding, S., Zhang, H., Xia, F., Wang, J., & Zhou, M. (2019). Utilization of red mud in geopolymer-based pervious concrete with function of adsorption of heavy metal ions. *Journal of Cleaner Production*, 207, 789–800. https://doi.org/10.1016/j.jclepro.2018.09.263

Chuang, C. W., Chen, T. A., & Huang, R. (2023). Effect of finely ground coal bottom ash as replacement for portland cement on the properties of ordinary concrete. *Applied Sciences (Switzerland)*, 13(24). https://doi.org/10.3390/app132413212

Davidovits, J. (1994). GEOPOLYMERS: Man-made rock geosynthesis and the resulting development of very early high strength cement. *Journal of Materials Education*, 16(3), 91–139.

Davidovits, J. (2002). 30 Years of successes and failures in geopolymer applications. Market trends and potential breakthroughs. *Geopolymer 2002 Conference*, 1–16. www.geopolymer.org

Epstein, M., Yogev, A., Yao, C., & Berman, A. (2001). Carbothermal reduction of alkali hydroxides using concentrated solar energy. *Energy*, 26(5), 441–455. https://doi.org/10.1016/S0360-5442(01)00010-X

Ergeshov, Z., Örklemez, E., Ketema, A. F., Kwami, M. I. A., Ilkentapar, S., Durak, U., Karahan, O., & Atis, C. D. (2025). Influence of silica fume on the mechanical and microstructural properties and life cycle assessment of fly ash-based geopolymer mortar. *Arabian Journal for Science and Engineering*. https://doi.org/10.1007/s13369-025-10142-9

Forcina, A., Petrillo, A., Travaglioni, M., di Chiara, S., & De Felice, F. (2023). A comparative life cycle assessment of different spent coffee ground reuse strategies and a sensitivity analysis for verifying the environmental convenience based on the location of sites. *Journal of Cleaner Production*, 385, 135727. https://doi.org/10.1016/j.jclepro.2022.135727

Gharehbaghi, K., Tee, K. F., Farnes, K., & Blackburne, L. (2023). Pertinency of geopolymer concrete for the Australian construction industry: A broader contextual appraisal. In *The Cement and Concrete: Design, Performance and Structure* (pp. 153–189). Nova Science Publishers.

Glukhovsky, V. D. (1952). *Грунтосиликаты* (Ground Silicates).

Gökçe, H. S. (2024). Durability of slag-based alkali-activated materials: A critical review. *Journal of the Australian Ceramic Society*, 60(3), 885–903. https://doi.org/10.1007/s41779-024-01011-z

Hijazi, D. A., Bibi, A., & Al-Ghouti, M. A. (2024). Sustainable waste utilization: Geopolymeric fly ash waste as an effective phenol adsorbent for environmental remediation. *Resources, Environment and Sustainability*, 15. https://doi.org/10.1016/j.resenv.2023.100142

Huang, W., & Wang, H. (2021). Geopolymer pervious concrete modified with granulated blast furnace slag: Microscale characterization and mechanical strength. *Journal of Cleaner Production*, 328. https://doi.org/10.1016/j.jclepro.2021.129469

Jeyasangar, P., Hazwan Hussin, M., & Ganesan, S. (2024). Analysis of efflorescence of spent bleaching clay based geopolymer using alkaline leachability methods. *Malaysian Journal of Analytical Sciences*, 28.

Khan, M. O., Klamerus-Iwan, A., Kupka, D., & Słowik-Opoka, E. (2023). Short-term impact of different doses of spent coffee grounds, salt, and sand on soil chemical and hydrological properties in an urban soil. *Environmental Science and Pollution Research*, 30(36), 86218–86231. https://doi.org/10.1007/s11356-023-28386-z

Khong, S. C., & Chin, S. C. (2024). Potential utilization of spent coffee waste in permeable concrete. *The Open Construction & Building Technology Journal*, 18(1). https://doi.org/10.2174/0118748368295411240206080606

Khong, S. C., Yee, J. J., Gimbun, J., Tee, K. F., & Chin, S. C. (2025). Spent coffee grounds enhanced strength of alkali-hydroxide free geopolymer concrete: An optimization study. *Journal of Materials Science: Materials in Engineering*, 20, 109.

Koay, H. Y., Azman, A. T., Mohd Zin, Z., Portman, K. L., Rusli, N. D., Hasmadi, M., Aidat, O., & Zainol, M. K. (2023). Assessing the impact of spent coffee ground (SCG) concentrations on shortbread: A study of physicochemical attributes and sensory acceptance. *Future Foods*, 8. https://doi.org/10.1016/j.fufo.2023.100245

Krivenko, P. (2017). Why alkaline activation – 60 years of the theory and practice of alkali-activated materials. *Journal of Ceramic Science and Technology*, 8(3), 323–333. Goller Verlag. https://doi.org/10.4416/JCST2017-00042

Kua, T. A., Arulrajah, A., Horpibulsuk, S., Du, Y. J., & Shen, S. L. (2016). Strength assessment of spent coffee grounds-geopolymer cement utilizing slag and fly ash precursors. *Construction and Building Materials*, 115, 565–575. https://doi.org/10.1016/j.conbuildmat.2016.04.021

Li, W., Lemougna, P. N., Wang, K., He, Y., Tong, Z., & Cui, X. (2017). Effect of vacuum dehydration on gel structure and properties of metakaolin-based geopolymers. *Ceramics International*, 43(16), 14340–14346. https://doi.org/10.1016/j.ceramint.2017.07.190

Longhi, M. A., Zhang, Z., Rodríguez, E. D., Kirchheim, A. P., & Wang, H. (2019). Efflorescence of alkali-activated cements (geopolymers) and the impacts on material structures: A critical analysis. *Frontiers in Materials*, 6. https://doi.org/10.3389/fmats.2019.00089

Luo, Y. P., Lv, Y., Wang, D., Jiang, Z., & Xue, G. (2023). The influence of coarse aggregate gradation on the mechanical properties, durability, and plantability of geopolymer pervious concrete. *Construction and Building Materials*, 382. https://doi.org/10.1016/j.conbuildmat.2023.131246

Marathe, S., Nieświec, M., & Gronostajska, B. (2025). Alkali-activated permeable concretes with agro-industrial wastes for a sustainable built environment. *Materials*, 18(1). https://doi.org/10.3390/ma18010087

Mohamed, R., Abd Razak, R., Abdullah, M. M. A. B., Abd Rahim, S. Z., Mohd Mortar, N. A., Chaiprapa, J., Tancharakorn, S., Othuman Mydin, M. A., & Vizureanu, P. (2024). Elucidation on alkali activation of slag with various solid-to-liquid ratios via heat evolution. *Journal of Building Engineering*, 84, 108639. https://doi.org/10.1016/j.jobe.2024.108639

Mostofizadeh, S., & Tee, K. F. (2024). Review of next-generation earthquake-resistant geopolymer concrete. *Discover Materials*, 4, 62.

Nehdi, M. L., & Yassine, A. (2020). Mitigating Portland cement CO_2 emissions using alkali-activated materials: System dynamics model. *Materials*, 13(20), 1–23. https://doi.org/10.3390/ma13204685

Neupane, K., & Hadigheh, S. A. (2021). Sodium hydroxide-free geopolymer binder for prestressed concrete applications. *Construction and Building Materials*, 293. https://doi.org/10.1016/j.conbuildmat.2021.123397

Ponomar, V., Luukkonen, T., & Yliniemi, J. (2023). Revisiting alkali-activated and sodium silicate-based materials in the early works of Glukhovsky. *Construction and Building Materials*, 398, 132474. https://doi.org/10.1016/j.conbuildmat.2023.132474

Provis, J. L. (2014). Geopolymers and other alkali activated materials: Why, how, and what? *Materials and Structures/Materiaux et Constructions*, 47(1–2), 11–25. https://doi.org/10.1617/s11527-013-0211-5

Ren, Q., Li, X., Ji, Y., Ding, X., Sun, Q., Zhao, P., Li, F., & Vandeginste, V. (2023). Thermal insulation of phenolic resin modified fly ash geopolymer. *Construction and Building Materials*, 409, 133840. https://doi.org/10.1016/j.conbuildmat.2023.133840

Roychand, R., Kilmartin-Lynch, S., Saberian, M., Li, J., Zhang, G., & Li, C. Q. (2023). Transforming spent coffee grounds into a valuable resource for the enhancement of concrete strength. *Journal of Cleaner Production*, 419. https://doi.org/10.1016/j.jclepro.2023.138205

Samoggia, A., & Riedel, B. (2019). Consumers' perceptions of coffee health benefits and motives for coffee consumption and purchasing. *Nutrients*, 11(3). https://doi.org/10.3390/nu11030653

Şenol, A. F. (2025). Performance of geopolymer mortar incorporating spent coffee grounds as a recycled building material: An experimental and predictive analysis. *Hybrid Advances*, 10. https://doi.org/10.1016/j.hybadv.2025.100479

Soffian, M. S., Abdul Halim, F. Z., Aziz, F., A. Rahman, M., Mohamed Amin, M. A., & Awang Chee, D. N. (2022). Carbon-based material derived from biomass waste for wastewater treatment. *Environmental Advances*, 9, 100259. https://doi.org/10.1016/j.envadv.2022.100259

Solomakou, N., Tsafrakidou, P., & Goula, A. M. (2022). Valorization of SCG through extraction of phenolic compounds and synthesis of new biosorbent. *Sustainability (Switzerland)*, 14(15). MDPI. https://doi.org/10.3390/su14159358

Tee, K. F., & Mostofizadeh, S. (2021). A mini review on properties of portland cement concrete with geopolymer materials as partial or entire replacement, *Infrastructures*, 6(2), 26.

Tee, K. F., & Mostofizadeh, S. (2024). Soil-pipe interaction and structure response under static and seismic loading for geopolymer concrete pipes. *Innovative Infrastructure Solutions*, 9, 49.

Tian, H., Zhou, T., Huang, Z., Wang, J., Cheng, H., & Yang, Y. (2021). Integration of spent coffee grounds valorization for co-production of biodiesel and activated carbon: An energy and techno-economic case assessment in China. *Journal of Cleaner Production*, 324, 129187. https://doi.org/10.1016/j.jclepro.2021.129187

Tofalo, R., Renda, G., De Caterina, R., & Suzzi, G. (2016). Coffee: Health effects. In B. Caballero, P. M. Finglas, & F. Toldrá (Eds.), *Encyclopedia of Food and Health* (pp. 237–243). Academic Press. https://doi.org/10.1016/B978-0-12-384947-2.00182-3

Wang, F., Sun, X., Tao, Z., & Pan, Z. (2022). Effect of silica fume on compressive strength of ultra-high-performance concrete made of calcium aluminate cement/fly ash based geopolymer. *Journal of Building Engineering*, 62, 105398. https://doi.org/10.1016/j.jobe.2022.105398

Whiting, J. (1895). *US544706: Manufacture of Cement* (Patent 544706). United States Patent Office.

Xu, L., Liu, Q., Ding, X., Sun, S., & Huang, Z. (2023). Mechanical properties of fiber-reinforced permeable geopolymer concrete. *Materials*, 16(17). https://doi.org/10.3390/ma16176030

Yee, J. J., Khong, S. C., Tee, K. F., & Chin, S. C. (2024a). Physical and mechanical properties of spent coffee grounds (SCG) in concrete. *Malaysian Journal of Analytical Sciences*, 28(4), 843–858.

Yee, J. J., Khong, S. C., Tee, K. F., Jolius, G., & Chin, S. C. (2024b). Spent coffee grounds enhanced compressive strength of cement mortar: An optimization study. *Discover Applied Sciences*, 6(7). https://doi.org/10.1007/s42452-024-06077-9

Zaetang, Y., Wongsa, A., Sata, V., & Chindaprasirt, P. (2015). Use of coal ash as geopolymer binder and coarse aggregate in pervious concrete. *Construction and Building Materials*, 96, 289–295. https://doi.org/10.1016/j.conbuildmat.2015.08.076

Zhang, Y., Nan, X., Zhang, S., Jia, L., Zhu, F., Yu, W., & Zheng, Q. (2025). Enhancing performance of mining phenolic filling materials by tailoring closed cell morphology with fly ash geopolymer. *International Journal of Mining Science and Technology*. https://doi.org/10.1016/j.ijmst.2025.06.008.

8 Alkali-Activated Binders in Infrastructure Development

Redefining Infrastructure with Smart AAB Binders

Arun Murugesan, Nidhya Rathinavel and Mohammed Jafer Sathik Ali Abdul Hameed

8.1 INTRODUCTION

The construction industry remains deeply entrenched in the use of Ordinary Portland Cement (OPC), the environmental impact of which is just as massive as its structural footprint. OPC cementing processes put out nearly 8% of global carbon dioxide emissions, outgassing more than many industrialized countries. Due to the production methods that include calcination of limestone, about 900 kg of CO_2 per ton of cement is directly released into the atmosphere; it is also energy-intensive and depletes the available resources. From an engineering perspective, there have been improvements in structural engineering, digital modeling, and project automation; yet, the actual cement chemistry of OPC has largely remained the same: chemically inert, heavy on energy, and heavy on pollution. As the population grows and urbanization increases, particularly in the Global South, reliance on older cementitious systems constitutes a great stumbling block with regard to climate mitigation, environmental resilience, and sustainable development[1, 2].

The environmental and functional standstill shows the acute necessity to take a serious hard look at the definition given to "building materials." Today, infrastructure simply cannot remain passive: materials have to become intelligent enough to sense damage, adapt to stress, or generate and secure energy for themselves. Materials of this nature would combine multiple functions: self-healing, self-monitoring, and energy interacting. The binder, a twin object in old terms, should thus become an active force in achieving structural health, energy efficiency, and environment-symbiosis. Strength, then, durability, responsiveness, resilience, and regenerative capacities form the new paradigm[2, 3].

This study introduces a new class of binders, that is, AABs based on alkali activation, derived from industrial waste such as fly ash, slags, and silica-rich by-products

DOI: 10.1201/9781003664284-8

to meet this ever-pressing need. These binders, in turn, offer a unique multifunctional platform above and beyond carbon reduction vis-à-vis an OPC. MnO_2 nanosheets for supercapacitive energy storage; reduced graphene oxides for electrical conductivity useful for real-time structural health monitoring; carbon nanotubes for self-sensing; and microencapsulated agents for autonomous crack healing combine to form conscious concrete-a low-carbon, energy-interactive, and self-healing material created in response to growing needs of smart cities, disaster-resilient structures, and off-grid intelligent infrastructure[4].

8.2 MATERIALS ALCHEMY: ECO-INDUSTRIAL WASTE TO FUNCTIONAL MATRIX

8.2.1 From Ashes to Architecture: Harnessing Industrial By-Products

Alkali-activated binders (AABs) represent the future of material science because they transform industrial waste into modern high-performance construction products. OPC production depends on limestone calcination and clinker grinding processes that generate substantial carbon dioxide emissions while AABs use alternative methods to avoid these carbon-intensive steps. The strength and reactivity of these materials originate from aluminosilicate by-products that include fly ash and Ground Granulated Blast-Furnace Slag (GGBS). The materials that people used to throw away now serve as fundamental components for production. The geopolymerization process uses alkaline conditions to transform these materials into three-dimensional aluminosilicate networks. The end product forms a compact durable chemically stable matrix that achieves mechanical performance equivalent or better than OPC systems [4, 5].

Table 8.1 displays essential performance metrics for OPC and AABs, which demonstrates AAB superiority in technological and environmental domains. The combination of AABs demonstrates improved thermal resistance and multi-functionality together with reduced carbon impact, but OPC produces high CO_2 emissions and shows weak performance in challenging environments. The environmentally friendly characteristics of AABs emerge from their ability to function as intelligent materials that adjust to present-day construction requirements.

Waste management experiences an entire revolution through this method that shifts from conventional disposal toward circular material regeneration systems. The process extends beyond landfill prevention because it allows material recovery that leads to the creation of upcoming building foundations. The combination of these binders with suitable design and curing approaches results in fast-setting concrete that creates initial strength along with sulfate resistance and thermal stability that protects environments conventional materials cannot handle. Through alkali-activation, waste materials transform into products with increased value. The solution does not require any concessions because it transforms waste into essential high-performance building materials that revolutionize sustainable construction under carbon constraints[6, 7].

TABLE 8.1
Comparative Performance Metrics of Ordinary Portland Cement (Opc) and Alkali-Activated Binders (Aabs)

Parameter	Ordinary Portland Cement (OPC)	Alkali-activated binder (AAB)
CO_2 emissions (kg/ton binder)	~900	~100 or net negative (with optimized waste usage)
Primary raw material	Limestone, clay	Fly ash, GGBS, silica fume, industrial by-products
Processing temperature (°C)	~1400	Ambient to ~80°C
Chemical durability	Moderate (vulnerable to sulfates, acids)	High resistance to acids, sulfates, and chlorides
Thermal resistance	Poor (spalling under high temperature)	Excellent (stable up to 1,000°C)
Electrical conductivity	Nonconductive	Conductive (with rGO/CNT integration)
Energy storage Capability	None	Yes (with MnO_2 & rGO integration)
Self-healing ability	Absent	Present (via microencapsulation)
Structural health monitoring	External sensors required	Intrinsic via CNT piezoresistivity
Lifecycle sustainability	High CO_2 footprint, short service life	Low-carbon, long lifespan, minimal maintenance

8.2.2 Chemistry with a Conscience: The Carbon-Negative Lifecycle

AABs serve as superior sustainable options that transform the entire lifecycle of cementitious materials. The production of AABs requires lower temperatures than OPC and produces less carbon dioxide because they activate through chemical processes at low temperatures. This procedure takes place under normal or slightly increased temperature conditions which leads to substantial energy savings. AABs combined with renewable power and by-product materials such as fly ash or red mud or slag can result in environmental impact reduction to zero or even negative carbon emissions[8].

Life cycle assessments (LCAs) demonstrate that AABs achieve net carbon sequestration through optimized sourcing and transport, which enables carbonation and CO_2-reactive precursors and biological integration to capture more carbon than they release during their lifespan. AABs stand as perfect choices for countries that aim to achieve net-zero or regenerative construction goals. The benefits of AABs extend beyond carbon advantages because they possess outstanding longevity properties that include acid resistance and sulfate and chloride ingress protection alongside excellent thermal stability and low shrinkage. The material characteristics make AABs perfect for usage in challenging environments including marine platforms and underground transit and fire-prone areas. The built environment potential receives a new definition because AABs represent sustainable construction methods that also bring engineered durability and environmental responsibility and endurance[9, 10].

8.2.3 Functional Symbiosis: MnO_2, rGO, and CNT Embedded

The path to infrastructure intelligence requires AABs to unite sensing capabilities with energy interaction and adaptive response mechanisms. Nanomaterials embedded within the binder enable electronic and mechanical and chemical functionality to operate. Pseudocapacitive manganese dioxide (MnO_2) introduces redox-active sites to the material, which enables solid-state energy storage. The material possesses the ability to take kinetic energy from foot traffic and vibrations and vehicle loads and store it as a decentralized embedded energy system within the structure[11].

The system benefits from reduced graphene oxide (rGO) through enhanced electron transport and nanoscale cohesion properties. The large surface area combined with high conductivity of rGO enables better electrical performance and improves material toughness and crack resistance. Carbon nanotubes (CNTs) exhibit remarkable strength and electromechanical capabilities that enable them to establish a sensing network inside the composite material. The nanotubes operate as an internal nervous system to detect minor strain and stress variations that signal structural failure risks before reaching critical levels[12].

These nanomaterials unite to create a binder that delivers functional as well as responsive and self-aware properties. Real-time diagnostics merge with load-bearing capacity and energy storage and adaptive feedback to form this "living" system. The development of materials that possess cognitive abilities and environmental adaptability represents the beginning of smart bridges alongside energy-independent buildings and self-monitoring pavements[3].

8.2.4 Microstructural Harmony: Integration Without Disruption

High-performance nanomaterials embedded in cementitious matrices create structural risks to the composite. The alkali-activated systems that contain amorphous C–A–S–H and N–A–S–H gels function as perfect hosts for nanoscale additives. The gel matrix allows MnO_2 and rGO to spread evenly throughout the material while preventing their clumping and preserving their functional properties. The functionalization or ultrasonic dispersion of CNTs creates continuous conductive pathways that do not affect gel formation. The microencapsulated healing agents stay passive until mechanical stress or environmental changes activate them, which leads to precise healing for extended periods[13, 14].

8.3 POWER FROM WITHIN: ENERGY HARVESTING THROUGH CONCRETE

8.3.1 Energizing the Inert: Embedding Pseudocapacitive Materials

People used to think concrete functioned purely as a structural element until recent developments have shown its potential to become an active interface for energy.

When pseudocapacitive materials manganese dioxide (MnO_2) and reduced graphene oxide (rGO) are incorporated in the AAB, the material becomes a fundamental part of decentralized energy networks. The materials allow infrastructure components to hold electrical charges thus making pavements and columns and wall panels capable of acting as supercapacitors or energy storage devices[15, 16].

MnO_2, which belongs to the family of transition metal oxides, operates as a necessary component to transform cement matrices into storage devices because it offers both high theoretical capacitance and broad redox properties. Through its faradaic charge transfer process, MnO_2 enables fast ion and electron exchange that makes it suitable for pseudocapacitive energy storage. The material's layered or tunnel-like structure allows for large surface area development together with easy electrolyte diffusion, and its ionic conductivity helps quicken redox reactions. The electrical conductivity problems of MnO_2 led researchers to introduce reduced graphene oxide (rGO) as a conductive carbon nanomaterial with high surface area. The dual role of rGO enables electron conduction while providing structural support to MnO_2 particles and preventing their aggregation as well as enhancing electrical conductivity. The supercapacitive network of MnO_2 together with rGO exists within the AAB to integrate faradaic and non-faradaic storage mechanisms. The developed binder functions as both structural load handler and energy storage medium that enables the construction of self-powered infrastructure[17].

The change extends beyond simple addition because it represents a fundamental transformation. Modern infrastructure functions as an active energy node that receives mechanical and solar input to store power for backup operations and grid interconnections and auxiliary functions. The building walls function as solar energy storage units while pavements capture motion power to illuminate sensors during nighttime hours. The material functions as an energy-embedded system that transforms conventional concrete into a "structural battery" to revolutionize smart cities and independent energy systems[18].

Industrial waste conversion into a multifunctional AAB appears through the flow diagram as showed in Figure 8.1. The process involves activation and nanomaterial integration with MnO_2 and rGO and CNTs and includes self-healing capsules. The final product is an intelligent concrete with low carbon emissions that can interact with energy and perform structural monitoring.

8.4 INTELLIGENT SKIN: STRUCTURAL HEALTH MONITORING VIA CNT NETWORKS

8.4.1 Carbon Nanotubes: Wiring the Material Nervous System

AABs that integrate carbon nanotubes (CNTs) transform concrete into a material that understands its structure and functions autonomously. The rolled graphene nanostructures deliver excellent electrical conductivity alongside high tensile strength and thermal stability. Microstructural gaps become bridged by their high aspect ratio that allows them to establish conductive pathways across the matrix. After dispersion, CNTs establish a percolation network that detects minimal mechanical stimuli thus transforming concrete into an electro-responsive system that monitors stress and strain and initial microcracking[11, 19].

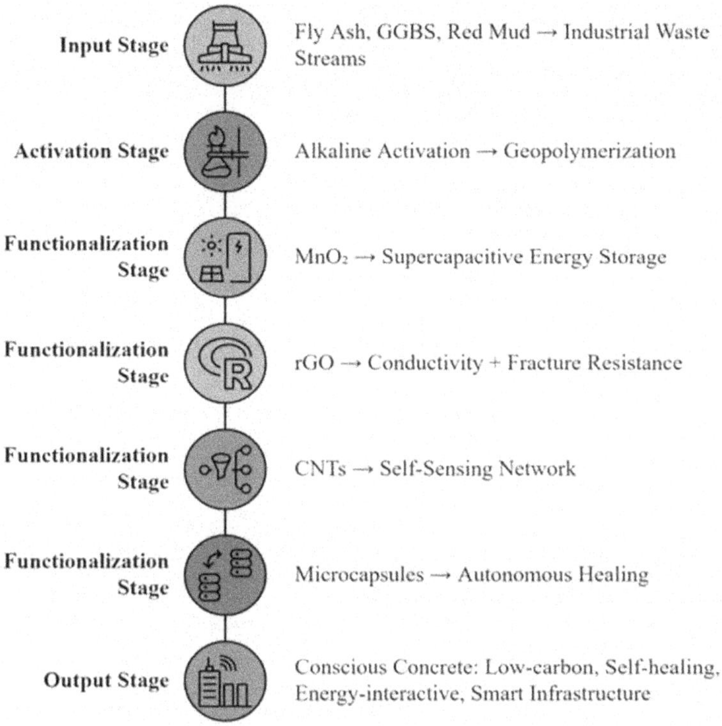

FIGURE 8.1 Transformation of industrial waste into multifunctional "conscious concrete" via nanomaterial-embedded alkali-activated binders.

Table 8.2 presents the electrical and sensing properties of AABs enhanced with CNTs, which demonstrate their structural health-monitoring abilities. The parameters of resistivity along with gauge factor and crack detection threshold show how CNT networks allow for instant detection of stress and microcracks. The integrated sensing function makes concrete an intelligent material with self-awareness that decreases the need for external monitoring systems.

Electro-responsive materials enable concrete to fulfill new functions within infrastructure systems. Traditional structural health monitoring (SHM) operates with external sensors and devices, but CNT-modified binders function as self-contained sensors that eliminate the requirement for these devices. The CNT network acts like sensory receptors in human skin by sensing internal changes and sending immediate responses directly to the system without needing additional equipment. The binder develops into a responsive layer that detects changes in stress and damage and environmental conditions while sending notifications. Concrete receives a new identity through this concept which enables it to monitor itself and perform maintenance tasks independently[20, 21].

In Figure 8.2, a conceptual illustration shows MnO_2 nanosheets and rGO layers integrated into concrete wall and pavement surfaces. The technology transforms collected solar and kinetic energy into electrochemical storage energy that can be

TABLE 8.2
Electrical and Sensing Properties of Cnt-Modified Alkali-Activated Binders

Parameter	Typical Range / Value	Significance
CNT content (wt.%)	0.1–1.0%	Optimum for conductivity without agglomeration
Initial resistivity (Ω·cm)	$10^2 - 10^4$	Tunable depending on dispersion and dosage
Gauge factor	5–15	Higher than traditional strain gauges
Strain detection range	0.001% – 1%	Suitable for micro to macro stress monitoring
Crack detection threshold (mm)	~0.05 – 0.10 mm	Early detection before visible failure
Signal type	Resistance change ($\Delta R/R_0$)	Easily measurable with low-cost electronics
Sensing mode	Embedded, passive	No need for external sensors

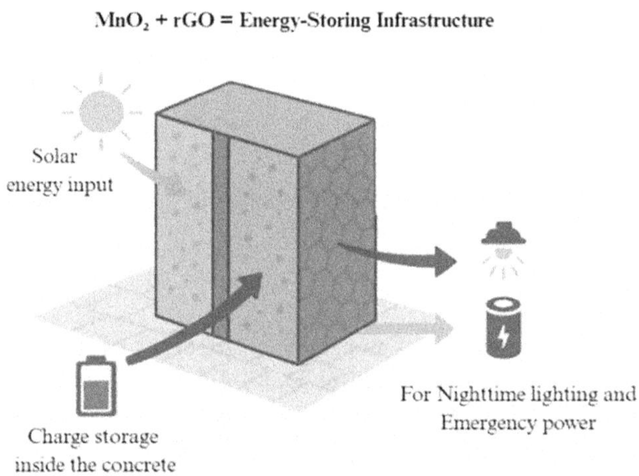

FIGURE 8.2 MnO_2 and rGO layers enable the binder to harvest, store, and supply energy for smart infrastructure applications.

accessed at a later time. Smart infrastructure uses this technology to provide electrical power for lighting systems and sensors together with emergency equipment.

8.4.2 Listening to Stress: Piezoresistivity in Action

Concrete's self-sensing ability via carbon nanotubes (CNTs) is based on piezoresistivity – the change in a material's electrical resistance under mechanical strain. In CNT-enhanced AABs, this effect is amplified. During tension, compression, or fatigue, the CNTs' distribution and contact points move, which leads to changes in the conductive network. The changes in current flow create measurable continuous resistance variations that directly correspond to stress levels. Through this method, it becomes possible to monitor structural performance in real time by using the material itself[22].

Standard SHM systems use external sensors or fiber optics, but CNT-based piezoresistive sensing integrates its sensing functionality straight into concrete materials. The technology removes the requirement for external sensors and wiring while also reducing maintenance operations. The material operates as a built-in diagnostic device that identifies stress and detects initial cracks while differentiating between damage that can be reversed versus permanent damage. The embedded noninvasive sensing system provides higher spatial and temporal resolution than conventional approaches by allowing infrastructure to send real-time actionable data directly to engineers[23, 24].

8.4.3 From Material to Messenger: Real-Time Wireless Data

The transition from self-sensing to self-reporting functionality depends on linking piezoresistive nanomaterial networks with embedded wireless electronics. The integration of ultra-low-power microcontrollers together with energy harvesters and IoT transceivers inside the concrete matrix enables autonomous data collection and transmission of performance information. Every structure transforms into a continuous self-monitoring sensor network that detects stress along with strain measurements and crack growth as well as temperature and vibration data in real time. The innovation transforms concrete into an active cyber-physical node that operates within intelligent infrastructure systems[21, 23].

The consequences are substantial. Such self-reporting functionality allows smart city infrastructure to communicate with cloud platforms and digital twins and AI-based asset management systems, which enables predictive analytics and adaptive planning[25, 26]. The system provides structural awareness in areas where inspections are hard to perform including tunnels and dams and offshore sites. Wireless alert systems enable structures to transmit direct messages to engineers and maintenance teams and network infrastructure through stress threshold detection and rapid crack growth and abnormal pattern recognition[21, 27].

8.4.4 Engineering Foresight: Predictive Maintenance Enabled

The greatest revolutionary benefit of self-sensing systems using CNT technology enables organizations to move their operation from reactive maintenance to predictive prevention. Regular inspections that depend on human visual checks detect problems after deterioration occurs. Real-time data collection from embedded CNT networks monitors stress and microcracks along with temperature and fatigue levels. Engineers can use pattern analysis to receive early warnings, which helps them predict material behavior before they schedule necessary interventions to prevent failure[28]. Such functionality transforms the financial model as well as safety standards of infrastructure systems. Maintenance activities become proactive and precisely targeted, which reduces the need for emergency repairs along with downtime.

Predictive intelligence plays a vital role for critical systems that include high-speed rails, long-span bridges, and offshore wind farms and nuclear reactors. Early diagnostics together with data-driven planning enable service life extension and resource optimization while stopping catastrophic failures. The new approach enables infrastructure to forecast its own future state while adapting and protecting itself through molecular-level resilience engineering[29].

8.4.5 Synergies and Overlaps Between SHM and Energy Storage

Normally, energy storage and structural health monitoring (SHM) are considered two different functionalities, whereas in AABs, the two share a common material platform. Conductive nanomaterials are incorporated to enhance electron transport: for electrochemical energy storage and similarly for piezoresistive sensing, providing a dual benefit. Likewise, MnO_2 inputing pseudocapacitive behavior in the device also modifies the electrochemical environment in a manner that can affect sensing responses[30]. The multifunctionality opens up a range of opportunities and challenges. It is advantageous as it integrates energy storage and SHM into one binder matrix, maximizing resource efficiency and minimizing the requirement for external devices, thereby granting structural systems the ability to become "power active" and "self-aware" within themselves. On the other hand, trade-offs may exist due to functional overlaps; for example, simultaneously optimizing the material for charge storage might affect sensing sensitivity, or tuning the dispersion for crack detection could reduce the energy density in one way or another. Defining these interdependencies will serve as the basis for the design of multifunctional concrete that can balance practical durability, sensing fidelity, and energy performance[31].

8.5 CRACKS THAT HEAL THEMSELVES: AUTONOMOUS RECOVERY OF STRUCTURAL INTEGRITY

8.5.1 Healing From Within: Microencapsulation of Repair Agents

Self-healing concrete functioned as science fiction until recent times when it became a scientific reality. AAB materials that contain microencapsulated healing agents have the ability to autonomously recognize and fix internal damage. The capsules contain repair agents that include sodium silicate and reactive polymers along with mineral catalysts and bio-based precursors that exist in dimensions between several hundred and a few microns. The particles exist in a state of dormancy under standard conditions while evenly distributed throughout the matrix to create a hidden internal restoration mechanism[32, 33].

The release of healing agents occurs when mechanical forces break the capsules near cracks and microfractures, which then initiate chemical reactions with the matrix to form polymerized C–S–H-like gels that restore structural integrity. The system functions independently without needing any sensors or triggers. Capsules function similarly to immune cells because they stay inactive until physical damage triggers their activation. The embedded network of resilience functions to increase

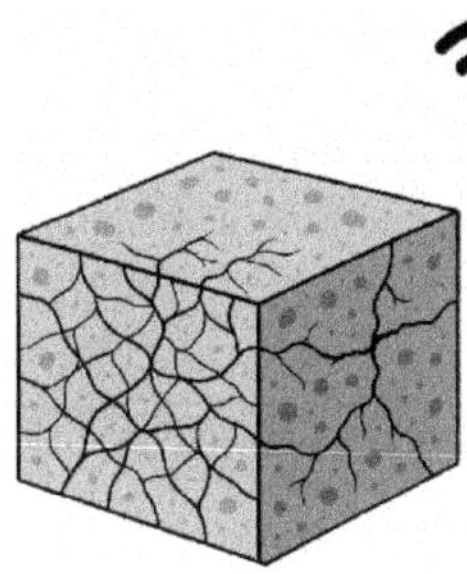

FIGURE 8.3 CNT network integrated into concrete, enabling internal strain, crack, and resistivity sensing with wireless real-time monitoring.

service life while reducing repair requirements and enhances structural durability to unprecedented levels[34, 35].

In Figure 8.3, the concrete matrix that contains CNTs allows for stress detection through its conductive network. Mechanical strain along with crack development and resistivity alterations causes various parts of the material to respond differently. The sensing system operates for real-time wireless monitoring through its built-in design without needing any external sensors.

8.5.2 Triggered by Damage: Release and React Mechanism

The fundamental power of microencapsulated self-healing systems exists in their inactivity until damage happens. The passive system functions autonomously without the need for monitoring equipment or control systems. Microcracks generated by cyclic loading, thermal stress, shrinkage, or corrosion trigger the breakage of nearby capsules because of local stress. The activation process targets capsules that are close to the damage thus minimizing material wastage and optimizing operational performance. The sealing agent exists as a liquid or gel that moves through small cracks by capillary action to repair damage that standard methods cannot reach[36].

The healing agent starts chemical reactions with surrounding matrix materials under moisture or high-pH conditions or embedded catalysts. Reaction mechanisms vary based on the agent because sodium silicate produces C–S–H-like gels with calcium ions, while mineralizing agents produce dense $CaCO_3$ deposits and polymeric systems from flexible seals. The reactions restore strength while simultaneously preventing fluid penetration. The sealing process happens within hours to prevent crack expansion and block dangerous materials from entering[37].

The intelligent damage-responsive system functions as biological repair systems by replicating the operations of clotting and scar tissue formation. The human body along with the self-healing binder uses matching injury response patterns that immediately stop bleeding and seal wounds through their autonomous internal defense systems. The bioinspired design allows an early detection of structural damage while it halts small cracks from turning into major structural failures. The repair system enhances structural strength at the same time as it extends operational lifespan resulting in infrastructure that requires minimal maintenance. The new material system shows adaptive resilience because it integrates repair functionality that outperforms conventional concrete technology standards[38].

8.5.3 The Healing Cycle: Impact on Durability and Strength

The integration of microencapsulated healing agents within AABs results in permanent changes to their deterioration patterns. Studies show that healed cracks can recover up to 80–90% of original mechanical strength, including flexural and tensile capacity. Self-healing mechanisms enhance both the impermeability of structures and their ability to resist water and chloride penetration that prevent steel corrosion inside the concrete. The passive defense mechanism proves vital for harsh environmental conditions such as coastal and freeze–thaw and chemically aggressive zones as it stops structural deterioration processes that cause long-term failure[35].

This innovative system presents groundbreaking capabilities through its capacity to sustain continuous cycles of damage followed by recovery. Modern capsule engineering creates multilayered and phase-tuned healing agents that trigger responses at various stress levels across multiple environmental conditions to allow multiple healing events over time. The ability to sustain multiple healing cycles functions as engineered resilience when applied to structures that experience cyclic loading such as bridges, pavements, seismic-prone buildings, and wind turbine foundations. The material strengthens itself through microcrack formation so it develops enhanced damage resistance for upcoming threats. Through its adaptive functionality, concrete evolves from being a disposable material to a self-healing substance that can lower maintenance demands and reduce expensive interventions while enhancing the lifespan of infrastructure. The binder demonstrates evolutionary responsiveness to stress environments through each healing cycle because it exceeds repair functions by achieving durability recalibration[39].

8.5.4 Built for Harsh Realities: Environmental Resistance and Lifecycle Gains

When exposed to tough conditions, infrastructure systems encounter significant operational difficulties. Chlorides in marine areas create corrosion that results in internal cracking inside concrete, while freeze–thaw conditions cause water expansion that damages concrete surfaces through spalling, and industrial environments with acids and sulfates and thermal shocks exceed Portland cement capabilities. The concrete barrier becomes compromised by cracks, which allows deterioration to speed up. AABs provide a revolutionary solution through autonomous healing because microcracks activate immediate sealing that prevents aggressive agents from damaging durability or safety features[40].

The environmental resistance of AABs naturally increases the advantages that self-healing provides. The dense microstructure with high chemical stability of geopolymerized matrices provides better protection against sulfate attack and acid erosion and chloride penetration than OPC-based systems. The combination of dense microstructure with internal healing abilities leads to a dynamic shield-like binder system that performs autonomously. This system goes beyond passive resistance of environmental effects because it responds to them directly in real time. The combination of chemical durability with automatic repair functions produces long-lasting performance in extreme conditions, which decreases the necessity of manual inspections particularly in offshore platforms and subway tunnels and nuclear containment structures[41, 42].

Strong materials convert into durable materials by showing adaptability to diverse conditions while preserving their structural integrity during prolonged time periods. Concrete design bases its approach on actual conditions rather than theoretical ones because it endures salt-laden winds together with seismic shocks and polluted atmospheres and temperature extremes. The infrastructure matures with intention as natural processes reduce maintenance requirements to extend its service life. Through this model, the material transitions from being a maintenance burden into a lifecycle asset that has been designed to survive multiple generations while adapting to its environmental surroundings.

8.5.5 Limitations and Challenges

There are still a few limitations that restrict the further industrial applications of self-healing AABs despite their promising performance. The mass production of microcapsules and their uniform dispersion within the matrix are often expensive and technically demanding. Healing may be limited to microcracks of certain widths, with wider cracks not entirely closing. In addition, concerns exist on the long-term durability of encapsulated agents subjected to aggressive environmental conditions as premature degradation of the capsules might reduce healing potential compromising the full-service life[43]. There is also a possibility that the addition of capsules locally weakens the matrix or might reduce compressive strength if not well optimized. Lastly, there might be trade-offs once the self-healing mechanism is implemented in multifunctioning systems (such as sensing and energy storage combined). Hence, experimental validations still need to be carried out. These challenges pointed at sustain Tunisia enhancements in scaling processes, cost-effectiveness, and durability[44, 45].

8.6 From Promise to Practice: Navigating the Barriers and Building the Future of Intelligent Binders

Functional AABs show promise as a future generation of materials that combine detection and healing capabilities with energy utilization. This progressive concept demands the resolution of various scientific as well as technical and economic hurdles. The commercial manufacturing of nanomaterials and pseudocapacitors together with healing agents requires processing methods that deliver precise results at a low cost. The stability of matrix and functional inclusion combinations after a long-term environmental stress exposure requires additional investigation. The stable integration of interference-free electronics into concrete faces significant challenges in wireless power transmission systems and signal filtering mechanisms

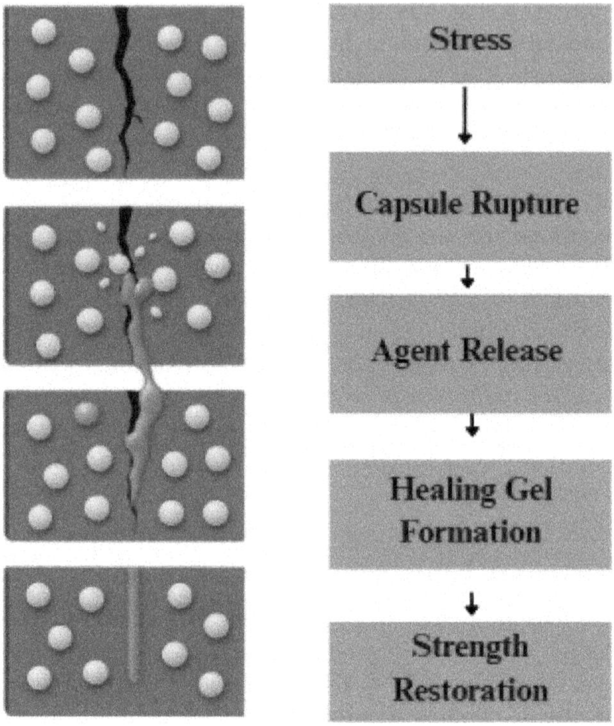

FIGURE 8.4 Microcapsules embedded within the binder release healing agents upon cracking, enabling autonomous repair.

and IoT connectivity solutions. Engineering teams must collaborate with nanotech specialists together with digital infrastructure experts to find solutions for these obstacles. The upcoming era will replace conventional multifunctionality by integrating programmed binders with bio-hybrid concrete while AI systems self-develop through continuous use[46]. The realization of these reachable milestones requires joint efforts between research and policy sectors and innovation sectors. The future infrastructure will unite adaptive qualities together with enduring longevity and living system characteristics[47, 48].

In Figure 8.4, the drawing demonstrates how capsules within the binder break open when cracks develop. The healing substances penetrate microcracks to repair the structure and block entry points to the damaged area. The self-healing mechanism draws inspiration from nature to allow multiple crack repairs, which lengthens the operational lifespan of the system.

8.7 CONCLUSION

Infrastructure advancement needs a complete transformation of the materials that form its basic structure. Binders need to evolve into active cores that enhance performance and sustainability and resilience. The chapter introduces AABs as both cement substitutes and intelligent materials that transform concrete. AABs transform industrial waste into intelligent matrices that possess sensing and healing functions and energy-harvesting abilities to serve as the central nervous system for upcoming infrastructure systems that interact with their environment. Three fundamental innovations enable this transformation.

- The integration of MnO_2 together with reduced graphene oxide (rGO) and carbon nanotubes (CNTs) transforms static concrete into an electrically active multifunctional matrix that can store ambient energy and sense mechanical stress and conduct signals throughout its internal framework. The technology enables infrastructure to function as both structural elements and sensors by eliminating the material-to-device barrier.
- The incorporation of self-healing microencapsulated agents enables concrete to develop an internal immune response that autonomously heals cracks and maintains performance while preventing degradation without human intervention. Concrete design receives both enhanced durability and self-healing capabilities through this approach.
- The combination of embedded sensors together with wireless communication technologies and IoT modules propels concrete into the digital age. The infrastructure system gains the ability to report itself while connecting to predictive maintenance systems and smart city networks where it provides data and adapts in real time to operational conditions.

The advancement represents a change in construction materials beyond standard concrete composition. Through the fusion of alkali-activated chemistry with industrial waste reuse and built-in digital components, these materials develop sensing capabilities while adapting themselves, repairing damage, and communicating in real time. The escalating challenges of climate change and population expansion and infrastructure degradation demand essential features of self-awareness along with energy independence and resilience. This development establishes pathways for infrastructures that persist while actively engaging with their surroundings by converting buildings, roads, and bridges into responsive living systems.

REFERENCES

1. M. Soomro, V. W. Y. Tam, and A. C. J. Evangelista, "Production of cement and its environmental impact," in *Recycled Concrete*, Elsevier, 2023, pp. 11–46.
2. N. Mahasenan, S. Smith, and K. Humphreys, "The cement industry and global climate change: Current and potential future cement industry CO_2 emissions," in *Greenhouse gas control technologies-6th international conference*, Elsevier, 2003, pp. 995–1000.
3. S. Nihalani, U. Joshi, and A. Meeruty, "Smart materials for sustainable and smart infrastructure," in *Materials Science Forum*, Trans Tech Publ, 2019, pp. 278–283.
4. K. Setlak, J. Mikuła, and M. Łach, "Application of industrial waste materials by alkaline activation for use as geopolymer binders," *Materials (Basel).*, vol. 16, no. 24, p. 7651, 2023.
5. E. Luga, E. Mustafaraj, M. Corradi, and C. D. Atiş, "Alkali-activated binders as sustainable alternatives to portland cement and their resistance to saline water," *Materials (Basel).*, vol. 17, no. 17, p. 4408, 2024.

6. B. A. Lanjewar, R. Chippagiri, V. A. Dakwale, and R. V Ralegaonkar, "Application of alkali-activated sustainable materials: A step towards net zero binder," *Energies*, vol. 16, no. 2, p. 969, 2023.
7. R. A. Robayo-Salazar, W. Valencia-Saavedra, and R. Mejía de Gutiérrez, "Construction and demolition waste (CDW) recycling – As both binder and aggregates – In alkali-activated materials: A novel re-use concept," *Sustainability*, vol. 12, no. 14, p. 5775, 2020.
8. S. Contessi, M. Bellotto, M. Chiara Dalconi, and G. Artioli, "Alkali-activated cements incorporating marble-cutting powders: Towards low GHG emissions, sustainable binders implementing circular economy and by-product valorization," EGU General Assembly Conference Abstracts, 2018, p. 6303.
9. J. He, Q. Meng, H. Xu, and Y. Liu, "Exploring material selection and applications for embedded carbon reduction in the built environment," in *The 1st International Scientific and Practical Conference "Technologies for Improving Old Methods, Theories and Hypotheses"(January 07–10, 2025) Sofia, Bulgaria*, International Science Group, 2025, 405 p., 2025, p. 62.
10. L. Mouton, K. Allacker, and M. Röck, "Bio-based building material solutions for environmental benefits over conventional construction products–life cycle assessment of regenerative design strategies (1/2)," *Energy Build.*, vol. 282, p. 112767, 2023.
11. A. D'Alessandro, D. Coffetti, E. Crotti, L. Coppola, A. Meoni, and F. Ubertini, "Self-sensing properties of green alkali-activated binders with carbon-based nanoinclusions," *Sustainability*, vol. 12, no. 23, p. 9916, 2020.
12. Y. Z. Huang, X. M. Chen, and D. X. Li, "Reduced graphene oxide/carbon nanotube composites for high electrochemical performance supercapacitors," in *Materials Science Forum*, Trans Tech Publ, 2018, pp. 25–29.
13. R. Vargas-Bernal, "Multifunctional contributions of nanomaterials to cementitious composite materials," in *Production, Properties, and Applications of Engineered Cementitious Composites*, IGI Global Scientific Publishing, 2024, pp. 20–52.
14. N. Bheel, B. S. Mohammed, M. S. Liew, and N. A. W. A. Zawawi, "Durability behaviours of engineered cementitious composites blended with carbon nanotubes against sulphate and acid attacks by applying RSM modelling and optimization," *Buildings*, vol. 13, no. 8, p. 2032, 2023.
15. D. N. Bangera, Y. N. Sudhakar, and R. A. Nazareth, "Concrete-based energy storage: Exploring electrode and electrolyte enhancements," *RSC Adv.*, vol. 14, no. 39, pp. 28854–28880, 2024.
16. S. I. Basha, S. S. Shah, S. Ahmad, M. Maslehuddin, M. M. Al-Zahrani, and M. A. Aziz, "Construction building materials as a potential for structural supercapacitor applications," *Chem. Rec.*, vol. 22, no. 11, p. e202200134, 2022.
17. H. J. Lee, N. Noor, C. Gumeci, and D. Higgins, "One-pot synthesis of MnO2/reduced graphene oxide via hydrothermal routes for supercapacitor applications," in *Electrochemical Society Meeting Abstracts 241*, The Electrochemical Society, Inc., 2022, p. 1686.
18. H. Qin, S. Ding, A. Ashour, Q. Zheng, and B. Han, "Revolutionizing infrastructure: The evolving landscape of electricity-based multifunctional concrete from concept to practice," *Prog. Mater. Sci.*, vol. 145, p. 101310, 2024.
19. M. Zhan, G. Pan, F. Zhou, R. Mi, and S. P. Shah, "In situ-grown carbon nanotubes enhanced cement-based materials with multifunctionality," *Cem. Concr. Compos.*, vol. 108, p. 103518, 2020.
20. 丁思齐, 韩宝国, and 欧进萍, "本征自感知混凝土及其智能结构.," *Eng. Mech. Lixue*, vol. 39, no. 3, 2022.
21. W. Li, W. Dong, A. Castel, and D. Sheng, "Self-sensing cement-based sensors for structural health monitoring toward smart infrastructure," in *Journal and Proceedings of the*

Royal Society of New South Wales, Royal Society of New South Wales Sydney, 2021, pp. 24–32.
22. J. Suchorzewski, M. Prieto, U. Mueller, and K. Malaga, "Damage and stress detection (self-sensing) in concrete with multi-walled carbon nanotubes," in *Proceedings*, MDPI, 2019, p. 17.
23. M. Civera, A. Naseem, and B. Chiaia, "Recent advances in embedded technologies and self - sensing concrete for structural health monitoring," *Struct. Concr.*, vol. 26, pp. 5300–5334, 2024.
24. C. Rainieri, Y. Song, G. Fabbrocino, M. J. Schulz, and V. Shanov, "CNT-cement based composites: Fabrication, self-sensing properties, and prospective applications to structural health monitoring," in *Fourth International Conference on Smart Materials and Nanotechnology in Engineering*, SPIE, 2013, pp. 228–237.
25. R. Joshi and R. Badola, "Digital twin: A transformative tool for smart cities," in *Smart Cities: Innovations, Challenges and Future Perspectives*, Springer, 2024, pp. 177–198.
26. S. S. Suryawanshi *et al.*, "Techno-cities: Artificial intelligence in the evolution of smart cities," in *Advanced Manufacturing Processes*, CRC Press, 2025, pp. 189–208.
27. M. B. Денисов and M. C. Проскуряков, "Цифровые технологии управления инфраструктурой 'Умного города'," Экономика и предпринимательство., vol. 9, no. 146, p. 467, 2022.
28. A. Hehr, Y. Song, B. Suberu, J. Sullivan, V. Shanov, and M. Schulz, "Embedded carbon nanotube sensor thread for structural health monitoring and strain sensing of composite materials," in *Nanotube Superfiber Materials*, Elsevier, 2014, pp. 671–712.
29. F. Olaoluwa and K. Potter, "Predictive maintenance for critical infrastructure security," 2024. https://doi.org/10.20944/preprints202409.0413.v1
30. C. Vlachakis, M. Perry, and L. Biondi, "Self-sensing alkali-activated materials: A review," *Minerals*, vol. 10, no. 10, p. 885, 2020.
31. B. Han, L. Zhang, and J. Ou, "Future developments and challenges of smart and multifunctional concrete," in *Smart and Multifunctional Concrete Toward Sustainable Infrastructures*, Springer, 2017, pp. 391–400.
32. S. Formagini, "Self-healing concrete-what is it good for?," *Mater. Construcción*, vol. 71, no. 341, p. e237, 2021.
33. K. Ramya and S. Hemavathi, "Experimental study on self-healing concrete using micro encapsulation," *IJERT*, vol. 5, pp. 1–4, 2017.
34. Y. Zhao, J. Fickert, K. Landfester, and D. Crespy, "Encapsulation of self-healing agents in polymer nanocapsules," *Small*, vol. 8, no. 19, pp. 2954–2958, 2012.
35. K. Van Tittelboom and N. De Belie, "Self-healing concrete by means of encapsulated polymers," in *13th International Congress on Polymers in Concrete (ICPIC-2010)*, Oficinas Graficas de Barbosa & Xavier, Lda, 2010, pp. 681–688.
36. S. A. McDonald, S. B. Coban, N. R. Sottos, and P. J. Withers, "Tracking capsule activation and crack healing in a microcapsule-based self-healing polymer," *Sci. Rep.*, vol. 9, no. 1, p. 17773, 2019.
37. H. M. Jonkers, "Self healing concrete: A biological approach," in *Self Healing Materials: An Alternative Approach to 20 Centuries of Materials Science*, Springer, 2007, pp. 195–204.
38. B. Ben-Nissan, G. Choi, A. H. Choi, I. Karacan, and L. Evans, "Natural and synthetic intelligent self-healing and adaptive materials for medical and engineering applications," *Innov. Bioceram. Transl. Med. I Fundam. Res.*, pp. 89–124, 2022.
39. G. Anglani, J.-M. Tulliani, and P. Antonaci, "Behaviour of pre-cracked self-healing cementitious materials under static and cyclic loading," *Materials (Basel).*, vol. 13, no. 5, p. 1149, 2020.
40. S. S. Sneha and K. P. Ramaswamy, "A comprehensive review on the mechanism of concrete deterioration in accelerated aggressive environment," *AIJR Proc.*, pp. 332–340, 2021.

41. R. Cornelis and I. Rustendi, "Review durability beton geopolymer berbasis fly ash," *Teodolita Media Komunkasi Ilm. di Bid. Tek.*, vol. 21, no. 2, 2020.
42. J. Ekeocha, C. Ellingford, M. Pan, A. M. Wemyss, C. Bowen, and C. Wan, "Challenges and opportunities of self-healing polymers and devices for extreme and hostile environments," *Adv. Mater.*, vol. 33, no. 33, p. 2008052, 2021.
43. K. Yin *et al.*, "Preparation and application of Na2SiO3@ EC microcapsules for self-healing alkali-activated slag," *Constr. Build. Mater.*, vol. 400, p. 132651, 2023.
44. H. Chang *et al.*, "Mechanical properties, durability and self-healing performance of mortar containing double-layers capsules," *Mater. Today Commun.*, vol. 38, p. 107880, 2024.
45. M. L. Durão, L. Nobre, C. Mota, J. Bessa, F. Cunha, and R. Fangueiro, "Self-healing composites: A path to redefining material resilience – a comprehensive recent review," *Materials (Basel).*, vol. 17, no. 19, p. 4681, 2024.
46. Z. Tian, D. Liu, W. Chen, W. Wang, S. Pan, and N. Guo, "Analysis and optimization of asymmetric wireless power transfer in concrete," *Wirel. Commun. Mob. Comput.*, vol. 2020, no. 1, p. 8841210, 2020.
47. C. K. Chau, X. Guan, B. Zhang, and M. L. Sham, "Nano-modified alkali-activated cementitious materials for rapid repair," Oct. 31, 2017, *Google Patents*.
48. M. Sun, X. Mao, X. Gao, Y. Lin, K. Li, and C. Zou, "Innovative encapsulation of alkali activators in alkali-activated slag concrete: A sustainable strategy for regulating setting time and durability," *Constr. Build. Mater.*, vol. 427, p. 136230, 2024.

9 Sustainable Innovation – Smart Strategies for Resilient Infrastructure

Shweta Goyal and Himanshu Guleria

9.1 INTRODUCTION

Greenhouse gases are essential for sustaining life on earth by regulating its temperature through the natural greenhouse effect. Among greenhouse gases, carbon dioxide (CO_2) is the predominant contributor, accounting for approximately 79.7%, with an atmospheric lifetime spanning thousands of years (USEPA 2020). The CO_2 levels are rising at an unprecedented rate of 2.2 ppm annually, with current concentration at the end of 2024 reaching 422.5 ppm, potentially responsible for increase in global temperatures with the projected rise of 2 to 3° Celsius by 2050 (IEA 2025). Such rapid emissions have intensified global warming, leading to glacial retreat, sea-level rise, and more frequent extreme weather events. This has created an urgent need to mitigate CO_2 emissions, now recognized as a defining global challenge.

The construction sector, principally driven by cement and steel production, two of the most energy-intensive industries, remains a major source of CO_2 emissions. The cement industry let alone accounts for nearly 8% of total global emissions (Gardarsdottir et al. 2019; IEA 2021; Meng et al. 2023), primarily due to limestone calcination, fossil fuels combustion, and the associated energy demands of transportation and electricity (Worrell et al. 2001). For perspective, production of 1 kg of cement generates more than 0.8 kg of CO_2 (Kaliyavaradhan and Ling 2017). Cement, being a key constituent of concrete, further amplifies its impact, as after water, concrete is the second-most used material. Currently, concrete usage in construction sector stands at approximately 30 billion tons annually worldwide which is thrice compared to four decades ago (York and Europe 2021). This escalating demand underscores the urgent need to reform construction practices to align with the Paris Agreement (COP21) targets and subsequently achieving net zero targets. Therefore, adopting sustainable technologies within the construction industry becomes crucial to mitigate climate change impacts and reduce environmental degradation. However, the journey towards sustainability in construction is complex, with construction firms facing stringent environmental regulations and operational challenges that make adopting sustainable practices more difficult. Despite these constraints, construction firms are increasingly required to innovate responsibly, developing solutions that align with evolving environmental standards and support sustainable development goals.

DOI: 10.1201/9781003664284-9

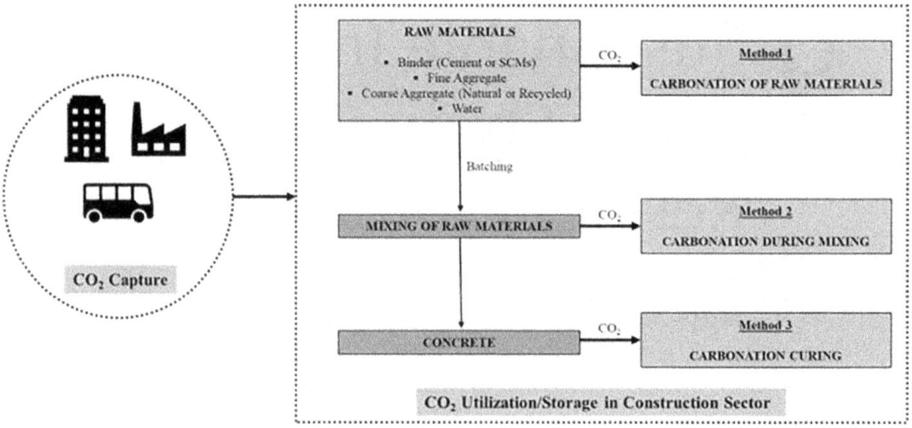

FIGURE 9.1 Pathways of CO_2 utilization in construction sector.

Carbon Capture, Utilization, and Storage (CCUS) technologies have emerged as pivotal strategies to mitigate emissions from hard-to-abate sectors like construction. The adoption of sustainable construction techniques and technological advancements in resilient civil infrastructure presents a viable pathway towards decarbonizing the construction industry while promoting durable, resource-efficient, and climate-resilient built environments. The CCUS in construction primarily focuses on integrating CO_2 directly into cement-based composites during the initial production process. The incorporation of CO_2 within the construction sector primarily occurs through three pathways: (i) carbonation of raw materials, (ii) direct injection of CO_2 during mixing, and (iii) CO_2 usage in the curing phase, commonly referred to as carbonation curing as presented in Figure 9.1. This integration not only helps CO_2 to sequester within the material, potentially reducing emissions significantly, but it may also enhance the material properties of the resulting construction products, such as improving material strength and durability.

In this chapter, three principal methods for CO_2 utilization pathways in the construction industry, carbonation of raw materials, incorporation of CO_2 during mixing, and carbonation curing, are comprehensively discussed. Each method is examined with reference to its commercial implementations, highlighting pioneering companies and their technological processes and the number of patents granted in pursuit of innovation and intellectual property protection. Furthermore, a critical evaluation of the advantages and disadvantages associated with each method is provided to establish a balanced understanding of their practical feasibility and limitations. These methods help in the integration of sustainable construction techniques and technological advancements in resilient civil infrastructure which offers a promising pathway towards decarbonizing the construction sector. These innovations are designed not only to mitigate greenhouse gas emissions but also to promote the development of durable, resource-efficient, and climate-resilient infrastructure that aligns with long-term sustainability goals.

9.1 PATHWAY 1: CARBONATION OF RAW MATERIALS

As concrete is produced through the blending of raw materials, the first pathway focuses on the utilization of CO_2 directly in these raw materials, enhancing their properties while enabling permanent carbon sequestration.

9.1.1 Concept and Mechanism

The raw material carbonation is a strategy in which raw materials are pre-treated with CO_2 before mixing, thereby initiating early carbonation reactions (Lu et al. 2024). This approach facilitates the formation of carbonated supplementary cementitious materials (SCMs) and fillers, where the precipitation of calcium carbonate can be integrated into binder systems to enhance performance. When processed to sufficient fineness, these carbonated powders not only act as reactive constituents but also contribute to the filler effect, improving particle packing density and optimizing the microstructure of cement composites.

A wide range of minerals and industrial by-products have demonstrated potential in this context due to their calcium and magnesium richness, including serpentine, olivine, wollastonite, basalt, steel slag, cement kiln dust, waste cement paste, and various types of fly ash (Chen, Bai, and Gao 2021; Jiang and Ling 2020; Kelemen et al. 2019; Lu et al. 2018; Sanna et al. 2014; Zajac et al. 2022; Zhang et al. 2015). Beyond fine powders, this technique has also been extended to the production of carbonated aggregates, developed by treating steel slag and Recycled Coarse Aggregates (RCA). Moreover, carbonated aggregates can also be formed using cementitious material by granulation method or press-moulding followed up by solidifying using CO_2 (Liu and Li 2023). The carbonation process improves the microstructure of aggregates; thereby decreasing the water absorption and porosity of aggregates (Poon et al. 2023). These carbonated aggregates not only provide stable sequestration of CO_2 but also encourage the sustainable reuse of industrial residues within construction applications. Additionally, the carbonation of RCA holds particular importance, as natural aggregates are generally inert and exhibit negligible reactivity towards carbonation (Poon et al. 2023). Overall, this methodology provides twin benefit of reducing the carbon footprint while advancing circularity in resource utilization.

9.1.2 Experimental Methodology

The experimental methodology for this process begins with the pre-treatment of raw materials using water or other suitable solutions, depending on the specific material selected. Such pre-treatment enhances the reactivity of the material by increasing the available surface area for subsequent CO_2 interaction. Subsequently, raw material is transferred into sealed carbonation chamber and exposed to carbon dioxide at concentrations ranging from 10% to 100% for a predetermined duration. Upon the completion of exposure, a post-treatment stage may be applied, which varies according to the nature of the material. This stage may involve techniques such as granulation and sieving, although in some cases, post-treatment can be omitted. The carbonated raw material is then blended with other concrete constituents, cast into specimens, and

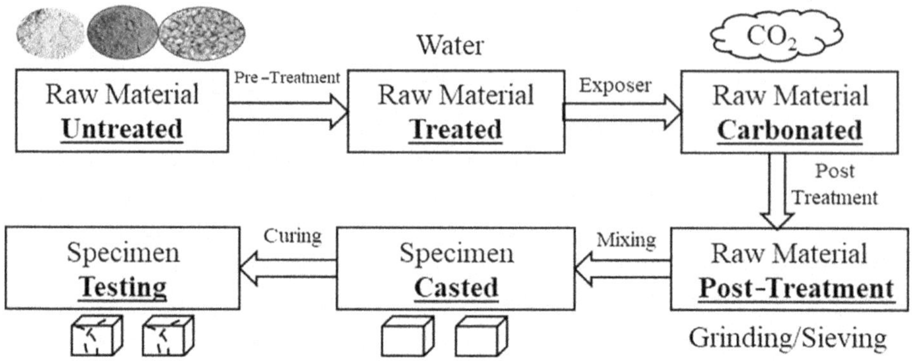

FIGURE 9.2 Flow chart for the process of carbonation of raw materials.

subjected to curing. Finally, the specimens are tested at designated ages to evaluate their performance. The overall experimental procedure is schematically illustrated in Figure 9.2.

9.1.3 Commercial Technologies

9.1.3.1 Carbon8

Carbon8 Systems, established in 2006 by Dr Colin Hills and Dr Paula Carey from the University of Greenwich, London, United Kingdom, have pioneered the commercialization of Accelerated Carbonation Technology (ACT). The research in this field was started way back in year 1994 at Imperial College and Queen Mary and Westfield College. This innovative process utilizes captured CO_2 in combination with industrial residues to produce CircaBuild – a carbon-negative alternative to natural aggregates. The technology embodies the principles of circular economy by transforming industrial waste streams into valuable construction materials. CircaBuild, produced through Carbon8's patented ACT process, not only sequesters CO_2 permanently within the material but also enhances performance characteristics, being over 20% lighter than conventional aggregates. By embedding CO_2 into solid mineral phases, the process ensures its safe and permanent storage, thereby converting otherwise harmful emissions into sustainable assets. To date, the company has secured multiple patents in this domain, reinforcing its role as a global leader in carbon utilization technologies (Carbon8 Systems 2024).

9.1.3.2 Blue Planet

Blue Planet technology established in 2012 in Los Gatos, USA, presents an innovative carbon capture and utilization approach that synthesizes limestone aggregates by combining waste carbon dioxide with calcium sourced from industrial and construction by-products. In this process, CO_2 is permanently and securely mineralized within the aggregate, thereby ensuring long-term carbon storage. This approach is referred to as geomimetic, as it mirrors the natural geological mechanisms responsible for carbonate rock formation. The calcium feedstock, referred to as geomass, is derived from a wide range of secondary materials, including demolished or returned

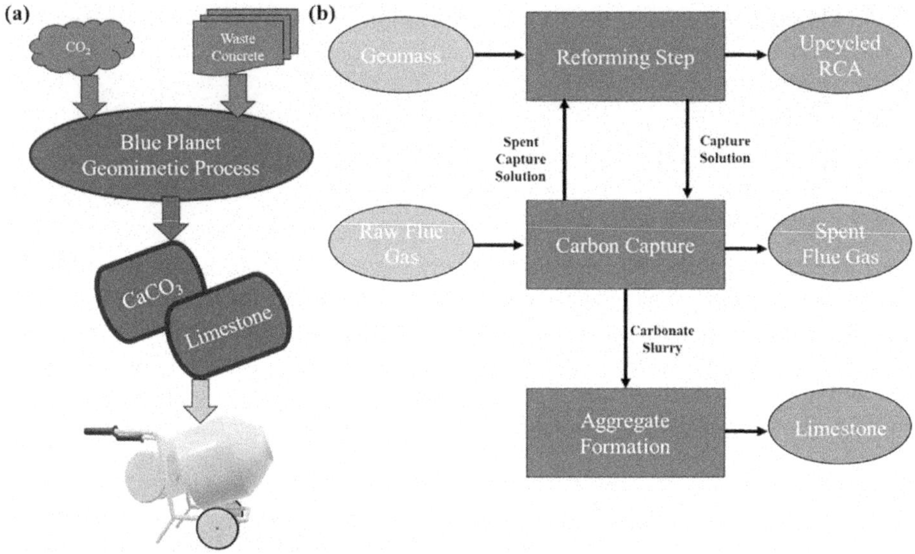

FIGURE 9.3 Blue Planet technology: (a) science and (b) circular solution.

Source: Blue Planet Systems (2024)

concrete, CKD, steel slag, fly ash and air pollution control residues (APCr), bauxite residue, and silicate rocks. More so, each ton of CO_2-sequestered limestone produced through this process can permanently lock away approximately 440 kilograms of CO_2, preventing its release into the atmosphere. By valorizing such waste materials, Blue Planet not only mitigates CO_2 emissions but also advances circular economy principles by transforming industrial residues into durable, CO_2-negative construction products (Blue Planet Systems 2024). Figure 9.3 represents the technology adopted by the Blue Planet.

9.1.4 Advantages and Disadvantages

The principal advantages and disadvantages of carbonation of raw materials, as identified through various literature, are outlined as follows:

1. The carbonation provides a sustainable route for valorizing waste streams such as steel slag, cement kiln dust, and fly ash, thereby reducing landfill disposal and supporting circular economy practices.
2. The carbonated RCA exhibits enhanced interfacial bonding with cement paste due to surface densification, which can improve the compressive and tensile strength of concrete composites.
3. By upgrading secondary raw materials through carbonation, the demand for virgin aggregates and natural resources is reduced, alleviating environmental pressures.
4. Some raw materials require grinding, sieving, or other pre-treatment steps before carbonation, which can increase processing costs and reduce efficiency.

5. Although beneficial, the CO_2 sequestration potential of raw materials is finite and depends on their chemical composition (CaO, MgO content) which limits its scalability.

9.2 PATHWAY 2: CARBONATION DURING MIXING

9.2.1 Concept and Mechanism

As the name suggests, this approach entails the introduction of gas CO_2 directly into the constituents during the ingredients' mixing stage, thereby exposing unhydrated cement phases to CO_2. This interaction results in the precipitation of nanoscale $CaCO_3$ particles, which are uniformly distributed across the cementitious matrix due to the homogeneous dispersion of CO_2 during mixing. Broadly, carbonation during mixing can be implemented through three pathways by carbonation of (Figure 9.4):

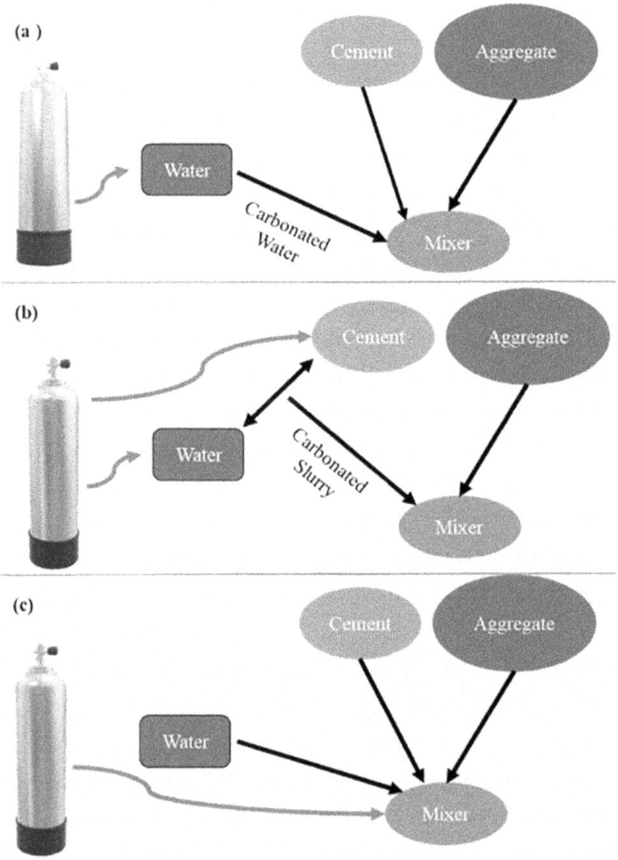

FIGURE 9.4 Different methods for carbonation during mixing: (a) carbonation of mixing water, (b) carbonation using cement slurry, and (c) direct injection during mixing.

Source: Davolio et al. (2025)

(a) mixing water, (b) cement slurry, and (c) direct injection of CO_2 during mixing. In the first approach, CO_2 is dissolved in the mixing water; however, as water alone lacks reactive compounds, calcium hydroxide is often introduced to facilitate the precipitation of nano-calcium carbonate, which serves both as nucleation sites and as a filler to accelerate hydration and enhance matrix densification (Davolio et al. 2025). The second approach, carbonation of cement slurry, involves the reaction of CO_2 with binder–water mixtures and is regarded as the most efficient and widely applied laboratory method, as it ensures targeted interaction of CO_2 with reactive cementitious phases. The third method, direct CO_2 injection, introduces carbon dioxide into the mixing chamber during ingredient blending through a controlled injection system, allowing carbonation to occur without compromising the workability or performance of fresh concrete. This latter technique has already been translated into commercial practice, with Canada-based CarbonCure Technologies pioneering its deployment at scale.

9.2.2 Experimental Methodology

In laboratory practice, carbonation of mixing water is generally carried out in three distinct stages. Initially, calcium hydroxide is dissolved in deionized water to form a calcium hydroxide solution. Subsequently, carbon dioxide is introduced into this solution using a gas injection system at a controlled flow rate of approximately 20 L/min for 2.5 minutes per litre of solution to produce nano-calcium carbonate. This step is followed by high-speed agitation to obtain a homogeneous suspension of fine calcium carbonate particles. Finally, cement is added to this suspension, blended for 5 minutes using a mechanical mixer, and then cast into specimens. The manufacturing flow chart for whole process is presented in Figure 9.5.

Slurry carbonation process is initiated with the placement of cement and water into a sealed mixing unit as shown in Figure 9.6. The mixer is first evacuated to create a vacuum prior to CO_2 injection, after which pressurized CO_2 is introduced until the desired pressure level is attained. The deionized water is then supplied through the mixer lids, followed by mixing at a predefined speed for approximately 5 minutes. Once carbonation is complete, the treated slurry is subsequently blended with other constituents, depending on the intended end-product requirements.

For direct CO_2 injection during mixing, the experimental procedure is comparatively straightforward. The carbonation chamber operates at pressures ranging from 1 bar to 15 bar, under 100% CO_2 concentration. The system has two main components: (i) a mixing bowl equipped with shafts and blades and (ii) a cover lid fitted with pressure gauges, an inlet, and an outlet as shown in Figure 9.7. To maintain a fully carbon-dioxide-saturated environment, the mixer is initially vacuumed prior to injection. Upon CO_2 introduction, its higher density ensures the displacement of residual air through the outlet, enabling a uniform CO_2 atmosphere throughout the mixing chamber (Saikia and Rajput 2025; Saikia et al. 2025).

9.2.3 Commercial Technologies

9.2.3.1 CarbonCure

CarbonCure is one of the companies that has focused on CO_2 utilization during the mixing stage of concrete production. The company established by Robert Nive in year

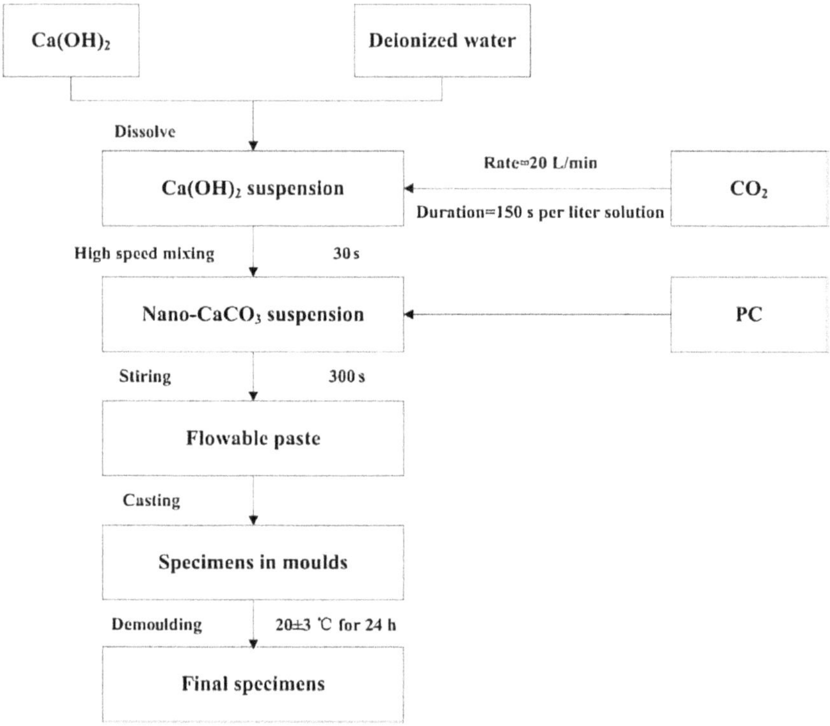

FIGURE 9.5 Flow chart for carbonation using mixing water.

Source: Reproduced from Qin et al. (2018)

2012 in Nova Scotia, Canada, developed a technology that can be retrofitted into existing concrete plants with minimal installation effort. The company has already filled above 25 patents in the field of CO_2 utilization (Hanifa et al. 2023). The system consists of a valve box connected to an on-site CO_2 storage tank, which automatically injects a controlled dosage of CO_2 into the mix. A control unit integrated with the batching software ensures precise synchronization making the process simple and efficient for the operators. In ready-mix applications, the technology has been shown to reduce approximately 7–11 kg of CO_2/m^3 of concrete, leading to substantial emission savings. To date, the implementation of CarbonCure's system has resulted in the mitigation of more than 648,000 metric tons of CO_2 emissions worldwide (Concrete Cure TM 2022).

9.2.4 Advantages and Disadvantages

The principal advantages and disadvantages of carbonation during mixing, as identified through various literature, are outlined as follows:

1. The incorporation of CO_2 directly into the mixing process addresses the challenge of slow gas diffusion in early-age concrete, which is commonly encountered in conventional carbonation curing.

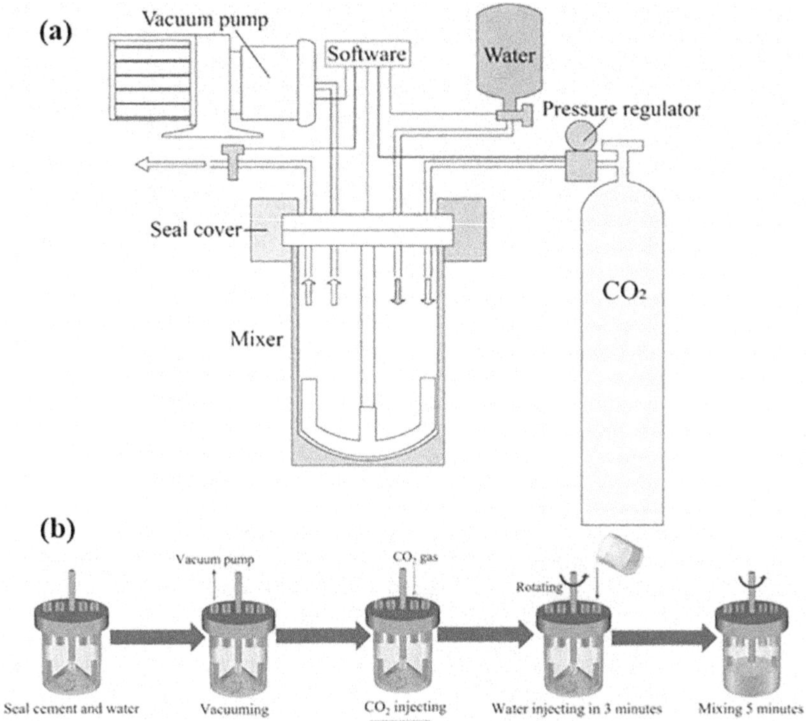

FIGURE 9.6 (a) Typical mixing setup and (b) CO_2 mixing procedure for carbonated slurry.

Source: Reproduced from Li et al. (2025)

2. The process eliminates the need for specialized curing chambers, thereby simplifying operational requirements.
3. This method allows the casting of concrete specimens of any size, as the procedure is not restricted by the dimensions of carbonation chambers.
4. Maintaining uniform CO_2 distribution within the mix can be technically challenging, potentially leading to inconsistent carbonation efficiency.
5. The dosage of CO_2 must be carefully controlled; excessive amounts may adversely affect workability and hydration, whereas insufficient amounts may reduce carbonation efficiency.

9.3 METHOD 3: CARBONATION CURING

9.3.1 Concept and Mechanism

Historically, carbonation curing first came into existence back in 1960s, when it was initially employed as a method to reduce concrete shrinkage (Shi et al. 2017). In the 1970s, CO_2 was further investigated for its potential to accelerate cement hydration in the presence of water (Berger and Klemm 1972; Berger et al. 1972). However, widespread adoption of the technique was hindered by limited scientific

FIGURE 9.7 Typical mixing setup for direct CO_2 injection during mixing.

Source: Reproduced from Saikia and Rajput (2025)

understanding and concerns regarding long-term durability. In recent years, driven by advancements in material science and mounting pressures to mitigate CO_2 emissions, carbonation curing has re-emerged as a promising accelerated curing technology, garnering significant research attention for sustainable construction practices. Carbonation curing involves subjecting young concrete to carbon dioxide under regulated pressure, temperature, and relative humidity within a carbonation chamber for a specified period. Unlike natural weathering carbonation, where CO_2 gradually penetrates hardened concrete over its service life, the carbonation curing process is characterized by a rapid chemical reaction, in contrast to the slow and progressive nature of weathering carbonation. During the exposure period in carbonation curing, CO_2 interacts with Bogue chemicals and hydrous cement compounds to produce stable calcium carbonate, efficiently sequestering carbon inside the young concrete. The Bogue compounds, including di- and tri-calcium silicate, react with CO_2 to produce $CaCO_3$ and calcium silicate, whereas the other Bogue compound, tri-calcium aluminate, reduces reactivity with CO_2 and generates ettringite, as illustrated in Equations 9.1 to 9.5. The hydrated cement compounds yield calcium carbonate, silica gel, and water as products, with CO_2, as illustrated in Equations 9.4 and 9.5 (Bukowski and Berger 1979; Klemm and Berger 1972; Young et al. 1974). This procedure is also occasionally referred to as accelerated carbonation curing (ACC). ACC is different from weathering carbonation as CO_2 is introduced to fresh concrete while specimen is exposed to CO_2 throughout its lifetime in weathering carbonation. Carbonation curing is a fast reaction, while weathering carbonation is a slow process.

$$3(3CaO.SiO_2)+(3-x)CO_2+yH_2O \rightarrow xCaO.SiO_2.yH_2O+(3-x)CaCO_3 \quad (9.1)$$

$$2(3CaO.SiO_2)+(2-x)CO_2+yH_2O \rightarrow xCaO.SiO_2.yH_2O+(2-x)CaCO_3 \quad (9.2)$$

$$3CaO.Al_2O_3.3CaSO_4.32H_2O+3CO_2 \rightarrow 3CaCO_3+3CaSO_4.H_20+Al_2O_3 \quad (9.3)$$

$$Ca(OH)_2+CO_2 \rightarrow CaCO_3+H_2O \quad (9.4)$$

$$C-S-H+CO_2 \rightarrow CaCO_3+H_2O+SiO_2 \quad (9.5)$$

9.3.2 Experimental Methodology

The laboratory implementation of carbonation curing generally involves four sequential stages: (i) mould-assisted curing, (ii) pre-curing, (iii) carbonation curing, and (iv) post-conditioning or post-curing. The mould-assisted curing is the period in which freshly cast concrete remains in moulds until demoulding, typically lasting between 5 to 24 hours (Sidhu et al. 2023; Zhang and Shao 2016), depending on the raw materials used and the prevailing environmental conditions during casting. The pre-curing is a critical preparatory stage prior to CO_2 exposure, aimed at reducing the moisture percentage within freshly prepared specimen to assist faster CO_2 ingress. The excess moisture can act as a barrier to gas diffusion; hence, partial drying is achieved either by storing the specimens in an environmental chamber or by fan-drying at room temperature. The duration of pre-conditioning typically ranges from 1 to 6 hours (Guleria and Goyal 2025a; Guleria et al. 2024; Xian and Shao 2021). The third stage carbonation curing involves subjecting the pre-conditioned specimens to CO_2 within a controlled chamber, where pressure, temperature, and relative humidity are carefully regulated. The CO_2 pressure may exceed up to 5 bar, while temperature and relative humidity generally range between 23 and 27° Celsius and 50 and 70%, respectively (Xian et al. 2024). The curing duration commonly extends up to 24 hours; however, some studies have explored extended periods of up to 28 days (Das et al. 2024; Sharma et al. 2023), depending on the specific experimental parameters and objectives. The fourth stage, post-curing, is strongly recommended, as the chemical reaction between CO_2 and the cementitious phases consumes calcium hydroxide, thereby reducing the alkalinity of the concrete matrix and potentially compromising the passive protective layer against steel reinforcement corrosion. To restore alkalinity, post-curing is performed by maintaining specimens in a moist environment such as immersion in water, wrapping with wet burlap, continuous water spraying, and storage in a moisture-controlled room (Guleria and Goyal 2025b; Sharma and Goyal 2018; Zhang and Shao 2018). The duration of post-curing may vary from three days to the intended testing age, ensuring adequate pH recovery and long-term durability of the concrete system.

9.3.3 Commercial Deployment

9.3.3.1 Solidia

Solidia Technologies founded in the United States in the year 2008 has been at the forefront of CO_2 utilization in the cement and concrete sector. The company operates with two primary product lines: Solidia Cement and Solidia Concrete. The production of Solidia Concrete involves a three-step process: (i) mixing Solidia cement powder with aggregates to create a loosely packed structure, (ii) filling the pore spaces by introducing water and CO_2, and (iii) initiating a reaction in which only Solidia cement interacts with CO_2 to form $CaCO_3$ and silica phases, thereby densifying the matrix and producing Solidia Concrete. Unlike Ordinary Portland Cement (OPC), which contains alite and belite phases and hydrates in the presence of water, Solidia cement is non-hydraulic in nature and is primarily composed of wollastonite and rankinite phases, reacting exclusively with CO_2. The adoption of Solidia products offers a significant reduction in carbon footprint, estimated at approximately 550 kg of CO_2/tons of cement. Moreover, the production of 30 Solidia Concrete blocks can sequester about 22 kg of CO_2, equivalent to the annual CO_2 uptake of a mature tree. In addition, this technology enables manufacturers to reduce water consumption by up to 100%, further enhancing its sustainability credentials (Solidia 2019).

9.3.3.2 Carbstone

Carbstone technology spearheaded by a Belgian-based concrete manufacturer represents an innovative approach to CO_2 utilization by transforming industrial residual streams, particularly steel slags, into durable and sustainable building materials. Through the chemical binding of CO_2 with calcium- and magnesium-rich constituents, this process eliminates the need for conventional cement while producing high-strength, carbon-negative construction products. The environmental benefits of Carbstone are two-fold: it ensures the permanent sequestration of CO_2 and reduces the reliance on primary raw materials, thereby lowering the ecological footprint. At the same time, it generates economic value by converting industrial waste into high-quality, circular products such as building blocks, facade bricks, paving stones, curbstones, tiles, roofing materials, and briquettes. The production process begins with the collection of steel slag from the steel industry, which is subjected to an upcycling process where metallic fractions are separated, and the remaining mineral fraction is utilized. This mineral component is combined with supplementary materials to form a cement-free, concrete-like composition, which is then pressed into various product shapes. These products are transferred to a controlled curing chamber, where they undergo a CO_2 curing cycle. Within 24 hours, the products are fully carbonated, cured, and ready for transport and on-site application. Figure 9.8 represents a typical schematic of Carbstone technology used in lab scale. The development and implementation of Carbstone technology have been advanced through collaborative efforts between the organizations VITO and Orbix, demonstrating its potential as a scalable and sustainable alternative to traditional construction materials (Carbstone 2025).

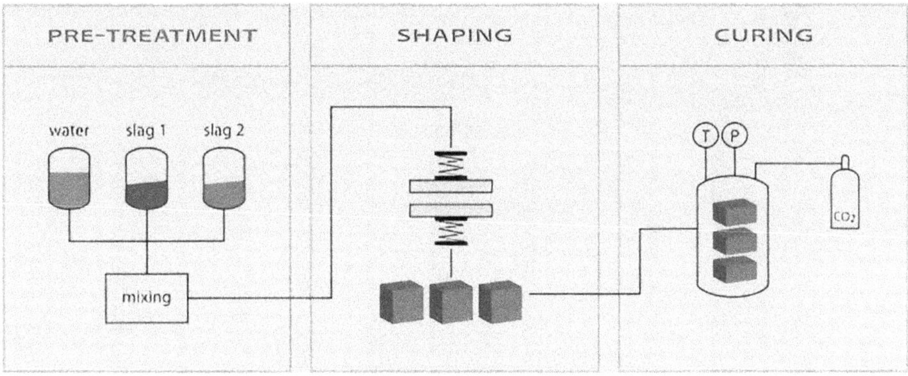

FIGURE 9.8 Schematic of Carbstone technology at lab scale.

Source: Reproduced from Quaghebeur et al. (2015)

9.3.4 Advantages and Disadvantages

The principal advantages and disadvantages of carbonation curing, as identified through various literature, are outlined as follows:

1. The operating conditions such as temperature, pressure, humidity are mild during carbonation curing unlike steam curing, which demands high energy input.
2. In addition to permanently sequestering CO_2, the process enhances both the mechanical properties and durability performance of concrete.
3. Compared to other accelerated curing techniques like steam curing, carbonation curing is economically advantageous due to reduced energy consumption and simpler operational requirements.
4. The rate of carbonation in concrete is inherently slow in carbonation curing, and the overall efficiency of the process is largely governed by the diffusion of CO_2 within the concrete matrix.
5. The process requires controlled curing environments, making it more suitable for precast elements rather than large-scale in-situ construction.

9.4 PATENT LANDSCAPE OF CO_2-BASED TECHNOLOGIES

In recent years, several companies such as Solidia Technologies, CarbonCure, Carbon8 Systems, Carbstone, and Blue Planet have been at the forefront of developing and commercializing CO_2 utilization technologies in the construction sector. These companies, along with others, have filed and secured numerous patents across key regions including the United States, Canada, Europe, and Japan, reflecting the growing global interest in this domain. To date (20th August, 2025), approximately 63 patents have been granted in this field for aforementioned companies, a

TABLE 9.1

Granted Patent Data of Commercial Working in Carbonation Field (as of 20 August 2025)

Sr. no	Patent ID	Title	Assignee	Year	Remarks	Aligned pathway
1	US-10668443-B2	Curing systems for materials that consume carbon dioxide and method of use thereof	Solidia	2020	CO_2-based curing system for carbonatable materials	3
2	US-12012365-B2	Mitigation of corrosion in carbonated concrete based on low-calcium silicate cement	Solidia	2024	Corrosion mitigation in carbonated low-calcium concrete	3
3	US-10662116-B2	Synthetic pozzolans	Solidia	2020	Production of engineered synthetic pozzolanic materials	1
4	US-10766819-B2	Compositions and methods for controlling setting of carbonatable calcium silicate cements containing hydrating materials	Solidia	2020	Setting control for carbonatable calcium-silicate cements	3
5	EP-3601197-B1	An aerated concrete product and a process of production thereof	Solidia	2024	Aerated concrete production using optimized processes	3
6	JP-6837836-B2	Hollow core articles and composites, and how to manufacture and use them	Solidia	2021	Manufacturing hollow-core composite structures	1, 3
7	US-11225438-B2	Pavers and block composite materials and methods of preparation thereof	Solidia	2022	CO_2-cured composite pavers and block materials	3
8	US-10611690-B2	Carbonatable calcium silicate compositions and methods thereof	Solidia	2020	Carbonatable calcium-silicate cement compositions	1
9	US-10787390-B2	Aerated composite materials, methods of production and uses thereof	Solidia	2020	Aerated composite materials and production methods	1, 3
10	US-10851022-B2	Aerated composite materials, methods of production and uses thereof	Solidia	2020	Aerated composite materials with enhanced properties	1, 3

Sr. no	Patent ID	Title	Assignee	Year	Remarks	Aligned pathway
11	EP-3515879-B1	Advanced curing equipment	Solidia	2023	Advanced CO_2 curing equipment for composites	3
12	US-10351478-B2	Advanced curing equipment and methods of using same	Solidia	2019	Advanced curing systems and operational methods	3
13	CA-3011251-C	Novel cement chemistries	Solidia	2023	Novel cement chemistries for improved performance	1
14	US-11352297-B2	Carbonatable calcium silicate-based cements and concretes having mineral additives and methods thereof	Solidia	2022	Mineral-additive carbonatable calcium-silicate cements	1, 3
15	EP-3097405-B1	Method and apparatus for curing CO_2 composite material objects at near ambient temperature and pressure	Solidia	2021	Near-ambient CO_2 curing of composite materials	3
16	US-9393715-B2	Composite railroad ties and methods of production and uses thereof	Solidia	2016	Composite railroad ties using CO_2 reactive materials	3
17	US-10781140-B2	Method and apparatus for the curing of composite material by control over rate-limiting steps in water removal	Solidia	2020	Curing via controlled moisture-removal mechanisms	3
18	US-10695949-B2	Lightweight composite materials produced from carbonatable calcium silicate and methods thereof	Solidia	2020	Lightweight composites from carbonatable silicate systems	1,3
19	US-10338053-B2	Curing-drying model and its applications	Solidia	2019	Curing drying modelling for optimized processing	3
20	US-10315335-B2	Rapid curing of thin composite material sections	Solidia	2019	Rapid curing for thin composite elements	3
21	EP-3004019-B1	Production of pervious composite materials	Solidia	2019	Production of pervious CO_2 reactive composites	3

(*Continued*)

TABLE 9.1 (*Continued*)
Granted Patent Data Of Commercial Working in Carbonation Field (as of 20 August 2025)

Sr. no	Patent ID	Title	Assignee	Year	Remarks	Aligned pathway
22	EP-3426619-B1	Hazing control for carbonatable calcium silicate-based cements and concretes	Solidia	2020	Hazing control in carbonatable calcium-silicate concretes	3
23	EP-3177384-B1	Controller for the curing of composite material by controlling over rate-limiting steps in water removal	Solidia	2023	Controller for moisture-limited curing processes	3
24	US-10752545-B2	Steam-assisted production of metal silicate cements, compositions, and methods thereof	Solidia	2020	Steam-assisted metal-silicate cement production	1
25	EP-3423421-B1	White carbonatable calcium silicate-based cements and method of preparation	Solidia	2020	White carbonatable calcium-silicate cement formulation	1
26	JP-7450605-B2	Multi-step curing of substrate	Solidia	2024	Multi-step CO_2 curing of substrates	3
27	US-11667573-B2	Compositions and method to improve the strength development of calcium silicate-based cements and concretes	Solidia	2023	Strength enhancement for calcium-silicate cement systems	1,3
28	US-12012362-B2	Compositions and method to improve the aesthetics of calcium silicate-based cements and concretes	Solidia	2024	Aesthetic improvement of calcium-silicate concretes	3
29	US-10457605-B2	Composite materials, methods of production, and uses thereof	Solidia	2019	Composite materials and their production processes	1,3
30	US-11597678-B2	Cement chemistries	Solidia	2023	Cement chemistries enabling improved performance	1
31	US-11034625-B2	Conditioned curing systems and processes thereof	Solidia	2021	Conditioned curing systems for CO_2 reactive materials	3
32	US-11518715-B2	Compositions and method to improve the durability of calcium silicate-based cements and concretes	Solidia	2022	Durability enhancement for calcium-silicate concretes	3

Sr. no	Patent ID	Title	Assignee	Year	Remarks	Aligned pathway
33	US-11618714-B2	Methods of forming cured composite materials with optimized pH and related compositions and systems	Solidia	2023	Optimized pH curing of composite materials	3
34	AU-2014244068-B2	Curing systems for materials that consume carbon dioxide	Solidia	2016	CO_2-consuming curing systems for composites	3
35	US-10927042-B2	Methods and compositions for concrete production	CarbonCure	2021	CO_2-injected concrete production methods	2
36	US-11958212-B2	Methods and compositions for treatment of concrete wash-water	CarbonCure	2024	Treatment of concrete wash-water using CO_2	2
37	US-9388072-B2	Methods and compositions for concrete production	CarbonCure	2016	Methods for CO_2-enhanced concrete manufacturing	2
38	US-9108883-B2	Apparatus for carbonation of a cement mix	CarbonCure	2015	Apparatus for carbonating cement mixtures	2
39	US-12319626-B2	Apparatus for delivery of carbon dioxide to a concrete mix in a mixer and determining flow rate	CarbonCure	2025	CO_2 delivery and flow-rate control in mixers	2
40	US-12330336-B2	Methods and compositions for treatment of concrete wash-water	CarbonCure	2025	Improved treatment of concrete wash-water	2
41	US-10683237-B2	System and method of applying carbon dioxide during the production of concrete	CarbonCure	2020	CO_2 application during concrete production processes	2
42	US-10654191-B2	Carbon dioxide treatment of concrete upstream from product mould	CarbonCure	2020	Upstream CO_2 treatment before concrete moulding	2
43	US-9738562-B2	Methods and compositions for concrete production	CarbonCure	2017	CO_2-based methods for producing concrete	2
44	US-10350787-B2	Carbonation of cement mixes	CarbonCure	2019	Carbonation of cement mixes for improved performance	2

(Continued)

TABLE 9.1 (*Continued*)
Granted Patent Data of Commercial Working in Carbonation Field (as of 20 August 2025)

Sr. no	Patent ID	Title	Assignee	Year	Remarks	Aligned pathway
45	US-11878948-B2	Integrated carbon dioxide capture	CarbonCure	2024	Integrated CO_2 capture for concrete production	2
46	CA-2821776-C	Carbon dioxide sequestration in concrete articles	CarbonCure	2016	Sequestration within concrete	3
47	EP-3224012-B1	A method for carbonating a concrete mix and an apparatus for the addition of carbon dioxide	CarbonCure	2024	Method and apparatus for carbonating concrete mixes	2
48	EP-3709118-B1	Network comprising a plurality of spatially separate cement mix operations	CarbonCure	2023	Networked multi-site cement-mix carbonation system	2
49	EP-3129753-B1	Compositions and methods for delivery of carbon dioxide	CarbonCure	2019	Methods for delivering CO_2 into concrete	2
50	EP-2276714-B1	Method of producing a mainly carbonate-bonded article by carbonation of alkaline materials	Carbstone	2018	Carbonation of alkaline materials into bonded articles	1
51	EP-2771305-B1	Method for producing a bonded article comprising a press-moulded, carbonated granular material	Carbstone	2019	Press-moulded carbonated granular material products	1,3
52	US-11840481-B2	Production of aggregates	Carbon8	2023	Production of CO_2-reactive manufactured aggregates	1
53	US-10203434-B2	Highly reflective microcrystalline / amorphous materials and methods for making and using the same	Blue Planet	2019	Reflective microcrystalline materials and fabrication methods	1
54	US-10197747-B2	Carbon sequestration methods and systems	Blue Planet	2019	CO_2 sequestration systems and operational methods	1

Sr. no	Patent ID	Title	Assignee	Year	Remarks	Aligned pathway
55	US-10322371-B2	Ammonia-mediated carbon dioxide (CO_2) sequestration methods and systems	Blue Planet	2019	Ammonia-mediated CO_2 sequestration processes	1
56	US-9707513-B2	Alkali-enrichment-mediated CO_2 sequestration methods and systems for practising the same	Blue Planet	2017	Alkali-assisted CO_2 sequestration methods	1
57	US-9714406-B2	Carbon sequestration methods and systems and compositions produced thereby	Blue Planet	2017	CO_2 sequestration and resulting carbonate materials	1
58	US-10766015-B2	Continuous carbon sequestration material production methods and systems for practising the same	Blue Planet	2020	Continuous production of CO_2-sequestered materials	1
59	US-11712654-B2	Direct air capture (DAC) carbon dioxide (CO_2) sequestration methods and systems	Blue Planet	2023	Direct-air-capture CO_2 mineralization systems	1
60	US-11946343-B2	Geomass-mediated carbon sequestration material production methods and systems for practising the same	Blue Planet	2024	Geomass-mediated CO_2 sequestration material production	1
61	US-11154813-B2	Modular CO_2 sequestration units and systems and methods for using the same	Blue Planet	2021	Modular CO_2 sequestration units and systems	1
62	US-12146130-B2	Carbon sequestration methods and systems and compositions produced thereby	Blue Planet	2024	CO_2 sequestration processes and carbonate products	1
63	US-10287439-B2	Carbonate pigments and methods for making and using the same	Blue Planet	2019	Carbonate pigment production using CO_2 mineralization	1

TABLE 9.2
Comparative Summary of CO_2 Utilization Approaches

	Pathways		
Parameters	**Carbonation of Raw Materials**	**Carbonation During Mixing**	**Carbonation Curing**
Stage of CO_2 application	Pre-treatment of material	During batching and mixing	During or after curing
Equipment	Pressurized chamber	Mixer with CO_2 dosing system	Sealed chamber with CO_2 supply control
Suitable product/ material	RCA, slag, fly ash, cement kiln dust	Fresh concrete	Pavers, blocks, precast elements
Effect on cement hydration	Reactive phases convert into carbonates before mixing	Alters early hydration kinetics	Accelerates early hydration
Potential of carbon reduction	Very high	Moderate	High
CO_2 uptake capacity	Moderate to high; depends upon type of raw material	Low to moderate	Moderate
Overall cost	Moderate to high	Low to moderate	Moderate to high
Scalability	Medium; depends upon raw materials	High	High
Technical challenges	Energy requirement due to raw material variability	Maintains uniform CO_2 distribution in mixer	Requires controlled environment

considerable number, given that most of these technologies have only been rediscovered and advanced in the past decade. Table 9.1 presents the granted patent data for aforementioned companies working in the field of carbonation. Notably, the volume of patent activity has shown a marked increase since the COP21 Paris Agreement, highlighting the rising emphasis on carbon mitigation strategies. This surge in patenting activity underscores the crucial role of such innovations in advancing sustainable technologies for the construction of resilient civil infrastructure.

A comparative summary of the three pathways of CO_2 utilization approaches is presented in Table 9.2 to highlight various parameters related to application, effect on hydration process, carbon reduction potential, and scalability.

9.5 CONCLUSIONS

The utilization of carbon dioxide in cementitious systems through carbonation offers a promising pathway towards sustainable and low-carbon construction resilient civil infrastructure. The three principal strategies carbonation of raw materials, carbonation during mixing, and carbonation curing have been studied, each demonstrating distinct benefits and limitations.

1. The carbonation of raw materials focuses on pre-treating RCA, steel slag, fly ash, CKD, and similar by-products with CO_2, enhancing their performance and permanently sequestering carbon in stable mineral phases. Such practices align with circular economy principles by converting industrial by-products into high-value aggregates, fillers, and building materials. Successful commercial examples include Blue Planet (USA) and carbon8 (Canada), which have demonstrated synthetic aggregate production at scale. Nevertheless, variability in feedstock composition and relatively high energy demands of the process remain critical barriers to broader adoption.
2. Carbonation during mixing introduces CO_2 directly into fresh concrete, ensuring uniform binder carbonation and overcoming slow diffusion issues. It eliminates the need for curing chambers and allows casting of elements of any size, offering greater flexibility. Although pilot projects and adaptations, particularly by CarbonCure in ready-mix concrete plants, have shown promise, widespread deployment remains limited due to challenges in process optimization and maintaining consistent performance under field conditions.
3. The carbonation curing is the most established approach where precast elements are exposed to CO_2 under controlled conditions, enhancing early-age strength, durability, and reducing curing time. Commercial adoption by companies such as Solidia (USA), and Carbstone (Belgium) demonstrates industrial feasibility, supported by numerous patents. However, the need for sealed curing chambers restricts its use largely to precast applications.

Overall, carbonation curing currently leads in commercialization, but advances in mixing-stage carbonation and raw material treatment are expanding deployment opportunities. The granted patent data suggests commercial moving towards the resilient construction. Collectively, these methods strengthen the pathway towards durable, resource-efficient, and climate-resilient infrastructure, aligning with global decarbonization goals.

ACKNOWLEDGEMENT

The authors would like to acknowledge the financial support from the AICTE under grant sanction no 8–183/RIFD/RPS(POLICY-1)/2018-19 and NBCC under grant sanction no NBCC/ED(R&D)/LOI/2020/01.

REFERENCES

Berger, R. L., and W. A. Klemm. 1972. Accelerated Curing of Cementitious Systems by Carbon Dioxide. Part II. Hydraulic Calcium Silicates and Aluminates. *Cement and Concrete Research* 2 (6):647–652. https://doi.org/10.1016/0008-8846(72)90002-6.

Berger, R. L., J. F. Young, and K. Leung. 1972. Acceleration of Hydration of Calcium Silicates by Carbon Dioxide Treatment. *Nature Physical Science* 240 (97):16–18. https://doi.org/10.1038/physci240016a0.

Blue Planet Systems. 2024. *Permanent Carbon Capture | Blue Planet Systems | Los Gatos.* https://www.blueplanetsystems.com/.

Bukowski, J. M., and R. L. Berger. 1979. Reactivity and Strength Development of CO_2 Activated Non-Hydraulic Calcium Silicates. *Cement and Concrete Research* 9 (1):57–68. https://doi.org/10.1016/0008-8846(79)90095-4.

Carbon8 Systems. 2024. *Carbon8*. https://www.carbon8.co.uk/.

Carbstone. 2025. *Technology | Circular Building Blocks without Cement*. https://www.carbstone.be/en/technology/.

Chen, Tiefeng, Mengjie Bai, and Xiaojian Gao. 2021. Carbonation Curing of Cement Mortars Incorporating Carbonated Fly Ash for Performance Improvement and CO_2 Sequestration. *Journal of CO_2 Utilization* 51 (April). Elsevier Ltd:101633. https://doi.org/10.1016/j.jcou.2021.101633.

Concrete Cure TM. 2022. *CarbonCure's Sustainable Concrete Solution – Concrete Technology Reducing Carbon Impact*. https://www.carboncure.com/.

Das, Kunal Krishna, Raju Sharma, Xuanru Wu, and Jeong Gook Jang. 2024. Influence of Hygiene Tissue Paper Used as an Additive on the Physicochemical and Mechanical Properties of CO_2-Cured Belite-Rich Cement Paste. *Cement and Concrete Composites* 145 (October 2023). Elsevier Ltd:105317. https://doi.org/10.1016/j.cemconcomp.2023.105317.

Davolio, Marco, Giovanni Muciaccia, and Liberato Ferrara. 2025. Concrete Carbon Mixing – A Systematic Review on the Processes and Their Effects on the Material Performance. *Cleaner Materials* 15 (September 2024). https://doi.org/10.1016/j.clema.2025.100292.

Gardarsdottir, Stefania Osk, Edoardo De Lena, Matteo Romano, Simon Roussanaly, Mari Voldsund, José Francisco Pérez-Calvo, David Berstad, Chao Fu, Rahul Anantharaman, Daniel Sutter, et al. 2019. Comparison of Technologies for CO_2 Capture from Cement Production – Part 2: Cost Analysis. *Energies* 12 (3). https://doi.org/10.3390/en12030542.

Guleria, Himanshu, Devender Sharma, and Shweta Goyal. 2024. Long Term Performance of Concrete Using Accelerated Carbonation Curing: An Effective CCUS Technique. *European Journal of Environmental and Civil Engineering* 28 (9). Taylor & Francis:2138–2159. https://doi.org/10.1080/19648189.2023.2295502.

Guleria, Himanshu, and Shweta Goyal. 2025a. *Accelerated Carbonation Curing as a CCUS Strategy: A Bibliometric and Technical Review for Sustainable Construction. Innovative Infrastructure Solutions*. Vol. 10. Springer International Publishing. https://doi.org/10.1007/s41062-025-02081-z.

Guleria, Himanshu, and Shweta Goyal. 2025b. Investigating the Effect of Accelerated Carbonation Curing Duration on Portland and Pozzolana Cement-based Concrete. *Structural Concrete* (September):1–20. https://doi.org/10.1002/suco.70326.

Hanifa, Mohd, R. Agarwal, U. Sharma, P. C. Thapliyal, and L. P. Singh. 2023. A Review on CO_2 capture and Sequestration in the Construction Industry: Emerging Approaches and Commercialised Technologies. *Journal of CO_2 Utilization* 67 (October 2022). Elsevier Ltd:102292. https://doi.org/10.1016/j.jcou.2022.102292.

IEA. 2021. *IEA (2020), Global Energy Review 2020, IEA, Paris. Global Energy Review 2020.* France. https://www.iea.org/reports/global-energy-review-2021/.

IEA. 2025. CO_2 *Emissions – Global Energy Review 2025 – Analysis – IEA*. IEA Publications. https://www.iea.org/reports/global-energy-review-2025/co2-emissions.

Jiang, Yi, and Tung Chai Ling. 2020. Production of Artificial Aggregates from Steel-Making Slag: Influences of Accelerated Carbonation during Granulation and/or Post-Curing. *Journal of CO_2 Utilization* 36 (June 2019). Elsevier:135–144. https://doi.org/10.1016/j.jcou.2019.11.009.

Kaliyavaradhan, Senthil Kumar, and Tung Chai Ling. 2017. Potential of CO_2 Sequestration through Construction and Demolition (C&D) Waste – An Overview. *Journal of CO_2 Utilization* 20 (May). Elsevier:234–242. https://doi.org/10.1016/j.jcou.2017.05.014.

Kelemen, Peter, Sally M. Benson, Hélène Pilorgé, Peter Psarras, and Jennifer Wilcox. 2019. An Overview of the Status and Challenges of CO_2 Storage in Minerals and Geological Formations. *Frontiers in Climate* 1 (November):1–20. https://doi.org/10.3389/fclim.2019.00009.

Klemm, W. A., and R. L. Berger. 1972. Accelerated Curing of Cementitious Systems by Carbon Dioxide. Part I. Portland Cement. *Cement and Concrete Research* 2 (5):567–576. https://doi.org/10.1016/0008-8846(72)90111-1.

Li, Chunjin, Yi Zhang, Qiang Ren, and Zhengwu Jiang. 2025. Early-Age Hydration of Cement Paste Prepared with CO_2 Mixing Under Varying Pressures. *Construction and Building Materials* 472:140948.

Liu, Haibao, and Qiuyi Li. 2023. Preparation of Artificial Aggregates from Concrete Slurry Waste and Waste Brick Masonry Powder: CO_2 Uptake and Performance Evaluation. *Construction and Building Materials* 382 (April). Elsevier Ltd:131356. https://doi.org/10.1016/j.conbuildmat.2023.131356.

Lu, Bao, Caijun Shi, Jiake Zhang, and Jiyun Wang. 2018. Effects of Carbonated Hardened Cement Paste Powder on Hydration and Microstructure of Portland Cement. *Construction and Building Materials* 186:699–708. https://doi.org/10.1016/j.conbuildmat.2018.07.159.

Lu, Dong, Fulin Qu, Chao Zhang, Yipu Guo, Zhiyu Luo, Lei Xu, and Wengui Li. 2024. Innovative Approaches, Challenges, and Future Directions for Utilizing Carbon Dioxide in Sustainable Concrete Production. *Journal of Building Engineering* 97 (May). Elsevier Ltd:110904. https://doi.org/10.1016/j.jobe.2024.110904.

Meng, Dan, Cise Unluer, En Hua Yang, and Shunzhi Qian. 2023. Recent Advances in Magnesium-Based Materials: CO_2 Sequestration and Utilization, Mechanical Properties and Environmental Impact. *Cement and Concrete Composites* 138 (March 2022). Elsevier Ltd:104983. https://doi.org/10.1016/j.cemconcomp.2023.104983.

Poon, Chi Sun, Peiliang Shen, Yi Jiang, Zihan Ma, and Dongxing Xuan. 2023. Total Recycling of Concrete Waste Using Accelerated Carbonation: A Review. *Cement and Concrete Research* 173 (February). Elsevier Ltd:107284. https://doi.org/10.1016/j.cemconres.2023.107284.

Qin, Ling, Xiaojian Gao, and Qiyan Li. 2018. Upcycling Carbon Dioxide to Improve Mechanical Strength of Portland Cement. *Journal of Cleaner Production* 196:726–738.

Quaghebeur, Mieke, Peter Nielsen, Liesbeth Horckmans, and Dirk Van Mechelen. 2015. Accelerated Carbonation of Steel Slag Compacts: Development of High-Strength Construction Materials. *Frontiers in Energy Research* 3:52.

Saikia, Saurav Kumar, and Aditya Singh Rajput. 2025. Impact of Carbon Sequestration Technologies on Corrosion Resistance of Various Concrete Types. *Construction and Building Materials* 490 (October 2024). Elsevier Ltd:142372. https://doi.org/10.1016/j.conbuildmat.2025.142372.

Saikia, Saurav Kumar, Aditya Singh Rajput, and Kartik Saini. 2025. Utilizing Ground Granulated Blast Furnace Slag during Carbon Sequestration in Concrete. *Journal of Materials in Civil Engineering* 37 (2):1–15. https://doi.org/10.1061/JMCEE7.MTENG-18542.

Sanna, A., M. Uibu, G. Caramanna, R. Kuusik, and M. M. Maroto-Valer. 2014. A Review of Mineral Carbonation Technologies to Sequester CO_2. *Chemical Society Reviews* 43 (23). Royal Society of Chemistry:8049–8080. https://doi.org/10.1039/c4cs00035h.

Sharma, Devender, and Shweta Goyal. 2018. Accelerated Carbonation Curing of Cement Mortars Containing Cement Kiln Dust: An Effective Way of CO_2 Sequestration and Carbon Footprint Reduction. *Journal of Cleaner Production* 192. Elsevier B.V.:844–854. https://doi.org/10.1016/j.jclepro.2018.05.027.

Sharma, Raju, Junjie Pei, and Jeong Gook Jang. 2023. Microstructural Evolution of Belite-Rich Cement Mortar Subjected to Water, Carbonation, and Hybrid Curing Regime.

Cement and Concrete Composites 139 (November 2022). Elsevier Ltd:105028. https://doi.org/10.1016/j.cemconcomp.2023.105028.

Shi, Caijun, Zhenjun Tu, Ming Zhi Guo, and Dehui Wang. 2017. *Accelerated Carbonation as a Fast Curing Technology for Concrete Blocks. Sustainable and Nonconventional Construction Materials Using Inorganic Bonded Fiber Composites*. Elsevier Ltd. https://doi.org/10.1016/B978-0-08-102001-2.00015-2.

Sidhu, Gaganjot Singh, Himanshu Guleria, Devender Sharma, and Shweta Goyal. 2023. Strength and Permeation Characteristics of Pervious Concrete Subjected to Accelerated Carbonation Curing. *Journal of Sustainable Cement-Based Materials* 12 (10). Taylor & Francis:1242–1254. https://doi.org/10.1080/21650373.2023.2213241.

Solidia. 2019. *The Science behind Solidia*, 1–2. https://assets.ctfassets.net/jv4d7wct8mc0/5DwEAeEYqsFAYA9UC53EF7/4f8b7566221a8d9cb38f970867003226/Solidia_Science_Backgrounder_11.21.19__5_.pdf.

USEPA. 2020. *Overview of Greenhouse Gases | US EPA. Earth System Science Data*. https://www.epa.gov/ghgemissions/overview-greenhouse-gases.

Worrell, Ernst, Lynn Price, Nathan Martin, Chris Hendriks, and Leticia Ozawa Meida. 2001. Carbon Dioxide Emissions from the Global Cement Industry. *Annual Review of Energy and the Environment* 26:303–329. https://doi.org/10.1146/annurev.energy.26.1.303.

Xian, Xiangping, Mehrdad Mahoutian, Duo Zhang, and Yixin Shao. 2024. Development of Wet-Cast Portland-Cement-Free Concrete Based on Steel Slag and Ambient-Pressure Carbonation Activation. *Resources, Conservation and Recycling* 203 (August 2023). Elsevier B.V.:107455. https://doi.org/10.1016/j.resconrec.2024.107455.

Xian, Xiangping, and Yixin Shao. 2021. Carbonation Curing of Concretes in a Flexible Enclosure under Ambient Pressure. *Journal of Materials in Civil Engineering* 33 (4):1–14. https://doi.org/10.1061/(asce)mt.1943-5533.0003648.

York, In New, and In Europe. 2021. Concrete Needs to Lose Its Colossal Carbon Footprint. *Nature*. https://doi.org/10.1038/d41586-021-02612-5.

Young, J. F., R. L. Berger, and J. Breese. 1974. Accelerated Curing of Compacted Calcium Silicate Mortars on Exposure to CO_2. *Journal of the American Ceramic Society* 57 (9):394–397. https://doi.org/10.1111/j.1151-2916.1974.tb11420.x.

Zajac, Maciej, Jan Skocek, Mohsen Ben Haha, and Jan Deja. 2022. CO_2 Mineralization Methods in Cement and Concrete Industry. *Energies* 15 (10):1–26. https://doi.org/10.3390/en15103597.

Zhang, Duo, and Yixin Shao. 2016. Early Age Carbonation Curing for Precast Reinforced Concretes. *Construction and Building Materials* 113. Elsevier Ltd:134–143. https://doi.org/10.1016/j.conbuildmat.2016.03.048.

Zhang, Duo, and Yixin Shao. 2018. Surface Scaling of CO_2-Cured Concrete Exposed to Freeze-Thaw Cycles. *Journal of CO_2 Utilization* 27 (June). Elsevier:137–144. https://doi.org/10.1016/j.jcou.2018.07.012.

Zhang, Jiake, Caijun Shi, Yake Li, Xiaoying Pan, Chi Sun Poon, and Zhaobin Xie. 2015. Influence of Carbonated Recycled Concrete Aggregate on Properties of Cement Mortar. *Construction and Building Materials* 98:1–7. https://doi.org/10.1016/j.conbuildmat.2015.08.087.

10 Enhancing the Mechanical and Durability Properties of Concrete Through the Valorization of Bamboo Aggregates and Fibers

Suvash Chandra Paul, Shahriar Hasan, Md. Maruf Hossain, Noor Md. Sadiqul Hasan and M. Shariful Islam

10.1 INTRODUCTION

Rapid industrialization and urbanization worldwide have made concrete the second-most utilized material after water. The primary component of concrete is aggregates, which constitute approximately 70–80% of its volume, with coarse aggregates accounting for 40–50% of the total. As these aggregates are sourced from natural materials, the construction industry significantly impacts environmental degradation. One solution is to seek alternative materials that can be recycled and reused. The use of new, naturally occurring materials is essential in this regard. Examples include bamboo,[1] agricultural products,[2] coconut shells,[3] rice husks,[4] ceramics,[5] and crumb rubber.[6] Bamboo, in particular, is a rapidly renewable resource that has historically been used extensively in construction. Due to its rapid growth, affordability, and dependable mechanical characteristics, bamboo stands out among agricultural by-products as a cost-effective and sustainable alternative to traditional construction materials.[7,8]

Recent research has focused on utilizing bamboo as a structural material and reinforcement in concrete, highlighting its numerous benefits and ecological advantages.[9,10] Plain concrete has low tensile strength, limited strain capacity, and is brittle. Since the late 1960s, the use of fiber-reinforced concrete (FRC) has increased to address these limitations. Fibers have traditionally been employed as stabilizing agents in various types of mortar and concrete, with common materials including steel, organic polymers, carbon fiber, and various natural fibers.[11,12] Bamboo fiber (BF), derived from the bamboo culm, can enhance concrete properties due to its

DOI: 10.1201/9781003664284-10

advantages, such as low cost, low density, ecological friendliness, renewability, and biodegradability.[13]

Bamboo chip aggregates (BCA) and BF have emerged as promising materials in concrete applications, offering structural and environmental benefits. Incorporating BF improves mechanical properties like compressive and tensile strength, as well as durability. This enhancement is primarily due to the fibers' ability to bridge cracks, distribute loads more effectively, and limit crack propagation within the concrete matrix.[13,14] Additionally, BCA can help reduce the reliance on conventional aggregates, contributing to sustainability by lowering the carbon footprint of concrete. Studies indicated that incorporating BF at various percentages increased strength compared to traditional concrete, with significant gains observed in both early and later curing stages. This method enhances concrete resilience, particularly in conditions that typically lead to cracking and surface damage.[15] Prioritizing sustainable materials in construction is essential, as the industry is a major source of global carbon dioxide (CO_2) emissions. Therefore, replacing conventional building materials with sustainable alternatives that still meet requirements is essential.[16,17]

Bamboo was reported to have a tensile strength of approximately 238 MPa, earning it the designation of "green steel."[18] However, its properties are influenced by various factors, including the type of bamboo and treatment conditions, before using them in concrete. BF effectively can reduce crack propagation in concrete, enhancing the material's overall durability and longevity.[19] In a study conducted by Noh et al., BF was used to replace varying percentages of natural fine aggregates in concrete.[20] The researchers compared the compressive strength of BF-reinforced concrete with that of conventional concrete without BF and found that an optimal BF content of 5% yielded the highest compressive strength. Similarly, Yusra et al. examined the effects of adding BF on the compressive and tensile strength of concrete.[21] They discovered that increasing BF content led to a decrease in compressive strength while enhancing tensile strength. Specifically, with 0.5% BF, the compressive strength dropped by 4.23%, whereas tensile strength increased by 23.37%. At a 1.5% BF content, the compressive strength declined notably by approximately 12.08%, while the tensile strength exhibited a substantial increase of 56.14%. These findings suggest that while BF reduces compressive strength, it significantly enhances tensile properties. Additionally, Park et al. studied the influence of BCA and curing age on the compressive strength of concrete.[22] Their results indicated that compressive strength consistently increased with curing time across all samples. The reference concrete without any BCA exhibited a 71% increase in strength from 7 to 28 days. For concrete mixes incorporating 10%, 20%, and 30% BCA, the strength increases over the same curing period were 44%, 35%, and 32%, respectively. These variations were influenced by the treatment conditions of the BCA, whether the chips were coated, dried, or wetted – with the wetted BCA achieving the highest strength gains among the treated samples. However, as BCA content increased, the positive effect of curing on strength diminished, particularly at 30% BCA, where the bamboo treatment negatively impacted compressive strength.

Dudhatra et al. explored the replacement of natural aggregates with BCA in self-compacting concrete, using substitution levels of 0%, 2%, 4%, and 5%.[9] Their results showed that as the BCA content increased, the concrete's strength decreased. After 56 days of curing, the compressive strength of concrete with 2%, 4%, and

5% BCA replacement declined by around 9%, 17%, and 23%, respectively, in comparison to the control mix. Similarly, the flexural strength at 56 days decreased by 9%, 13%, and 16% for the same BCA levels. The study also found that the slump value of fresh concrete mixes reduced by a maximum of 3%, 5%, and 7% for BCA contents of 2%, 4%, and 5%, respectively. Another study corroborated these findings, observing a strength reduction in concrete when natural coarse aggregates were replaced with BCA at 5%, 10%, 15%, 20%, and 25%.[23] However, after 28 days, a slight increase in compressive strength was noted for concrete with 5% BCA. For splitting tensile strength, mixes with up to 10% BCA exhibited strength comparable to that of the reference concrete. Additionally, a 3% increase in flexural strength was recorded for concrete samples containing 15% BCA.[23] Overall, higher BCA content adversely affects workability and strength, underscoring its negative impact on concrete properties.

The inclusion of BF in concrete samples increases pore volume due to challenges in compaction as fiber content rises. Additionally, the crack-bridging ability of BF can enhance the tensile and flexural strength of concrete, as supported by various studies.[24,25] Typically, conventional concrete samples fracture immediately upon reaching their ultimate load capacity. In contrast, BF acts as a crack resistor in concrete, enabling the material to support greater loads and remain intact even after failure.[26] A study by Dewi and Wijaya also reported reduced crack widths and improved post-cracking load-carrying capacity in reinforced concrete beams containing BF.[27] While bamboo enhances certain properties, achieving consistent performance necessitates careful regulation of fiber replacement levels and treatment methods.[23,28]

In another study, Kavitha and Kala examined the effectiveness of BF as a strength enhancer in concrete.[24] They tested BF content levels of 0%, 0.5%, 0.75%, 1%, and 1.25%. The results indicated that 1% BF was optimal, yielding the highest compressive and tensile strengths, while 1.25% BF provided the greatest flexural strength. Notably, all strength measurements improved as the curing duration extended from 7 days to 28 days.[24] Collectively, these studies demonstrate that BF enhances the tensile strength, flexural toughness, and crack resistance of concrete, particularly at lower fiber content levels of 0.4% to 2%.[8] However, factors such as optimal fiber content, size, aspect ratio, and weak bonding due to higher fiber volume, along with the lower mechanical properties of BF, significantly influence concrete properties. While BF generally enhances the flexural and tensile behavior of concrete samples, its effect on compressive strength can be inconsistent and may even be negative at higher dosages.[29]

The workability of fresh concrete was noticeably reduced with the increasing addition of BFs. Compared to the control mix, the slump values for concrete with 0.5%, 0.75%, 1%, and 1.25% BF decreased by approximately 14%, 27%, 32%, and 45%, respectively.[24] The addition of fibers tends to enhance cohesion and internal resistance in the concrete mixes, resulting in reduced workability when compared to fiber-free mixes.[30] Higher fiber content also increases the overall fiber surface area, leading to a greater water demand from the mix. This reduction in available water may impair the cement paste's ability to provide adequate lubrication, which is vital for workability. Furthermore, the inclusion of BF influenced the air content in the concrete mixes, with increases of 26%, 74%, and 105% observed when 1%, 2%, and 3% BF were added by volume.[10] The elevated air content is attributed to the difficulties in compacting fresh concrete mixes with higher fiber levels. However, natural bamboo has a higher tensile strength than wood, making it an appealing material for FRC. This is particularly relevant in emerging nations, where the demand for reinforced concrete is rapidly growing.[31]

The ecological impact of traditional construction materials can be partially alleviated by utilizing renewable resources like bamboo as a concrete substitute. While research has examined the use of natural fibers and aggregates in concrete, there has been limited investigation into the combined effects of BF and BCA on the mechanical properties and crack resistance of concrete. Other properties, such as electrical resistivity, water absorption, and pore volume in bamboo-reinforced concrete, have been thoroughly studied. This study aims to develop concrete using BF and BCA sourced from waste bamboo after its use on construction sites. These materials were incorporated into the concrete at varying percentages to replace natural coarse aggregates by volume. The concrete composite was tested for compressive strength, splitting tensile strength, flexural strength, electrical resistivity, crack properties, and absorption. Although the addition of waste BF and BCA may slightly reduce strength compared to conventional concrete, this approach is crucial for advancing eco-friendly construction practices. It provides sustainable and affordable solutions for low-cost housing, paver blocks, temporary structures, and nonstructural elements in developing regions.

10.2 MATERIALS AND METHODS

10.2.1 MATERIALS AND MIX COMPOSITIONS

Seven mix compositions were used in the experiments, labelled as M0, M1, M2, M3, M4, M5, and M6, as detailed in Table 10.1. Ordinary Portland cement (CEM-I, 52.5N) and locally sourced natural coarse and fine aggregates were utilized. A sieve analysis was conducted on the aggregates in accordance with ASTM C136 standard to verify their grading for concrete, as illustrated in Figure 10.1.[32] The maximum

TABLE 10.1
Mix Design of Different Concrete Mixes in Kg/M³

Mix ID	Cement	FA	CA	Water	BCA	BF
M0	482.09	642.38	1162.70	207.30	-	-
M1	482.09	642.38	1046.43	207.30	10.38	-
M2	479.68	639.17	1041.20	206.26	10.33	1.54
M3	477.27	635.96	1035.97	205.22	10.25	3.08
M4	482.09	642.38	930.16	207.30	20.76	-
M5	479.68	639.17	925.51	206.26	20.65	1.54
M6	477.27	635.96	920.86	205.22	20.55	3.08

Abbreviations: FA = Fine aggregate, CA = Coarse aggregate, BCA = Bamboo chip aggregates, BF = Bamboo fiber

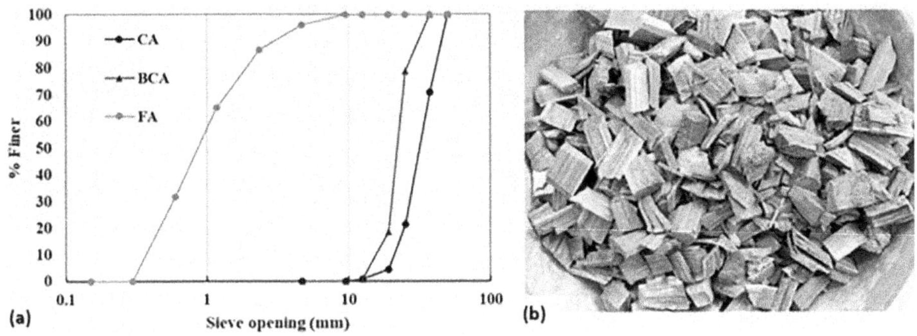

FIGURE 10.1 (a) Gradation curve of BCA, CA, and FA and (b) waste-recycled BCA used in the study.

particle sizes for the coarse and fine aggregates passed through 19 mm and 4.75 mm sieves, respectively. The BCA and BF were sourced from waste bamboo collected at a construction site, used at two different percentage levels. Mix M0 serves as the reference or control mix, containing no BCA or BF. Mixes M1, M2, and M3 incorporate 10% BCA, 10% BCA with 0.5% BF, and 10% BCA with 1% BF, respectively. Similarly, mixes M4, M5, and M6 contain 20% BCA, 20% BCA with 0.5% BF, and 20% BCA with 1% BF, respectively. In these mixes, BCA replaces natural coarse aggregates by volume, while BF replaces a portion of the total volume of the concrete mix. All ingredients were dry mixed for approximately 3 to 4 minutes before adding water. To ensure better consistency, an additional 2 to 3 minutes of continuous mixing was performed before measuring the slump using a conventional slump cone with a height of 300 mm, a base diameter of 200 mm, and a top opening of 100 mm.

10.2.2 Preparing of BCA and BF

For the experiment, waste bamboo was collected from a construction site and cut into pieces to create bamboo strips, followed by BCA and BF. The largest size of the BCA was below 25 mm, as confirmed by sieve analysis, shown in Figure 10.1a. Figure 10.1b displays the recycled BCA used in this study. To produce the BF, bamboo strips measuring approximately 25 to 35 mm were soaked in a 2% NaOH solution (by weight of water) for 72 hours. The BF was then manually extracted, as illustrated in Figure 10.2. All fibers were air-dried in sunlight for about seven days. The approximate diameter of the BF ranged from 0.50 mm to 0.75 mm, with an average aspect ratio of 69 and a water absorption of around 49%. The specific gravity of the BCA was approximately 0.31.

10.2.3 Sample Preparation and Methods of Testing

In this study, we examined various properties of concrete mixes, including compressive strength, tensile strength, three-point bending strength, water absorption, pore volume, and electrical resistivity. After mixing, we assessed the workability of

FIGURE 10.2 Bamboo fiber (BF) extraction process followed in this study.

the concrete using a slump cone test, alongside measuring the density of the fresh concrete mixes. Samples in the form of 100 mm cubes were prepared to conduct the compressive strength test. The splitting tensile and flexural behavior tests were conducted on cylindrical and beam samples, with cylindrical specimens of 100 mm diameter and 200 mm length and beam specimens measuring 100 mm × 100 mm × 500 mm. All tests adhered to the guidelines outlined in the BS-EN 12390 standards.[33] The loading rate for both compressive and tensile tests was set at 2 to 3 kN/s, while for the beam samples, the loading rate was approximately 0.20 kN/s with a span of 330 mm. A minimum of three samples were tested for each curing age, and average values were reported.

Additionally, we determined the water absorption capacity and permeable pore volume of the concrete samples after 28 days of curing, following ASTM-C642.[34] For this analysis, we created disc samples with a diameter of 100 mm and a thickness of 50 mm for all concrete mixes. Detailed sample preparation methods are discussed in previous studies and the ASTM standard.[34,35] We also performed an electrical resistivity test to evaluate the resistance of the concrete samples made from BCA and BF to chloride ion penetration. This nondestructive test was conducted on the same cylindrical samples used for the splitting tensile strength test. We followed AASHTO TP95 for this resistivity test, using an LCR meter to apply resistance at a frequency of 1 kHz by connecting the two ends of the cylinder.[36] The resistance (R value) recorded from the LCR meter allowed us to calculate the resistivity in kΩ-cm of the concrete samples, using the cylinder's cross-sectional area and length. Furthermore, we measured the crack width in the flexural samples post-bending testing with a crack width ruler. The details of these experimental outcomes are discussed in the following sections.

10.3 RESULTS AND DISCUSSION

10.3.1 Workability and Density of Concrete

The slump test was performed to evaluate the consistency and workability of concrete, which reflects the balance between water content and mix composition. Workability is a critical factor influencing both the strength and durability of concrete. Figure 10.3

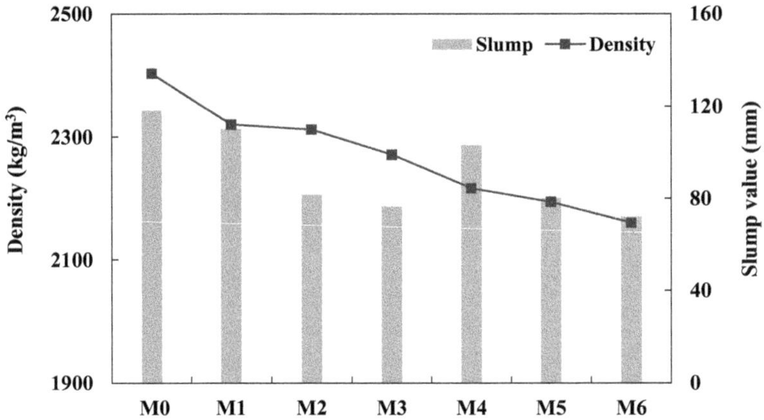

FIGURE 10.3 Variation of slump value and density of different concrete mix.

presents the slump values for different concrete mixes. A noticeable reduction in slump was observed when natural coarse aggregates were partially replaced with BCA and BF, with the most significant decrease recorded in the mix containing 20% BCA and 1% BF. Compared to the reference mix, the slump values declined by 6.8%, 30.9%, 35.2%, 12.7%, 32.2%, and 39% for mixes M1 through M6, respectively. This reduction in workability is likely due to the hygroscopic nature of bamboo, which absorbs water from the concrete mix, thereby reducing the effective water content. Similar declines in workability with unmodified BCA have been reported in previous studies.[37] However, research has shown that slump values can be improved by using epoxy-coated BCA[37] or BCA treated with hot water at 80°C for a specified duration.[38]

Figure 10.3 also illustrates the dry density of the concrete mixes. Generally, higher concrete density is associated with greater compressive strength. As observed, the density decreased progressively with the partial replacement of natural aggregates by BCA. The control mix (M0) exhibited the highest density, while the densities of mixes M1 through M6 were reduced by 3.6%, 4.0%, 5.8%, 8.4%, 9.5%, and 11.2%, respectively. The reduction in density can be attributed to the lower specific gravity of bamboo compared to conventional aggregates. Concrete density influences the dead load of a structure, and its optimal value depends on the structural application. In certain cases, lower density can be advantageous by reducing dead load, thereby contributing to more economical and sustainable construction practices.

10.3.2 Compressive Strength of BCA and BF Mixed Concrete

Figure 10.4 illustrates the development of concrete compressive strength at 28 as well as 56 days. Similar to density, the compressive strength of concrete decreases with the introduction of BCA and BF into the mixes. At 28 days, the strength of mixes M1 to M6 shows reductions of approximately 5.5%, 9.52%, 16.15%, 26.56%, 35.67%, and 38.12%, respectively, compared to the reference mix. After 56 days, these reductions

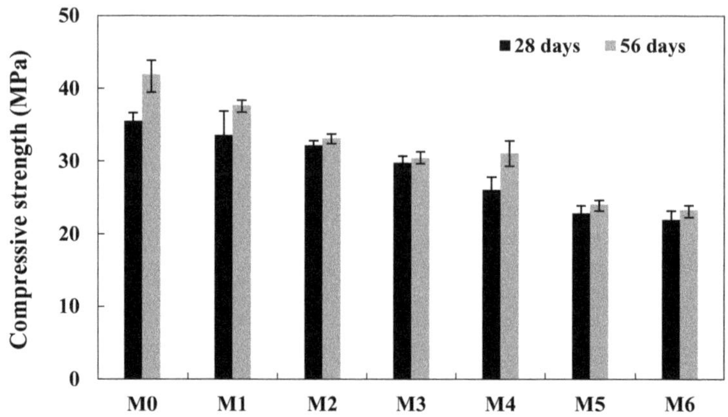

FIGURE 10.4 Compressive strength of different concrete mixes after 28 days and 56 days of curing.

increase to about 10.1%, 21.10%, 27.5%, 30.05%, 42.70%, and 44.39%. This lower strength can be attributed to insufficient water availability for the cement paste, as some water is absorbed by the BCA and BF in the mixes. The lack of adequate water may result in voids and unhydrated cement particles, negatively impacting the strength development of the concrete samples. Additionally, the lower density of bamboo aggregates compared to natural aggregates could contribute to this strength reduction. Furthermore, the agglomeration and improper distribution of BF within the concrete may create weak zones, disrupting stress distribution under compression and further diminishing the strength of the samples.[39]

10.3.3 Split Tensile Strength of BCA and BF Concrete Samples

Figure 10.5 illustrates the split tensile strength of concrete samples from various mixes after 28 and 56 days of curing. The results indicate that the introduction of fiber in mixes M2 and M3 enhances tensile strength. Additionally, longer curing periods lead to greater strength compared to shorter ones. When comparing the results to the reference mix M0, tensile strengths for M2 and M3 increased by approximately 5.34% and 10.34% after 28 days, respectively. In contrast, the tensile strengths of M1, M4, M5, and M6 decreased by about 0.93%, 11.85%, 10.66%, and 6.58%, respectively, at the same time point. Consistent results were noted after 56 days, with M2 and M3 showing increases of approximately 3.48% and 7.04%, while M1, M4, M5, and M6 exhibited decreases of 6.31%, 11.86%, 12.62%, and 7.88%, respectively. Overall, the combination of 10% BCA and 1% BF demonstrated the best performance in tensile strength. Like compressive strength, the concrete mix containing 20% BCA negatively impacted tensile strength as well. In FRC, the concrete matrix and fibers work together to resist external loads before cracking and failure. The properties of the fibers, such as length, diameter, and type, along with their content, are crucial for the performance of FRC structures.

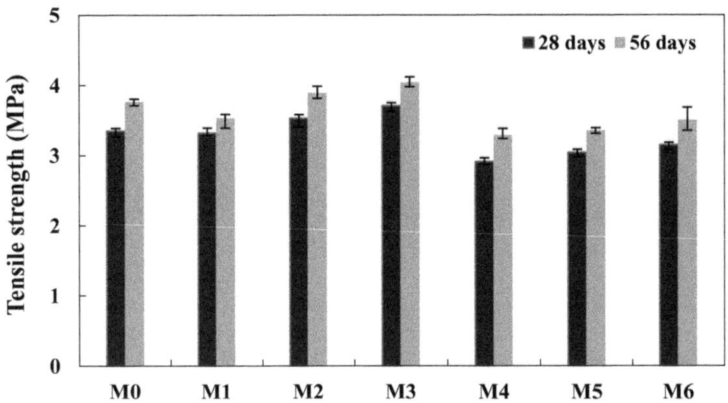

FIGURE 10.5 Split tensile strength of different concrete mixes after 28 days and 56 days of curing.

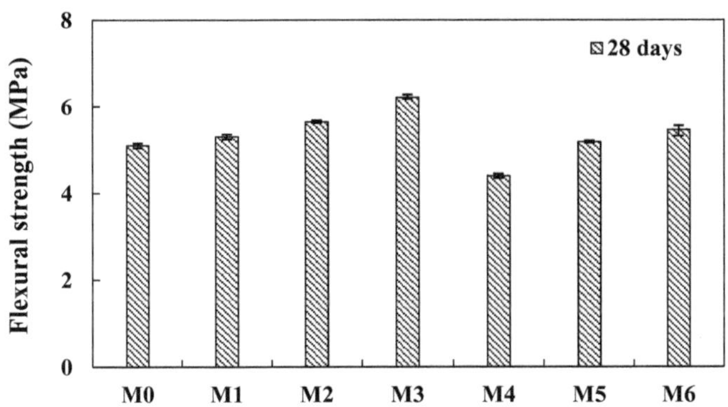

FIGURE 10.6 Flexural strength at 28 days of curing age.

10.3.4 Flexural Strength of BCA and BF Concrete Samples

Figure 10.6 illustrates the flexural strength behavior of beam samples under three-point bending testing at 28 days. The flexural strength data indicates a variety of mechanical responses among the experimental samples compared to M0, with increases ranging from 2.15% for M5 to 23.13% for M3 and a decrease of 12.75% for M4. Sample M6 shows a notable increase of 9.02%, while M2 demonstrates an improvement of 11.37%. These results suggest that bamboo has the potential to enhance flexural strength. The significant improvement observed in M3 indicates that this sample may possess an optimized composition or structural arrangement, making it a strong candidate for applications requiring higher flexural strength.

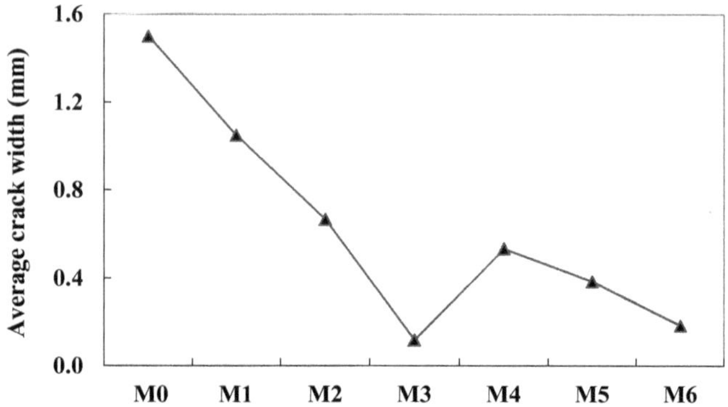

FIGURE 10.7 Crack width of different concrete mixes due to flexural strength.

10.3.5 Crack Width due to Flexural Loading

Figure 10.7 shows the average crack width of beams subjected to flexural loading in a three-point bending test. After unloading the samples from the testing machine, cracks were measured at three different points along their lengths, and the average widths were recorded. The average crack widths for all types of concrete mixes ranged from 1.5 mm to 0.15 mm, with the maximum crack width of 1.5 mm observed in mix M0 and the minimum of 0.15 mm in mix M3. The average reduction in crack width from sample M0 to M6 was 30%, 55.56%, 92.22%, 64.44%, 74.44%, and 87.78%, respectively. These reductions highlight the effectiveness of BF in mitigating crack formation. Previous studies have also reported that the inclusion of bamboo, particularly in fiber form, enhances the crack resistance and shrinkage performance of concrete by bridging microcracks and limiting their propagation.

10.3.6 Absorption Capacity, Pore Volume, and Electric Resistivity of Concrete

Figure 10.8 illustrates the permeable pore volume and water absorption capacities of various concrete types after 28 days of curing. The data shows that the inclusion of both BCA and BF resulted in increased pore volume and absorption in the concrete mixes. Compared to the reference mix (M0), the pore volume in mixes M1 through M6 was found to be 22%, 28%, 36%, 39%, 44%, and 49% higher, respectively. In terms of absorption, these values were 26%, 31%, 38%, 45%, 50%, and 54%.

Figure 10.9 presents the electrical resistivity of the different concrete mixes measured over 56 days. In contrast to pore volume and absorption, a decreasing trend in resistivity was observed across all mixes compared to the reference mix. After 28 days, mixes M1 (10% BCA) and M2 (20% BCA) exhibited resistivity values that were 7.6% and 8.8% lower, respectively, than the reference mix. The presence of BF had a significant impact on resistivity, with a mix of M6 (containing 20% BCA and 1% BF) showing a reduction of up to 73.8%. Despite these reductions, all mixes still met the AASHTO TP95 recommended resistivity threshold, where a value below 12

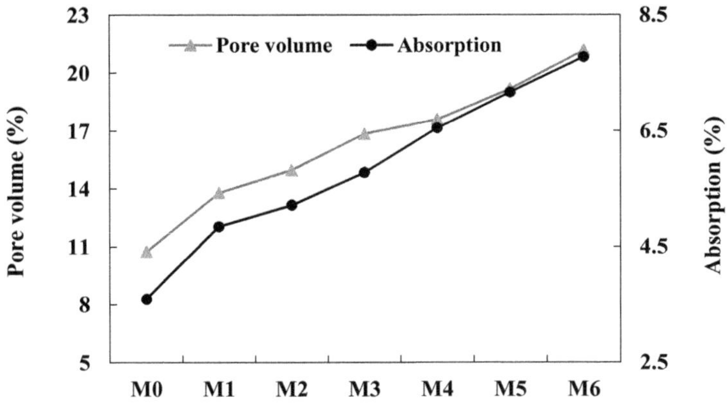

FIGURE 10.8 Pore volume and water absorption capacity of different concrete mixes.

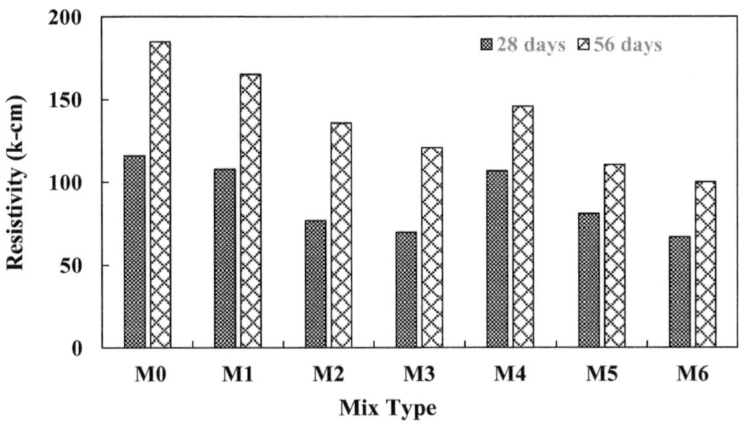

FIGURE 10.9 Electrical resistivity of different concrete mixes at different curing ages.

indicates a higher risk of corrosion in concrete.[36] Overall, pore volume, absorption, and resistivity serve as indicators of the durability properties of concrete. Higher pore volume and absorption, along with lower resistivity, facilitate the penetration of chemical substances such as chloride, oxygen, and carbon dioxide into the concrete, particularly impacting the steel surface and initiating corrosion activities. Although this process is gradual and influenced by various factors including concrete cover depth, crack characteristics, and environmental conditions, these considerations have been extensively discussed in existing literature.

The relationship between pore volume, water absorption, and electrical resistivity after 28 days of curing is illustrated in Figure 10.10. A clear correlation was identified between pore volume and resistivity, as well as between absorption and resistivity. Specifically, concrete samples exhibiting higher electrical resistivity tended to have lower water absorption, indicating a reduction in interconnected pore networks. This

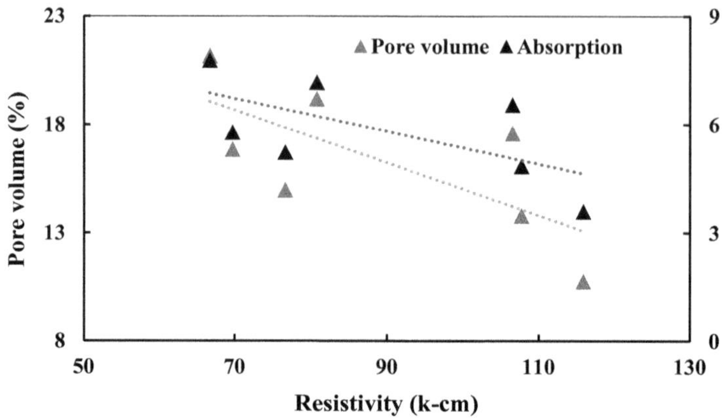

FIGURE 10.10 Correlation among electrical resistivity, pore volume, and absorption of different concrete mixes.

relationship highlights the significance of pore structure in governing moisture transport and, consequently, the durability of concrete. Reduced permeability due to lower pore connectivity contributes to enhanced resistance against aggressive environmental conditions, underscoring the importance of minimizing porosity for long-term performance.

10.4 CONCLUSION

This research examined how the combined effects of bamboo chip aggregates (BCA) and BF at various replacement levels influence the fresh and hardened properties of concrete. Based on the experiments conducted, the following conclusions are drawn.

I. The incorporation of BCA and BF gradually decreased both the density and slump values of the concrete. The reduction in slump value was more pronounced when BCA and BF were used together in the concrete mixes. This combination enhanced the interlocking forces, restricted the movement of the cement paste, and ultimately reduced the workability of the concrete.

II. The replacement of natural coarse aggregate with BCA resulted in a reduction in compressive strength relative to the control concrete sample. A 10% replacement of natural aggregate with BCA achieved a strength close to that of the reference sample at both 28 and 56 days. Even with a 20% BCA inclusion in the concrete mix, the strength reached 25 MPa, which remains acceptable for slightly low-grade structural applications.

III. Concrete samples incorporating both BCA and BF demonstrated positive results in tensile and flexural strength testing. A mix containing 0.5% BF and 10% BCA emerged as the optimal composite, exhibiting a well-balanced composition that positions it as a strong candidate for applications demanding higher flexural strength. Additionally, the crack resistance of concrete with BCA and BF improved, as bamboo enhances interconnection within the matrix, significantly reducing crack formation.

IV. However, negative effects were noted in terms of pore volume, water absorption, and electrical resistivity. The presence of BCA and BF increased pore volume and absorption capacity while decreasing the resistivity of the concrete samples. A linear relationship was established among pore volume, absorption, and resistivity.

V. Future studies should explore treatment methods (such as using organic or inorganic agents) for BCA and BF to assess their potential impact on concrete properties. Additionally, further research is needed on the size and shape of both BCA and BF to ensure that these substitutes can be used in concrete without compromising essential properties.

REFERENCES

1. Rajesh Kumar, K., Ramamohan, R. & Murthi, P. Shear resistance of portal frame reinforced with bamboo and steel rebar: Experimental and numerical evaluation. *Int. J. Recent Technol. Eng*, 2019. 8: p. 445–452.
2. Chandra Paul, S., Mbewe, P.B., Kong, S.Y., & Šavija, B. Agricultural solid waste as source of supplementary cementitious materials in developing countries. *Materials*, 2019. 12(7): p. 1112.
3. Palanisamy, M., et al., Permeability properties of lightweight self-consolidating concrete made with coconut shell aggregate. *Journal of Materials Research and Technology*, 2020. 9(3): p. 3547–3557.
4. Paul, S.C., Basit, M.A., Hasan, N.M.S., & Islam, M.S. Sustainable cement mortar production using rice husk and eggshell powder: A study of strength, electrical resistivity, and microstructure. *Smart Construction and Sustainable Cities*, 2024. 2(1): p. 13.
5. Amin, M., Tayeh, B.A. & Agwa, I.S. Effect of using mineral admixtures and ceramic wastes as coarse aggregates on properties of ultrahigh-performance concrete. *Journal of Cleaner Production*, 2020. 273: p. 123073.
6. Rahman, M.H., Kong, S.Y., Lam, B.Y.B., Murali, G., Wong, L.S., & Paul, S.C. Effects of palm oil fuel ash and crumb rubber on mechanical and thermal properties of sustainable engineered cementitious composites. *Journal of Building Engineering*, 2024. 97: p. 110795.
7. Sun, X., He, M. & Li, Z. Novel engineered wood and bamboo composites for structural applications: State-of-art of manufacturing technology and mechanical performance evaluation. *Construction and Building Materials*, 2020. 249: p. 118751.
8. Li, H., et al., Characterization of local compressive behavior for bamboo scrimber loaded perpendicular to the grain. *Construction and Building Materials*, 2023. 397: p. 132421.
9. Dudhatra, B., Parmar, D. & Patel, P. A study on bamboo as a replacement of aggregates in self compacting concrete. *Int. J. Eng. Res. Technol*, 2017. 6: p. 429–432.
10. Terai, M. & Minami, K. Basic study on mechanical properties of bamboo fiber reinforced concrete. *Global Thinking in Structural Engineering: Recent Achievements*, 2012. 8: p. 17–24.
11. Phong, N.T., et al., Study on how to effectively extract bamboo fibers from raw bamboo and wastewater treatment. *Journal of Materials Science Research*, 2012. 1(1): p. 144.
12. Liu, G.J., Bai, E.L., Xu, J.Y., Yang, N., & Wang, T.J. Dynamic compressive mechanical properties of carbon fiber-reinforced polymer concrete with different polymer-cement ratios at high strain rates. *Construction and Building Materials*, 2020. 261: p. 119995.

13. Osmi, S.K.C., et al., Performance of bamboo fibre on modifying mechanical properties of concrete. *The Open Civil Engineering Journal*, 2024. 18.
14. Xu, B., et al., Preparation and properties of natural bamboo fiber-reinforced recycled aggregate concrete. *Materials*, 2024. 17(12): p. 2972.
15. Li, L., Yang, C., Shu, W., Xu, H., Zeng, Y., Zhu, Z.,. . .& Jia, S. Study on the deformation and cracking characteristics of bamboo fiber reinforced concrete anti-slide pile considering the aspect ratio, fiber dosage and its coupling effect with rebars. *Engineering Structures*, 2023. 294: p. 116750.
16. Yeo, D., & Potra, F.A. Sustainable design of reinforced concrete structures through CO_2 emission optimization. *Journal of Structural Engineering*, 2015. 141(3): p. B4014002.
17. Latawiec, R., Woyciechowski, P., & Kowalski, K.J. Sustainable concrete performance CO_2-emission. *Environments*, 2018. 5(2): p. 27.
18. Ming, C.Y.T., Jye, W.K., & Ahmad, H.A.I. Mechanical properties of bamboo and bamboo composites: A review. *J. Adv. Res. Mater. Sci*, 2017. 35(1): p. 7–26.
19. Marrero, R., et al., Study of high-strength concrete reinforced with bamboo fibers. *Materials for Energy, Efficiency and Sustainability: TechConnect Briefs*, 2017. 2: p. 301–304.
20. Noh, H.M., Mohamad, N.A., Idris, N.A., Kasim, N., Zainal, R., & Musa, S.M.S. The performance of bamboo fiber as fine aggregate replacement in concrete. *International Journal of Integrated Engineering*, 2021. 13(5): p. 187–199.
21. Yusra, A., Triwulan, T., Safriani, M., & Ikhsan, M. Use of bamboo fiber on the relationship between compressive strength and split tensile strength of high strength concrete. *IOP Conference Series: Materials Science and Engineering*, 2020. 933(1): p. 012010.
22. Park, S.S., Hou, Y.L., Lee, J.C., & Jeong, S.W. Mechanical properties of concrete with bamboo chips. *Applied Sciences*, 2019. 9(16): p. 3367.
23. Manimaran, A., Somasundaram M., & Ravichandran, P. Experimental study on partial replacement of coarse aggregate by bamboo and fine aggregate by quarry dust in concrete. *International Journal of Civil Engineering and Technology*, 2017. 8(8): p. 1019–1027.
24. Kavitha, S. & Kala, T.F. Effectiveness of bamboo fiber as a strength enhancer in concrete. *Int J Earth Sci Eng*, 2016. 9(03): p. 192–196.
25. Sajjala, K. Study on structure, extraction and prevention of bamboo fibre as strength enhancer in concrete. *International Journal of Advances in Mechanical and Civil Engineering*, 2017: p. 16–20.
26. Bindu, M., Narendra, B. & Manjunatha, J. A study on characteristic strength of bamboo fiber reinforced concrete. *International Journal of Engineering Development and Research*, 2016. 4(2): p. 1464–1469.
27. Dewi, S.M. & Wijaya, M.N. The use of bamboo fiber in reinforced concrete beam to reduce crack. In *AIP conference proceedings*. 2017. AIP Publishing.
28. Javadian, A. *Composite bamboo and its application as reinforcement in structural concrete*. PhD Thesis, ETH Zurich, Switzerland, 2017.
29. Gupta, D.K. & Singh, R. An experimental evaluation of compressive strength and flexural strength of bamboo fiber reinforced concrete. *Carbon*, 2018. 1(3500): p. 200.
30. Ahmad, J., Zhou, Z., & Deifalla, A.F. Structural properties of concrete reinforced with bamboo fibers: A review. *Journal of Materials Research and Technology*, 2023. 24: p. 844–865.
31. Hebel, D.E., et al., Process-controlled optimization of the tensile strength of bamboo fiber composites for structural applications. *Composites Part B: Engineering*, 2014. 67: p. 125–131.
32. ASTM C136–06. *Standard test method for sieve analysis of fine and coarse aggregates*. West Conshohocken, PA, USA, 2006.
33. BS EN 12390–3:2019. *Testing hardened concrete – Part 3: Compressive strength of test specimens*. UK, 2019.

34. ASTM C642. *Standard test method for density, absorption, and voids in hardened concrete*. West Conshohocken, PA, 2018.
35. Shakil, M., Paul, S.C., Islam, S., Hasan, N.M.S., Babafemi, A.J., & Islam, M.S. Corrosion-induced mass loss of steel in concrete containing recycled coarse and fine concrete aggregates. *Multiscale and Multidisciplinary Modeling, Experiments and Design*, 2025. 8(1): p. 1–13.
36. AASHTO TP 95:2014. *Standard method of test for surface resistivity indication of concrete's ability to resist chloride ion penetration*. Washington, DC, USA, 2014.
37. Wang, G., Wei, Y., Ding, M., Wang, J., & Zhu, B. Physical and mechanical properties of sustainable bamboo coarse aggregate concrete. *Construction and Building Materials*, 2024. 431: p. 136605.
38. Andreola, V.M., da Gloria, M.H.Y., Pepe, M., & Toledo Filho, R.D.A comprehensive experimental study on the physical performance and durability of bamboo bio-concrete. *Sustainability*, 2024. 16(13): p. 5334.
39. Zhu, H., Zhou, H., & Gou, H. Evaluation of carbon fiber dispersion in cement-based materials using mechanical properties, conductivity, mass variation coefficient, and microstructure. *Construction and Building Materials*, 2021. 266: p. 120891.

11 Performance Assessment of Concrete with Partial Replacement of Fine and Coarse Aggregates Using Ceramic Waste

Samreen Bano, Abhishek Kumar, Vikash Singh and Zishan Raza Khan

11.1 INTRODUCTION

Concrete is universally recognized as the most extensively used construction material due to its adaptability, long service life, and affordability (Ray et al., 2021). It is a fundamental component in infrastructure development, playing a central role in both residential and commercial constructions worldwide. However, its production relies heavily on natural fine and coarse aggregates such as river sand and crushed stone. This over-dependence places immense strain on finite natural resources, leading to a cascade of environmental issues, including extensive land degradation, disruption of ecological habitats, and increased carbon emissions linked to the extraction and transportation of these materials (Blasius, 2025; Ray et al., 2021). With the pace of global urbanization accelerating rapidly, the demand for aggregates is expected to surpass 50 billion tonnes annually, which not only underscores the unsustainable trajectory of current practices but also intensifies the need to explore alternative, environmentally responsible materials (IEA, 2020; Tam et al., 2021). In parallel, the construction and demolition sector contributes significantly to solid waste generation, a substantial portion of which comprises ceramic waste – originating from rejected tiles, broken sanitary ware, and bricks that are either manufacturing rejects or demolition debris (Ray et al., 2021). These ceramic wastes are notably persistent in the environment due to their non-biodegradable and chemically inert nature, and their resistance to weathering further complicates landfill management (Sánchez de Juan & Gutiérrez, 2009). Nonetheless, ceramics exhibit several favorable physical and mechanical properties such as high compressive strength, abrasion resistance, chemical stability, and low thermal expansion, making them highly suitable for reutilization in concrete as recycled aggregates (Medina et al., 2017). In light of these attributes, the incorporation of ceramic waste as a partial replacement for both fine and coarse aggregates in concrete has emerged as a sustainable and innovative

DOI: 10.1201/9781003664284-11

solution. This dual-purpose approach not only mitigates the ecological footprint of concrete production by conserving natural resources but also provides an effective means of diverting large volumes of ceramic waste from landfills, thereby addressing critical environmental challenges faced by the construction industry (Blasius, 2025; Letelier et al., 2017).

Numerous experimental investigations have demonstrated that concrete incorporating ceramic waste as an aggregate – either fine or coarse – can deliver mechanical properties that are comparable to or even superior to those of conventional concrete, particularly when the replacement level is maintained within the optimal threshold of up to 20% (Silvestre et al., 2017; Senthamarai & Manoharan, 2005). This improvement in performance is primarily attributed to the inherent pozzolanic activity of finely ground ceramic particles. When introduced into the cementitious matrix, these particles react with calcium hydroxide ($Ca(OH)_2$), a by-product of cement hydration, to produce additional cementitious compounds such as calcium silicate hydrate (C–S–H` and calcium aluminate hydrate (C–S–H`. These secondary gels enhance the microstructure of concrete by filling the capillary pores, reducing porosity, and strengthening the interfacial transition zones (ITZs) between aggregates and the cement paste (Nair et al., 2013; Limbachiya et al., 2004). Beyond their technical contributions, ceramic wastes also offer clear environmental and economic benefits. Their utilization in concrete aligns with circular economy principles by promoting resource efficiency, reducing the demand for virgin natural aggregates, and transforming industrial waste into a valuable construction input – thereby lowering raw material costs and landfill burdens (Debieb & Kenai, 2008; Bignozzi et al., 2015). Various studies confirm that after suitable processing such as cleaning, crushing, and gradation, ceramic tile waste and sanitary ceramic debris can be effectively used as both coarse and fine aggregates, enhancing not only compressive and tensile strength but also the workability of concrete mixes, especially when replacement levels are carefully optimized (Kumar et al., 2018; Poon & Chan, 2007). However, despite these encouraging developments, the use of ceramic aggregates in concrete is not without challenges. The performance outcomes – particularly with regard to durability indicators like shrinkage, permeability, and long-term strength retention – show variability across studies due to differences in the mineralogical properties of ceramic waste, particle shapes, and mix design proportions (Senthamarai et al., 2011; Pedro et al., 2018). Notably, at higher replacement levels exceeding 25–30%, mechanical properties such as compressive strength and density tend to decline. This degradation is commonly linked to the formation of weak or poorly bonded ITZs, increased porosity due to poor particle packing, and heightened water absorption that compromises the integrity of the hardened concrete matrix (Blasius, 2025; Ray et al., 2021; Siddique & Kaur, 2016). These limitations underscore the need for a careful control of replacement ratios and an in-depth understanding of the physical characteristics of the ceramic source material. Furthermore, most of the existing research has treated ceramic waste as a replacement for either fine or coarse aggregates in isolation, with minimal emphasis on their combined application. As a result, there remains a significant knowledge gap regarding the synergistic effects of simultaneously incorporating ceramic waste in both fine and coarse fractions – an approach that may yield optimized concrete properties if appropriately engineered (Pacheco-Torgal &

Jalali, 2010). Compounding this issue is the absence of standardized design codes and guidelines, particularly in developing countries where ceramic waste is abundantly available but remains underutilized due to a lack of awareness, regulatory frameworks, and practical implementation strategies (Ikponmwosa & Ehikhuenmen, 2015). Addressing these research gaps is essential for enabling the widespread adoption of ceramic waste as a viable and sustainable alternative in concrete production.

Although substantial progress has been made in evaluating the individual use of ceramic waste as either fine or coarse aggregate in concrete, integrated studies that simultaneously consider both fractions under varying replacement levels are limited. This gap constrains our understanding of the combined influence of ceramic waste morphology, particle size distribution, and pozzolanic interaction on the overall concrete behavior. Furthermore, the durability aspects and long-term performance of such blended ceramic waste concretes under different curing regimes and environmental exposures remain underexplored. Hence, this study aims to bridge these gaps by systematically investigating the mechanical and physical performance of concrete incorporating ceramic waste as a partial replacement for both fine and coarse aggregates. The findings are expected to contribute significantly to sustainable construction practices, waste management strategies, and material circularity within the concrete industry.

11.2 MATERIALS AND METHODOLOGY

11.2.1 Materials

The materials employed in the present experimental study were selected on the basis of their compliance with the relevant Indian Standard specifications to ensure the production of durable and high-performance concrete incorporating ceramic waste as partial aggregate replacement. A detailed description of each material and its respective physical and chemical characteristics is provided below.

11.2.1.1 Cement

Ordinary Portland Pozzolanic Cement (OPC) was consistently used as the primary binder throughout the experimental program. The selected cement conformed to the physical requirements outlined in IS: 269–2015, ensuring its suitability for structural concrete applications. The measured physical properties, including fineness modulus, consistency, initial and final setting times, and specific gravity, were within permissible limits, validating its standard quality. The cement exhibited a fineness of 3.5%, a consistency of 35%, an initial setting time of 42 minutes, and a final setting time of 293 minutes, with a specific gravity of 3.12. These parameters are indicative of good workability and adequate setting behavior essential for optimal hydration and strength development.

11.2.1.2 Fine aggregate

Natural river sand, classified as Zone I per IS: 383–2016, was utilized as the fine aggregate. Fine aggregates play a vital role in concrete by filling voids between coarse particles, thereby enhancing the packing density, reducing porosity, and improving both the workability and finish of the concrete. The selected river sand exhibited a relatively coarse gradation with a fineness modulus of 4.12, aligning with the demands of structural concrete. The physical properties, including specific gravity (2.50), water absorption (1.76%), and free moisture content (nil), were thoroughly assessed to ensure consistency and quality. A well-graded particle size distribution was maintained to optimize the rheological properties of the mix and ensure superior mechanical performance and durability in the hardened state.

11.2.1.3 Coarse Aggregate

Coarse aggregates constitute the bulk of concrete and significantly influence its mechanical integrity and durability. In this study, crushed natural aggregates of 20 mm and 10 mm nominal sizes were used, conforming to IS: 2386 (Part I and III) – 1963. The sieve analysis confirmed appropriate gradation for structural applications. Specific gravity values of 2.60 (20 mm) and 2.65 (10 mm), along with water absorption values of 0.9% and 0.44%, respectively, reflected desirable density and porosity characteristics. Free surface moisture was negligible. Mechanical strength was evaluated through the Aggregate Crushing Value (24.35%) and Aggregate Impact Value (18.23%) tests in accordance with IS: 2386 (Part IV) – 1963, indicating acceptable compressive strength and toughness under impact, making the aggregates suitable for load-bearing concrete structures.

11.2.1.4 Water

Potable water, free from harmful salts and impurities, was used for both mixing and curing processes in accordance with IS: 456–2000, Clause 5.4.3. The pH of the water was maintained above 6.0 to avoid any deleterious reactions with cementitious materials. The use of clean, drinkable water ensures effective hydration and prevents chemical attacks on the steel reinforcement and cement matrix, thereby contributing to the durability and long-term performance of the concrete.

11.2.1.5 Superplasticizer

To enhance the workability of the concrete while maintaining a low water-to-cement (w/c) ratio, a high-range water-reducing admixture (HRWRA), commonly known as superplasticizer, was incorporated. These admixtures improve dispersion of cement particles through electrostatic or steric repulsion mechanisms, thereby reducing interparticle attraction and enhancing fluidity. Their use facilitates the production of high-strength and self-compacting concrete with improved compressive strength, lower permeability, and greater resistance to chemical attack. Superplasticizers are especially beneficial in concretes with complex reinforcement arrangements where adequate compaction is otherwise challenging.

11.2.1.6 Ceramic Waste

Ceramic waste, as shown in Figure 11.1, was sourced from demolition debris, defective ceramic tiles, discarded sanitary ware, and waste generated during industrial ceramic manufacturing. The material, predominantly composed of

FIGURE 11.1 Fine ceramic waste (a) and coarse ceramic waste (b).

alumino-silicate compounds, was processed into both fine and coarse fractions. The ceramic aggregates exhibit superior physical attributes such as high compressive strength, low thermal expansion, and resistance to chemical and thermal shock. These properties support their suitability as alternative aggregates in concrete. The ceramic waste had a specific gravity of 2.45, a bulk density of 2.45 g/cm^3, water absorption of 2.34%, and demonstrated acceptable crushing and impact values of 22% and 19%, respectively. The chemical composition revealed high silica (65%), alumina (18.34%), and calcium oxide (5.75%), reflecting its potential pozzolanic behavior. No free moisture or ignition loss was observed. In this study, ceramic waste was investigated as a partial replacement for both fine and coarse natural aggregates, a strategy less commonly explored in prior literature. This dual replacement approach aims to determine the optimal substitution level to achieve desired workability and strength while addressing sustainability goals by reducing reliance on virgin aggregate sources and promoting the valorization of industrial ceramic by-products.

11.2.2 Methodology

To evaluate the mechanical performance and workability of M25-grade concrete incorporating ceramic waste as a partial replacement for both fine and coarse aggregates, a comprehensive experimental methodology was adopted in accordance with relevant Indian Standard specifications. OPC, conforming to IS: 269–2015, served as the binding material throughout the study. Natural river sand, categorized under Zone I of IS: 383–2016, was employed as the fine aggregate, while crushed natural coarse aggregates of nominal sizes 10 mm and 20 mm, tested as per IS: 2386 (Parts I, III, and IV), were used as the coarse aggregates. Ceramic waste materials, sourced from discarded construction ceramics and rejected industrial tile and sanitary ware products, were crushed, cleaned, sieved, and categorized into fine and coarse fractions to match the gradation requirements of their natural counterparts. Their physical and chemical characteristics were analyzed to ensure suitability, including specific gravity, water absorption, impact and crushing values, and oxide composition indicative of pozzolanic activity.

Concrete mixes were designed for a target characteristic compressive strength of 25 MPa (M25 grade) with a constant water-to-cement (w/c) ratio of 0.45. An HRWRA (superplasticizer) was incorporated to maintain workability across mixes with varying ceramic content. The ceramic waste was introduced into the mix design by partially replacing the natural fine and coarse aggregates in varying proportions (5%, 10%, 15%, 20%, 25%, and 30%), both individually and in combination. For each mix, standard cube specimens (150 mm × 150 mm × 150 mm) were prepared to assess compressive strength, and slump tests were conducted on fresh concrete to evaluate workability as per IS: 1199–1959. The specimens were demolded after 24 hours and cured in potable water for 7, 14, and 28 days to monitor strength development over time. Compressive strength tests were conducted in accordance with IS: 516–1959 using a calibrated compression testing machine. The experimental data were systematically analyzed to determine the influence of ceramic waste on both the fresh and hardened properties of concrete, identifying optimal replacement levels that ensure structural integrity while enhancing sustainability.

11.3 RESULT AND DISCUSSION

11.3.1 Workability

The workability of concrete, measured through slump values, consistently decreased with increasing replacement of natural aggregates (fine, coarse, or both) by ceramic waste, as summarized in Tables 11.1 to 11.3. The control mix exhibited a slump of 62 mm, which progressively reduced as the replacement ratio increased. For fine aggregate substitution, slump reduction reached 21%, while coarse aggregate replacement showed a decline of 23%. When both fine and coarse aggregates were replaced simultaneously, the slump reduction intensified to 35%, highlighting the combined adverse effect on flowability.

This decline can be attributed to the angularity, rough surface texture, and higher porosity of ceramic waste particles compared with the smooth and rounded morphology of natural aggregates. The increased surface area and internal friction among particles restrict the free movement of the mix, while the higher absorption capacity of ceramic particles reduces the amount of available mixing water. These characteristics lead to a stiffer and less workable concrete mix. From a microstructural perspective, the rough ceramic surface enhances particle interlock but simultaneously impedes lubrication within the cement paste. Therefore, achieving adequate workability in such mixes may require the incorporation of water-reducing admixtures or superplasticizers to compensate for the loss of consistency.

TABLE 11.1
Effect of Fine Aggregate Replaced by Fine Ceramic Waste on Workability

Sr. no.	% replacement	Slump (mm)	% Change in Slump
1	0%	62	0%
2	5%	60	3%
3	10%	58	6%
4	15%	55	11%
5	20%	52	16%
6	25%	49	21%

TABLE 11.2
Effect of Coarse Aggregate Replaced By Coarse Ceramic Waste on Workability

Sr. no.	% replacement	Slump (mm)	% Change in Slump
1	0%	62	0%
2	5%	60	3%
3	10%	57	8%
4	15%	54	13%
5	20%	51	18%
6	25%	48	23%

TABLE 11.3
Effect of Both Fine and Coarse Aggregates Replaced by Ceramic Waste on Workability

Sr. no.	% replacement	Slump (mm)	% Change in Slump
1	0%	62	0%
2	5%	57	8%
3	10%	52	16%
4	15%	48	23%
5	20%	44	29%
6	25%	40	35%

11.3.2 Compressive Strength

The compressive strength exhibited distinct trends depending on whether fine, coarse, or both aggregates were replaced with ceramic waste (Figures 11.2, 11.3, and 11.4). For fine aggregate replacement, strength decreased gradually from 29.5 MPa

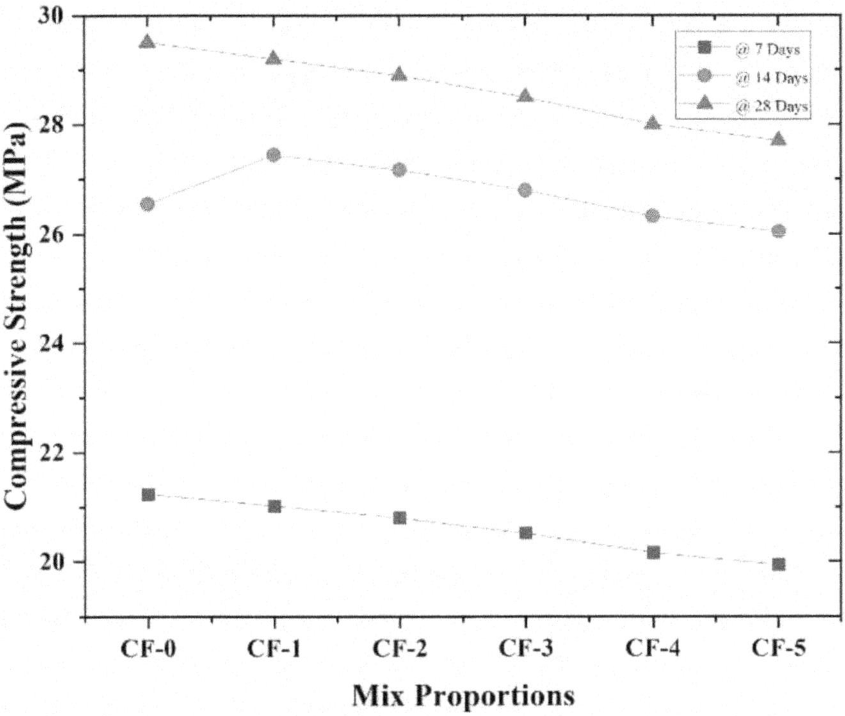

FIGURE 11.2 Effect of fine ceramic waste on compressive strength of concrete.

(control) to 27.7 MPa at 25% substitution – about a 6% reduction – due to the partial dilution of the cementitious matrix and weaker bonding at the aggregate–paste interface. In contrast, coarse aggregate replacement produced a mild enhancement in strength, peaking at 30.5 MPa (approximately 3–4% increase) at 10–15% substitution, before slightly declining beyond that level. This improvement is likely due to better mechanical interlocking and a stronger ITZ provided by the angular and high-strength ceramic aggregates.

When both fine and coarse aggregates were replaced simultaneously, compressive strength decreased more sharply – by nearly 10% at 25% replacement – indicating a cumulative negative effect.

The simultaneous presence of angular particles of both sizes increased void content and reduced compactness, resulting in a less dense and less continuous ITZ. Although fine ceramic particles may provide a micro-filler effect at lower replacement levels (≤15%), excessive substitution led to increased brittleness, micro-voids, and nonuniform stress transfer across the matrix, reducing overall strength.

In summary, the results suggest that partial incorporation (up to ~15%) of ceramic waste – particularly as coarse aggregate – can be beneficial for strength without compromising structural performance. However, higher replacement levels or combined

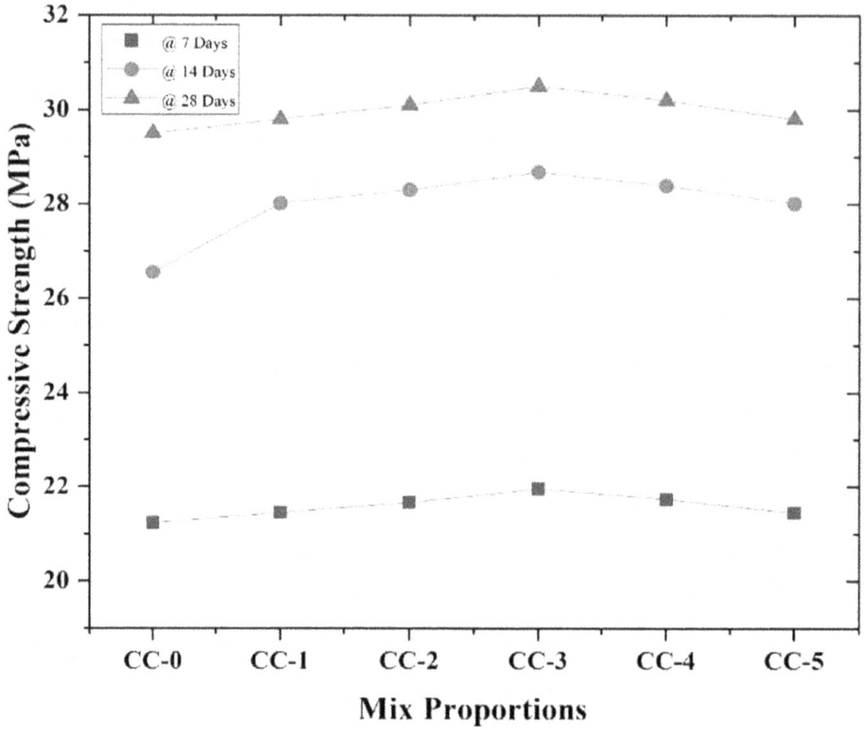

FIGURE 11.3 Effect of coarse ceramic waste on compressive strength of concrete.

substitution of both aggregates substantially reduces workability and compressive strength due to microstructural discontinuities, elevated internal friction, and diminished cement paste cohesion.

11.3.3 Durability Considerations

In addition to workability and strength characteristics, the durability of ceramic waste–modified concrete plays a crucial role in assessing its practical viability for structural applications. The inherently dense and inert nature of ceramic particles, especially those derived from fired tiles and sanitary ware, contributes positively to the chemical stability and impermeability of the concrete matrix. At moderate replacement levels (≤15%), the micro-filler effect of fine ceramic particles helps refine pore structure, thereby reducing capillary porosity and improving resistance to water ingress and chloride penetration. This refinement enhances durability against common deterioration mechanisms such as sulfate attack and carbonation.

However, at higher replacement ratios, excessive angularity and porosity of ceramic aggregates may increase internal voids and water absorption, potentially leading to higher permeability and reduced freeze–thaw resistance. The poor

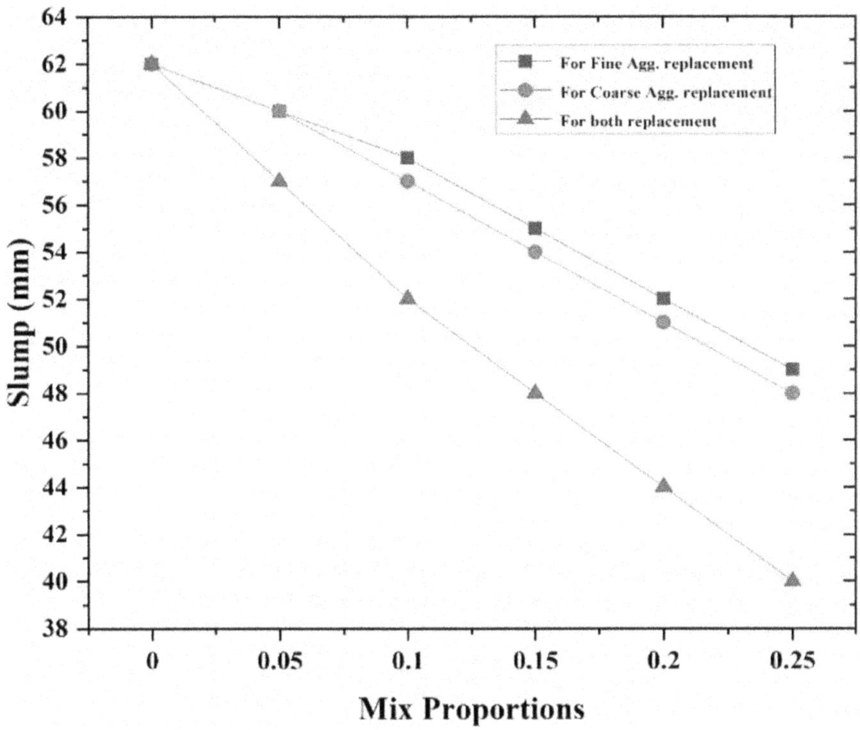

FIGURE 11.4 Effect of both ceramic waste on workability of concrete.

interfacial bonding observed beyond the optimal replacement level can also facilitate the ingress of aggressive agents, accelerating microcrack propagation under cyclic environmental stresses. Therefore, from a durability standpoint, controlled substitution up to 10–15% – particularly in the coarse aggregate fraction – appears most suitable. At this range, the material effectively balances sustainability benefits with long-term performance, aligning with the requirements for durable, low-maintenance, and eco-efficient concrete in modern infrastructure applications.

11.3.4 Environmental and Economic Benefits

The incorporation of ceramic waste in concrete offers substantial environmental and economic advantages, aligning with the principles of Life Cycle Assessment (LCA) and circular economy. Utilizing ceramic waste – typically sourced from tile and sanitary ware industries – diverts large volumes of nonbiodegradable material from landfills, thereby reducing solid waste generation and environmental pollution. It also conserves natural river sand and crushed stone, mitigating the ecological impacts of quarrying and dredging. From an energy perspective, recycling ceramic waste demands far less embodied energy compared to the extraction and processing of virgin aggregates, leading to lower CO_2 emissions per cubic meter of concrete. Economically, the partial substitution of aggregates by locally available ceramic waste can yield cost savings of 10–20%, depending on transportation logistics and waste availability. Hence, ceramic waste-based concrete contributes to sustainable construction practices by combining resource efficiency, reduced environmental footprint, and economic viability across its life cycle.

11.4 CONCLUSION

The study confirms that incorporating ceramic waste as a partial replacement for natural aggregates in concrete is both scientifically feasible and technically viable.

- Fine aggregate replacement up to 15–20% shows only a moderate reduction in workability and compressive strength, which can be mitigated through optimized mix design or superplasticizer use.
- Coarse aggregate replacement up to 15–20% improves strength slightly due to better interlocking and the high intrinsic strength of ceramic particles, with manageable workability loss.
- Combined replacement of fine and coarse aggregates significantly affects workability but maintains acceptable strength up to 10–15% substitution.

Overall, the controlled use of ceramic waste in concrete supports resource conservation, waste valorization, landfill reduction, and the advancement of green and circular construction practices without major compromise in structural performance.

REFERENCES

Bignozzi, M. C., Saccani, A., Sandrolini, F. (2015). New blended mortars with ceramic wastes as pozzolanic materials. *Construction and Building Materials*, 22(8), 167–174.

Blasius, H. N. (2025). Sustainable concrete production: The role of ceramic waste as a partial coarse aggregate substitute. *Discover Civil Engineering*, 2, 32. https://doi.org/10.1007/s44290-025-00198-9

Debieb, F., & Kenai, S. (2008). The use of coarse and fine crushed bricks as aggregate in concrete. *Construction and Building Materials*, 22(5), 886–893.

Ikponmwosa, E. E., & Ehikhuenmen, P. (2015). Suitability of ceramic waste as a replacement for fine aggregate for concrete production. *International Journal of Civil Engineering and Technology*, 6(1), 29–40.

International Energy Agency (IEA). (2020). *Technology Roadmap: Low-Carbon Transition in the Cement Industry*. IEA.

Kumar, R., Khan, S., & Rana, A. (2018). Utilization of ceramic waste in concrete: A review. *Journal of Cleaner Production*, 189, 26–39.

Letelier, V., Tarela, E., & Moriconi, G. (2017). Mechanical performance of concrete with recycled ceramic coarse aggregates. *Construction and Building Materials*, 134, 190–199.

Limbachiya, M., Leelawat, T., & Dhir, R. K. (2004). Use of recycled concrete aggregate in high-strength concrete. *Materials and Structures*, 33(9), 574–580.

Medina, C., Saez del Bosque, I. F., & Frias, M. (2017). Influence of ceramic waste on the properties of recycled aggregate concrete. *Construction and Building Materials*, 148, 746–755.

Nair, D. G., Fraaij, A. L., Klaassen, A. A. K., & Kentgens, A. P. M. (2013). A structural investigation relating to the pozzolanic activity of rice husk ashes. *Cement and Concrete Research*, 38(6), 861–869.

Pacheco-Torgal, F., & Jalali, S. (2010). Reusing ceramic wastes in concrete. *Construction and Building Materials*, 24(5), 832–838.

Pedro, D., de Brito, J., & Evangelista, L. (2018). Structural concrete with simultaneous incorporation of ceramic and waste glass aggregates. *Construction and Building Materials*, 154, 1137–1147.

Poon, C. S., & Chan, D. (2007). The use of recycled aggregate in concrete in Hong Kong. *Resources, Conservation and Recycling*, 50(3), 293–305.

Ray, S., Haque, M., Sakib, M. N., Mita, A. F., Rahman, M. D. M., & Tanmoy, B. B. (2021). Use of ceramic wastes as aggregates in concrete production: A review. *Journal of Building Engineering*, 43, 102567. https://doi.org/10.1016/j.jobe.2021.102567

Sánchez de Juan, M., & Gutiérrez, P. A. (2009). Influence of recycled aggregate quality on concrete properties. *Construction and Building Materials*, 23(2), 622–628.

Senthamarai, R. M., Devadas Manoharan, P., & Kalaiarasan, C. (2011). Concrete with ceramic waste aggregate – A review. *Construction and Building Materials*, 35, 874–881.

Senthamarai, R. M., & Manoharan, P. D. (2005). Concrete with ceramic waste aggregate. *Cement and Concrete Composites*, 27(9–10), 910–913.

Siddique, R., & Kaur, D. (2016). Properties of concrete containing ground waste ceramic powder as partial cement replacement. *Construction and Building Materials*, 102, 78–85.

Silvestre, J. D., Brito, J. D., & Pinheiro, M. D. (2017). Environmental impacts and benefits of the end-of-life of building materials. *Journal of Cleaner Production*, 143, 847–859.

Tam, V. W. Y., Soomro, M., & Evangelista, A. C. J. (2021). A review of recycled aggregate in concrete applications. *Construction and Building Materials*, 272, 121942.

12 Sustainable Utilization of Coconut Ash and Coir Fibre in Ferrocement Wall Panels

Deepana G., Chockalingam T., Harshani R. and Jeya Arthi A.J.

12.1 INTRODUCTION

The construction industry has long been perceived as a pressing grantor to global resource consumption and environmental derogation. The United Nations Environment Programme (UNEP) estimates that the building and construction sector accounts for approximately 39% of global carbon dioxide emissions, with a significant share arising from the production of conventional building materials such as cement and steel. The research community and industrial sector have maintained continuous interest in developing environmentally friendly construction material enhancements throughout the last 40 years [1, 2]. The inclusion of agricultural waste in cement-based composites has become an interesting discovery for sustainable construction material development. The combination of coir fibre with coconut ash produces positive outcomes in testing. The materials exist as waste products in nearby farmlands where they can be obtained quickly at affordable prices without compromising structural strength.

The burning of coconut husks produces coconut ash which contains convincing pozzolanic compounds that work as cement alternative in concrete matrix. The integration of classic cement with this sustainable material reduces environmental harm from classic cement production which provokes substantial global carbon emissions [3]. The organic fibre, coir fibre, synthesized from coconut husks exhibits exceptional strength attributes. The addition of coir fibre reduces microcracks in the concrete which strengthen the matrix [4].

Ferrocement is an ideal medium for evaluating the performance of such green materials, characterized by a thin mortar matrix reinforced with wire mesh, which exhibits a high surface-area-to-volume ratio, low self-weight, and good flexibility for fibre-based or lightweight materials [5, 6]. The use of coconut-based materials in ferrocement construction also supports the United Nations Sustainable Development Goals (SDGs), specifically SDGs 9 and 12, which focus on sustainable industrial innovation and responsible consumption and production of natural resources [7, 8].

 DOI: 10.1201/9781003664284-12

Although several studies have investigated the utilization of natural fibres in conventional concrete and mortar, little research has been conducted on the synergistic behaviour of coconut ash and coir fibre in ferrocement systems, which is the reason for this study. Therefore, this study investigates the feasibility of incorporating coconut ash and coir fibre, two by-products of coconut farming, into ferrocement wall panels for evaluation of the effect on structural efficiency, environmental sustainability, and climate adaptability through experimental analysis. The aim is to demonstrate that the use of agricultural waste as a high-performance construction material is in line with the principles of green infrastructure and the circular economy.

12.2 MATERIALS AND SUSTAINABILITY RATIONALE

12.2.1 Coconut Ash: A Pozzolanic Cement Substitute

Figure 12.1 exhibits coconut ash which is produced through the thermal decomposition of coconut shells. The ash contains high levels of silicon dioxide and aluminium oxide, which enable pozzolanic reactions when amalgamated with Portland cement [9]. These reactions generate more calcium silicate hydrate component, which improves the strength and reduces the porosity of the cementitious matrix. Due to the inclusion of coconut ash, the required amount of Portland cement is decreased, and, consequently, the embodied carbon is also reduced. Mainly, this work integrates agro-waste that would otherwise be scrapped.

12.2.2 Coir Fibre: A Renewable Reinforcing Agent

Coir fibre, extracted from the coconut husk (Figure 12.2) of the coconut plant (Cocos nucifera), is a natural lignocellulosic fibre with high tensile strength and durability, high elasticity, and high lignin content (40–45%) which results in slow degradation rates, which make it more suitable for cementitious composites [10]. It has been reported that the addition of coir fibre in low volume fractions increases the crack resistance, impact strength, and post-yield toughness of ferrocement panels [11]. Furthermore, it is a renewable and biodegradable material which is readily available in coconut-producing regions and has a lower environmental footprint than synthetic fibres [12].

FIGURE 12.1 Coconut ash.

FIGURE 12.2 Coir fibre.

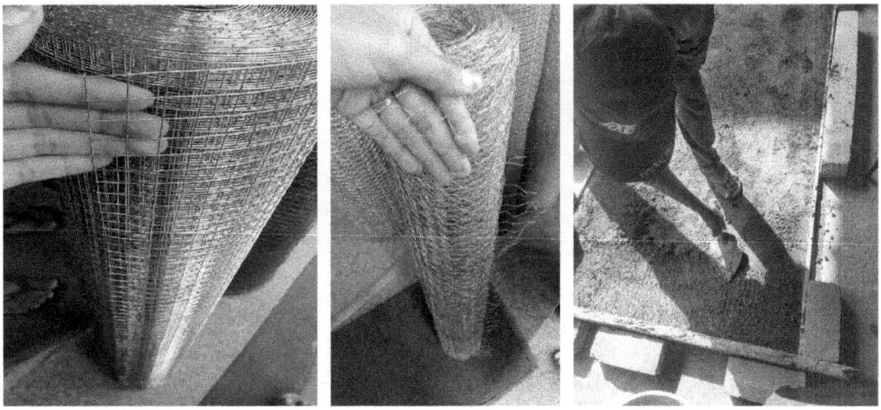

FIGURE 12.3 Constituents of ferrocement panel.

12.2.3 Ferrocement: A Versatile Construction Medium

Ferrocement is a thin wall made of mortar and different meshes (Figure 12.3) and is used for wall panels, roofing, and prefabricated structures [5, 6] because of its light weight, ductility, and ease of fabrication. This medium is particularly suitable for studying sustainable material composites because the material volume is relatively low, the surface-area-to-volume ratio is high and increases structural efficiency, and modular and prefabricated construction systems are compatible.

12.3 EXPERIMENTAL METHODOLOGY

In this study, the experimental methodology (Figure 12.4) and evaluation focused on ferrocement panels were enhanced with coconut ash and coir fibre. A total of six different mix combinations were designed to explore the effects of these natural additives. The coconut ash was incorporated at three different levels – 5%, 10%, and 15% – by weight of cement, while coir fibre was added at proportions of 0.5%, 1.0%, and 1.5%, also by the weight of cement. These varying concentrations were used to finalize the optimal performance characteristics of the modified panels.

For material preparation, the coconut shells from the nearby houses and agricultural farm were collected, then they were dried in the open air to avoid consuming more amount of heat while burning, and the dried coconut shells were burnt to ash. The collected ash was carefully sieved to ensure a fine particle size of less than 90 micrometres (Figure 12.5). Coir fibres were collected from a local coir processing facility and cut into uniform lengths of 25–30 mm to ensure consistent distribution within the cementitious matrix. The mortar used in all panel specimens was prepared using a standard cement-to-M-sand ratio of 1:2, in accordance with the specifications of IS 383:2016, and the ratio of water to cement in weight was kept constant at 0.45 to ensure uniform workability and hydration across all samples. Each panel was cast in moulds measuring 600 mm × 600 mm with a thickness of 25 mm.

To provide reinforcement, welded steel mesh with a 6 mm diameter was used along with galvanized chicken mesh with a 1 mm diameter. The spacing of the mesh

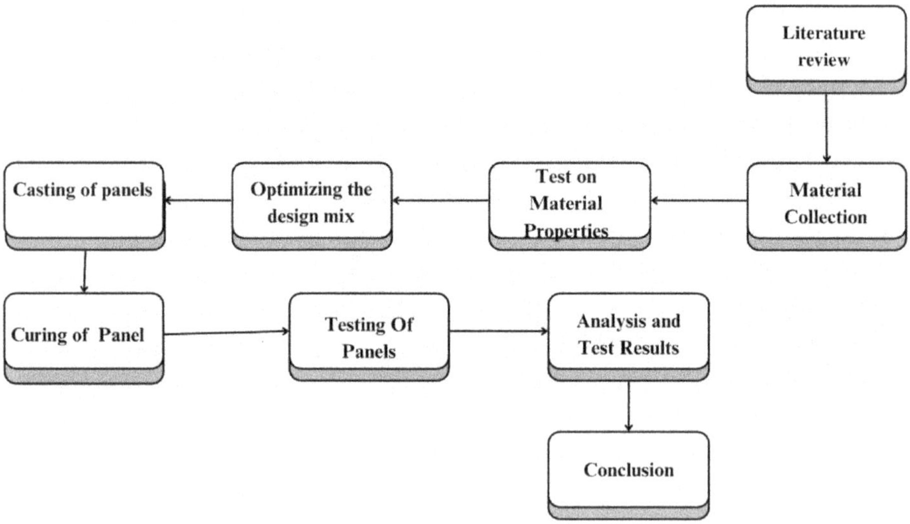

FIGURE 12.4 Methodology.

was standardized in all panels to eliminate variability due to reinforcement and to provide a uniform mechanical support and load distribution during testing. All specimens were then cured in water for 28 days to allow for proper hydration and strength development (Figure 12.6). After the completion of curing period, the panels were tested in accordance with the IS specification. The result are enhanced mechanical and durability properties.

12.3.1 Mix and Proportion

Table 12.1 exhibits the six distinct mix combinations used in the experimental investigation of ferrocement panels modified with coconut ash and coir fibre. Each mix is identified with a unique Mix ID (M1 to M6) shown in (Figure 12.7), and the variables under consideration are the percentage of coconut ash and coir fibre used, both measured by weight of cement. The control mix, labelled as M1, contains neither coconut ash nor coir fibre, serving as the baseline for evaluating the effects of the added natural materials. This mix provides reference values for mechanical and durability performance, against which all other modified mixes are compared.

Mix M2 introduces 5% coconut ash and 0.5% coir fibre, representing a lower-bound combination for both additives. This mix helps estimate the initial bounce of minimal reinstatement of cement with ash and the basic reinforcement provided by a small amount of coir fibre. The M3 blend, containing 10% coconut ash and 1% coir fibre, achieved an optimal balance between strength and durability. The ash likely helped things stick together and last, meanwhile the fibre seemed to prevent cracking without making the mix hard to use making the blend more good in mechanical and durability attribute. However, Mix M4 increased the amount of coconut ash and coir fibre to 15% and 1.5%, respectively, the highest proportions used, to see whether an

FIGURE 12.5 Coconut ash preparation.

increased quantity would lead to better performance, but in practice, this version did not quite attain the strength of M3. However, due to the slight weak results, there may be uneven fibre dispersion or reduced mix workability, which sometimes leads to micro-voids forming in the material structural integrity, even if the ingredients seem promising on paper.

FIGURE 12.6 Ferrocement panel specimen.

TABLE 12.1
Mix Proportion

Mix ID	Coconut Ash (%)	Coir Fibre (%)
M1 (Control)	0	0
M2	5	0.5
M3	10	1.0
M4	15	1.5
M5	10	0.5
M6	5	1.0

In order to accurately evaluate the performance of the trial mixes and determine their specific effects, systematic testing was conducted. In M5 mix, the ash content is similar to that of M3, but there is a reduction in the coir fibre content of 0.5%. By this, the effect of the fibre alone was studied. On the other hand, the trial mix M6 was the opposite, with the coconut ash content limited to 5% and the fibre content remaining at 1%. By this way of changing the trial mix, the results can be directly understood without any alterations in the fibre dosage. These variations in the trial mixes helped in the insights of the material behaviour. These mix designs also provide a matrix for analysing the interaction and main effects of coconut ash and coir fibre on ferrocement properties selected to determine not only the optimal proportion but also threshold levels above which performance may be compromised, enabling a systematic comparison of the results and providing valuable information about how sustainable materials can be used in construction.

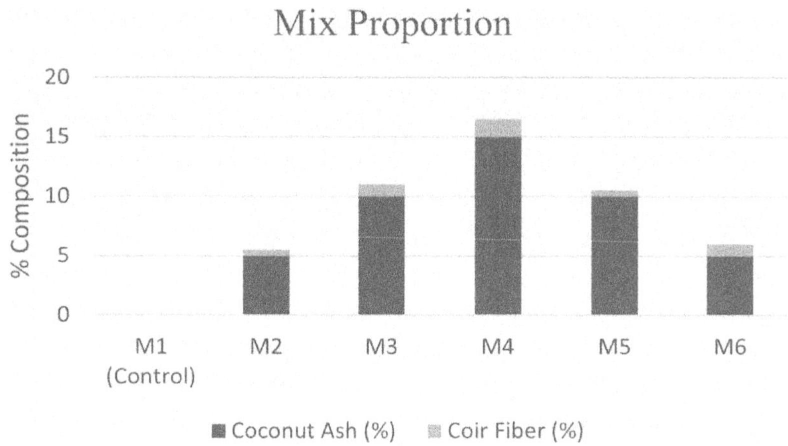

FIGURE 12.7 Mix proportion.

12.4. RESULTS AND DISCUSSION

12.4.1 Compressive Strength

The results in Table 12.2 and (Figure 12.8) show that the addition of coconut ash and coir fibre leads to an improvement in performance as the following compressive strength values indicate: the control mix (M1), with no additives, registered a strength of 32.0 MPa, whereas all modified mixes outperformed the control; the mix with the highest compressive strength was M3 (10% coconut ash and 1% coir fibre) with 38.4 MPa, about 20% higher than the control. The mechanisms involved in the enhancement are as follows:

a) **Pozzolanic Reaction:** During the hydration, silica and alumina react with the coconut ash. Due to the presence of these pozzolanic reactors, the calcium hydroxide compounds are converted into calcium silicate hydrates [12], therein reducing the microlevel pores and producing a dense compound structure.
b) **Micro-Crack Control by Coir Fibres:** Addition of 1% coir fibre improved the ductility of the matrix by bridging internal micro-cracks, reduced crack propagation, and enhanced post-peak strength behaviour [13]. The other mixes, M5 and M6, also showed significant enhancement in compressive strengths of 36.5 MPa and 35.8 MPa, respectively, indicating that even at lower fibre or ash contents, the material shows a synergistic strength enhancement. Nevertheless, the highest quantity of additives (15% coconut ash and 1.5% coir fibre) in M4 recorded a lower compressive strength, which could be attributed to reduced workability and fibre agglomeration, resulting in void formation and poor compaction [11].

Results from this study align with the results from previous studies. The inclusion of 0.5–1.5% coconut fibre in concrete mixtures has been reported to increase the compressive strength by 20–25% [14], and the incorporation of agricultural ash

TABLE 12.2
Compressive Strength

Mix ID	3 days (MPa)	7 days (MPa)	28 days (MPa)
M1 (Control)	18.5	25.7	32.0
M2	20.2	27.9	34.2
M3	22.4	30.5	38.4
M4	20.8	28.3	35.1
M5	21.6	29.7	36.5
M6	21.0	28.9	35.8

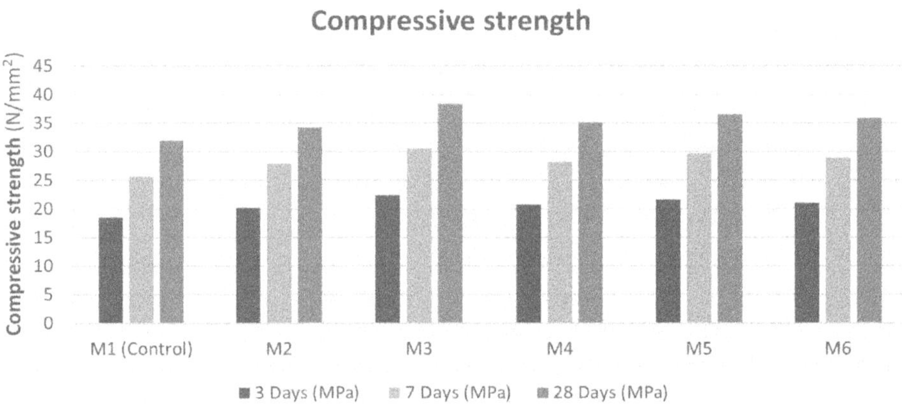

FIGURE 12.8 Compressive strength.

with high silica content has been reported to increase the strength and durability of cementitious composites [12]. In addition, the better energy absorption capacity of coir-fibre-reinforced composites has been reported to increase the compressive and flexural strength [10]. In summary, the increase in compressive strength observed in this study confirms the synergistic effect of coconut ash and coir fibre incorporation, which is in line with established literature and confirms the viability and potential of these sustainable materials for use in ferrocement and other cement-based applications.

12.4.2 Flexural Strength

The flexural strength results in Table 12.3 and Figure 12.9 show that all fibre-reinforced mixes perform better than the control mix (6.5 MPa), with mix M3 achieving a maximum flexural strength of 8.3 MPa, an increase of 28% compared to the control specimen. The enhanced flexural behaviour in M3 can be attributed to the combined effect of pozzolanic reactivity and fibre reinforcement mechanisms. Adding a small

TABLE 12.3
Flexural Strength

Mix ID	3 days (MPa)	7 days (MPa)	28 days (MPa)
M1 (Control)	3.9	5.2	6.5
M2	4.5	6.1	7.4
M3	5.0	6.9	**8.3**
M4	4.7	6.4	7.6
M5	4.8	6.6	7.9
M6	4.7	6.5	7.8

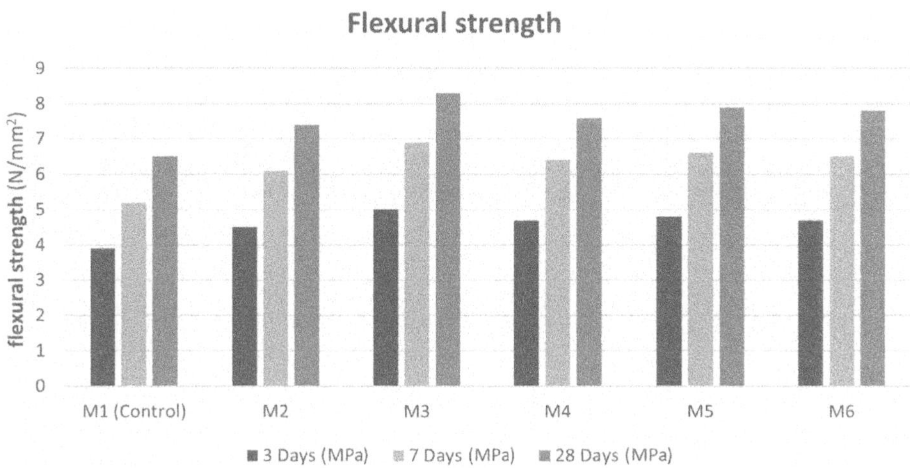

FIGURE 12.9 Flexural strength.

bit – just 1% – of coir fibre helped mend fissures while shifting pressure, so cracks grew slower plus the material became more durable after cracking [12]. Replacing a tenth of the cement with coconut ash actually made the blend more robust – it seems to generate additional binding material during combination [15].

The addition of fibre in the M3 seemed ideal. M5 and M6 achieved 7.9 MPa and 7.8 MPa respectively. However, this strength can be achieved even with a 1.5% addition of fibre (M4). This is mainly due to the clustering behaviour and improper mixing of the fibre. Al-Naghi [14] and Khan [10] also stated that there is an increase in strength when adding up to 1% of coir fibre; however, adding more than 2% of fibres results in a loss of compressive strength.

12.4.3 Water Absorption

The water absorption results (Table 12.4 and Figure 12.10) show that the moisture resistance of all coconut ash-modified panels is better than the control mix. The

TABLE 12.4
Water Absorption

Mix ID	3 days (%)	7 days (%)	28 days (%)
M1 (Control)	9.1	8.6	8.2
M2	8.3	7.8	7.5
M3	7.7	7.3	**7.0**
M4	8.0	7.5	7.3
M5	7.9	7.4	7.2
M6	8.1	7.6	7.4

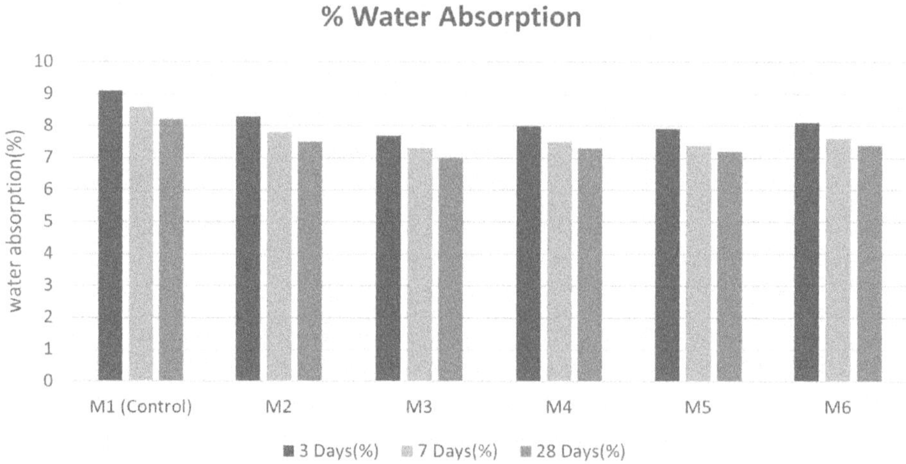

FIGURE 12.10 Water absorption.

highest water absorption (8.2%) was recorded in the control specimen while the lowest (7.0%) was found in Mix M3, which is about 15% less absorption than that of the control mix, indicating increased impermeability and better durability. The improvement in the M3 mix can be ascribed to the filler effect and pozzolanic activity of the finely ground coconut ash, which fills the capillary voids and micro-pores to give a denser internal structure with less permeability pathways [16]. Additionally, the pozzolanic reaction of the amorphous silica in the coconut ash with calcium hydroxide ($Ca(OH)_2$) to form additional calcium silicate hydrate (C–S–H) gel further refines the pore structure and enhances the strength of the mixture.

Similarly, mixes M2, M5, and M6 showed favourable reductions in water absorption (7.2–7.5%) indicating that even moderate amounts of coconut ash (20%) enhance the resistance to moisture penetration. On the other hand, higher amounts of ash (M4) that also had higher fibre content did not perform as well as M3 (7.4% absorption) possibly due to poor workability and reduced compaction efficiency (voids created by fibre agglomeration) [17].

12.5 CONTRIBUTION TO RESILIENT AND SUSTAINABLE INFRASTRUCTURE

12.5.1 Environmental and Economic Sustainability

Replacing the cement by coconut ash and inclusion of fibre allow buildings to be more sustainable and minimize the environmental impact of cement production, one of the largest sources of global carbon dioxide, and help to reduce suffering, promote more sustainable building practices, and conserve resources [15, 16]. This effort corresponds with the United Nations SDGs, particularly SDG 1 (No Poverty), SDG 9 (Industry, Innovation, and Infrastructure), and SDG 12 (Responsible Consumption and Production) [18]. Singh and his colleagues determined that replacing up to 15% of the cement with coconut shell ash did not lead to a decrease in durability and reduced cement use [15]. Meanwhile, Thomas and Gupta [16] demonstrated that replacing similar cementitious mixes with coir fibres reduced water demand by up to 35%.

12.5.2 Climate Resilience

Coconut-based ferrocement composites also have significant climate-resilient properties due to the improved material performance under adverse environmental conditions; the tensile ductility and crack resistance are improved by the addition of coir fibres, and permeability is reduced by the pozzolanic activity of coconut ash, all of which increase resistance to moisture ingress, freeze–thaw cycles, and chloride penetration [17], making it suitable for applications in coastal, humid, and cyclone-prone areas where conventional materials often deteriorate quickly.

Additionally, their durability enhances the need for minimal maintenance and repair during the life cycle, as well as provides resistance to long-term climatic degradation. These results are in agreement with the observations of Islam et al. [19], who demonstrated that agro-waste-ash-modified fibre-reinforced concrete showed better dimensional stability and less micro-cracking under varying humidity conditions, and with Ahmad et al. [13], who showed that coconut-ash-modified concretes retained higher residual strength after aggressive environmental cycles, proving superior climatic adaptability. Thus, the synergistic use of coconut ash and coir fibre provides a technically viable and environmentally responsible alternative for the construction of sustainable and climate-adaptive built environments.

12.6 TECHNOLOGICAL INTEGRATION

12.6.1 Prefabrication

Ferrocement composites reinforced with coconut ash and coir fibre have shown considerable potential for prefabricated construction systems, as the modified panels have a better strength-to-weight ratio, which enables reduced self-weight and adequate load-bearing capacity, which makes them more suitable for modular and precast applications where material efficiency, ease of transportation, and rapid assembly are key [4, 20].

These findings are in line with those of Ahmad et al. [13] who reported that agro-waste-based cementitious composites demonstrated 20–25% weight reduction with similar strength properties. They also resist cracking and have high bending strength – ideal for pre-fabricated houses or larger projects. Therefore, our new material presents a reasonable and sustainable solution to the future of building, perhaps increasing both environmental responsibility and cost-efficiency in construction.

12.7 IMPLEMENTATION CHALLENGES AND RECOMMENDATIONS

Although coconut waste, especially ash combined with coir fibre, could be an environmentally friendly alternative to traditional building materials, its use is not widespread because there are no clear guidelines on how to handle these materials, their properties vary widely, and builders have not fully embraced them yet [16, 21]; they are hesitant without firm benchmarks or ways to test consistency [21], and factors such as fibre size, wetness levels, and ash makeup influence strength and longevity [10]. In addition, the building industry is very cautious to adopt a new material; they require a significant level of real-world evidence [22]. Additionally, policies that provide benefits for using agricultural waste composites will help to encourage the industry to adopt new methods and technologies available in the research field.

12.8 CONCLUSION

The results demonstrate that combining 10% coconut ash with 1% coir fibre yields enhanced ferrocement specimens. The reduced cement content promotes sustainable, environmentally friendly projects.

- The 10% ash and 1% fibre mix showed the highest compressive and flexural strength and reduced water absorption.
- Synergistic interaction of ash and fibre enhances microstructure and crack control.
- Agro-waste utilization reduces environmental impact and supports sustainable construction practices.
- Future research should focus on standardized mix designs, testing protocols, and long-term field studies for large-scale implementation.

REFERENCES

1. Almohana, A. I., Abdulwahid, M. Y., Galobardes, I., Mushtaq, J., & Almojil, S. F. (2022). Producing sustainable concrete with plastic waste: A review. *Environmental Challenges*, 9, 100626. https://doi.org/10.1016/j.envc.2022.100626
2. Mushtaq, J., Danish, P., Salem, I. B., Shahmir, N. G., Gilani, T. A., Wani, T., & Gull, I. (2025). Analyzing the influence of eggshell powder (ESP) as partial replacement with cement in concrete. *ERT*, 8(3), 593–602. https://doi.org/10.35208/ert.1549955
3. Kadam, K. B., & Patil, S. S. (2016). Influence of coconut shell ash on concrete characteristics. *Int. J. Res. Eng. Technol.*, 5(5), 208–211. https://doi.org/10.15623/ijret.2016.0505038

4. Ali, M., Liu, A., Sou, H., & Chouw, N. (2012). Mechanical and dynamic behaviour of concrete incorporating coconut fibres. *Construction Building Materials*, 30, 814–825. https://doi.org/10.1016/j.conbuildmat.2011.12.068
5. Gopal, D., & Shobarajkumar, D. (2024). Comparative analysis on ferrocement wall panel structural behavior. *Earthq. Eng. Struct. Dyn.*, 53, 1727–1741. https://doi.org/10.1002/eqe.4081
6. Naaman, A. E. (2000). *Ferrocement and Laminated Cementitious Composites*. Techno Press 3000.
7. United Nations (2015). "Transforming our world: The 2030 Agenda for Sustainable Development," United Nations. [Online]. https://sdgs.un.org/goals
8. Pavoski, G., et al. (2025). Sustainable extraction of critical metals from phosphogypsum using nanohydrometallurgy. *J. Sustain. Metall.*, 11, 1689–1703. https://doi.org/10.1007/s40831-025-01070-x
9. Tural, H., Ozarisoy, B., Derogar, S., & Ince, C. (2024). Investigating the governing factors influencing the pozzolanic activity through a database approach for the development of sustainable cementitious materials. *Construction and Building Materials*, 411, 134253. https://doi.org/10.1016/j.conbuildmat.2023.134253
10. Khan, M. I., Alam, S., & Patel, H. (2020). Influence of natural fibres on mechanical behaviour of ferrocement panels. *Construction and Building Materials*, 260, 120132. https://doi.org/10.1016/j.conbuildmat.2020.120132
11. Krishnan, S., Gupta, R., & Das, M. K. (2022). Sustainability assessment of coir fibre reinforced composites. *Sustainable Materials and Technologies*, 31, e00325. https://doi.org/10.1016/j.susmat.2022.e00325
12. Singh, R., Kumar, A., & Sharma, P. (2021). Mechanical and durability performance of coir fibre reinforced cement composites. *Journal of Natural Fibre Reinforced Composites*, 15(3), 245–256.
13. Ahmad, J., et al. (2022). Effect of natural coir fibres on the mechanical performance of concrete composites. *Construction and Building Materials*, 314, 125653.
14. Al-Naghi, M., et al. (2025). Performance evaluation of coconut fibre reinforced concrete for sustainable infrastructure applications. *Scientific Reports*, 15(1), 234–245.
15. Singh, R., et al. (2021). Influence of pozzolanic materials on the microstructural and mechanical properties of blended cement composites. *Journal of Building Engineering*, 44, 102–118.
16. Thomas, B. S., & Gupta, R. C. (2021). A comprehensive review on the durability performance of sustainable concrete incorporating agricultural wastes. *Construction and Building Materials*, 288, 123103.
17. Rahman, M. M., & Khan, M. N. (2022). Effect of fibre dispersion on the water absorption and porosity of natural fibre reinforced concretes. *Materials Today: Proceedings*, 62, 185–192.
18. Geissdoerfer, M., Savaget, P., Bocken, N. M. P., & Hultink, E. J. (2017). Circular economy: Redefining sustainability in production and consumption systems. *J. Clean. Prod.*, 143, 757–768, https://doi.org/10.1016/j.jclepro.2016.12.048
19. Islam, M. S., et al. (2021). Effect of agro-based pozzolanic materials on permeability and strength characteristics of fibre-reinforced concrete. *Cleaner Engineering and Technology*, 5, 100326.
20. Azhar, S. (2011). BIM in the AEC sector: Emerging trends, opportunities, risks, and implementation challenges. *Leadership Manage. Eng.*, 11(3), 241–252. https://doi.org/10.1061/(ASCE)LM.1943-5630.0000127
21. Ali, M., & Chouw, N. (2020). Experimental investigations on reinforced natural fibre composite panels for prefabricated construction. *Construction and Building Materials*, 252, 119078.
22. Li, Y., & Zhang, L. (2022). Barriers to adoption of green materials in the construction industry. *Resources, Conservation and Recycling*, 180, 106197.

13 Stabilization of Sand Using Sugarcane Bagasse, Yeast, Human Hair Fibers, and Lime

Shambhavi Sinha, Mritunjay Kumar and Akhil Avchar

13.1 INTRODUCTION

Liquefaction-induced failures in cohesionless soils remain a critical challenge in earthquake-prone regions. Medium to fine sands are especially susceptible to strength degradation due to elevated pore water pressures generated under undrained cyclic loading, resulting in a loss of effective stress and shear resistance. Stabilization of such soils becomes essential to mitigate infrastructural failures in seismic zones.

Among conventional stabilization strategies ranging from compaction to chemical additives, bio-geotechnical methods have gained attention for their sustainability and environmental safety. One such technique is Microbially Induced Calcite Precipitation (MICP), where microbial activity is harnessed to promote the precipitation of calcium carbonate ($CaCO_3$), effectively binding soil particles and enhancing strength.

Sandy soils, valued for drainage, are prone to seismic liquefaction due to low cohesion, necessitating stabilization to enhance shear strength and reduce permeability. Traditional methods, such as chemical stabilization with cement or lime [1, 2] and reinforcement with synthetic fibers [3], improve soil properties but incur high environmental costs, with cement production contributing approximately 8% of global CO_2 emissions [4, 5]. MICP, a bio-mediated process, precipitates calcium carbonate ($CaCO_3$) to bind soil particles, offering a sustainable alternative [6]. Recent studies have further demonstrated the critical role of strength parameters, particularly the angle of internal friction, in governing the stability and performance of geotechnical systems under varying loading and boundary conditions [7].

Conventional MICP uses urease-producing bacteria like *Sporosarcina pasteurii* to hydrolyze urea, forming carbonate ions for $CaCO_3$ precipitation [8]. Effective for soil stabilization and self-healing concrete [6, 9], this method produces ammonia, posing environmental risks [10]. Specialized microbial cultures and nutrients increase costs [11], and sensitivity to pH and temperature affects efficiency [12].

DOI: 10.1201/9781003664284-13

Alternative pathways, including denitrification [13], sulfate reduction [14], and yeast-driven fermentation [15], mitigate these issues. Fermentation with *Saccharomyces cerevisiae* produces CO_2 from substrates like sugarcane bagasse (SCB), eliminating ammonia [16, 17]. The non-ureolytic approach, leveraging precipitation, is directly applicable to the development of self-healing concrete, which autonomously repairs cracks and enhances durability using both ureolytic and non-ureolytic MICP systems [18]. This pivot toward eco-friendly pathways, driven by the need to eliminate hazardous by-products and mitigate the high emissions associated with conventional cement production, aligns with the construction sector's broader mandate to incorporate diverse waste streams. For instance, the use of waste plastic as aggregate substitutes and fiber reinforcement in concrete is a parallel innovation that validates the use of waste materials for mechanical and environmental improvement across civil engineering [19].

Agricultural wastes like SCB, rich in sucrose, support fermentation-based MICP [20, 21], while rice husk and coir enhance soil strength [22]. Organic waste, such as wastepaper, improves MICP-treated soil strength [23]. Human hair fibers, a biodegradable alternative to synthetic fibers, increase CBR and shear strength in soils [24–27]. Lime provides calcium for MICP and enhances stiffness [28].

Fermentation-based MICP, using SCB and yeast, supports CO_2 production for carbonate formation [15, 29, 30], with applications in liquefaction mitigation [31], erosion control [32], and self-healing concrete [9]. Challenges include slow fermentation kinetics [33], uneven $CaCO_3$ distribution [34], and untested durability [22]. Future research should optimize fermentation, analyze $CaCO_3$ via Scanning Electron Microscopy (SEM)/X-Ray Diffraction (XRD), and test cyclic loading performance [31].

While the traditional MICP process involves ureolytic bacteria such as Sporosarcina pasteurii, it poses challenges, including slow kinetics, ammonia release, and the need for specific microbial strains. This study proposes a novel, urea-free MICP methodology utilizing yeast-driven alcoholic fermentation of SCB, a high-sucrose agricultural by-product, in saturated sand matrices. CO_2 generated in situ through fermentation is converted into carbonate ions under alkaline conditions, promoting calcite formation in the presence of lime. The inclusion of human hair fibers provides additional mechanical stability.

The primary objective is to evaluate the feasibility and effectiveness of this integrated bio-mechanical stabilization strategy using accessible, biodegradable materials under laboratory conditions.

13.2 MATERIALS AND METHODS

The experimental investigation involved the use of medium- to fine-grained sand collected locally, along with SCB, yeast cultures, human hair fibers, and lime. Sand was classified using sieve analysis (IS 2720–4, 1985) [35] and its specific gravity determined (IS 2720 Part 3, 1980) [36]. The poorly graded sand (SP) exhibited D10, D30, and D60 values of 0.331 mm, 0.552 mm, and 1.225 mm, respectively, with a coefficient of uniformity (Cu) of 3.70 and curvature (Cc) of 0.751.

The geotechnical properties of the sand used in the investigation are summarized in Table 13.1. The material was classified as poorly graded sand (SP) according to IS 1498, with effective grain size (D_{10}), intermediate grain size (D_{30}), and maximum grain size (D_{60}) values of 0.328 mm, 0.558 mm, and 1.230 mm, respectively. The soil exhibited a uniformity coefficient (Cu) of 3.75 and a curvature coefficient (Cc) of 0.76, indicating a relatively narrow particle-size distribution. The specific gravity of the sand was 2.53, while the hydraulic conductivity was determined as 9.2×10^{-3} cm/s, reflecting its permeable nature. The maximum and minimum void ratios were found to be 0.815 and 0.643, respectively, confirming the suitability of the material for evaluating the effectiveness of the proposed bio-mediated stabilization technique.

SCB was sourced from local vendors, with outer rind removed and inner pith chopped to 10–20 mm lengths. Yeast used included both commercial baker's yeast and *Saccharomyces cerevisiae* strains from a biotechnology lab. Human hair fibers were cleaned, air-dried, and incorporated at 1.5% by weight. Lime (CaO) was used in both powdered and limewater form for mechanical stabilization and as a source of calcium ions. Sand samples were prepared in cylindrical molds (38 mm diameter × 76 mm height) for unconfined compressive strength (UCS) testing (IS 2720 Part 10, 1991) [37]. Two methods were employed: mixed and layered placement of the yeast SCB mixture. After initial saturation, samples were cured under anaerobic conditions for 15 days. Starting on day 6, limewater solution was introduced on alternate days. This was done to facilitate the biocementation process by promoting the precipitation of calcium carbonate ($CaCO_3$). This mineral cementation serves to bind the sand particles, thereby increasing the sample's strength. To minimize the loss of yeast cells – which are essential for the metabolic processes leading to carbonate formation – the filtrate was recycled. The liquid drained from the samples was collected and reused during the preparation of the upper layers. This method ensured a higher

TABLE 13.1
Geotechnical Characteristics of Sand Employed in the Investigation

Parameter	Measured Value
Effective grain size (D_{10})	0.328 mm
Intermediate grain size (D_{30})	0.558 mm
Maximum grain size (D_{60})	1.230 mm
Uniformity coefficient (Cu)	3.75
Curvature coefficient (Cc)	0.76
Soil classification (IS:1498/ISSC System)	SP Poorly Graded Sand
Specific gravity (Gs)	2.53
Hydraulic conductivity (k)	9.2×10^{-3} cm/s
Maximum void ratio (emax)	0.815
Minimum void ratio (emin)	0.643

concentration of the biocatalyst within the samples, thereby enhancing the efficiency of the carbonate precipitation reaction.

13.2.1 Sample Preparation

To prepare stabilized sand samples for UCS testing as per IS 2720 Part-10 [37], 1 kg of oven-dried sand was uniformly mixed with 15 g (1.5% by weight) of cleaned human hair fibers to enhance tensile strength. Subsequently, 50 g (5% by weight) of powdered quicklime (CaO) was incorporated to provide initial stiffness and calcium ions for bio-cementation. A separate mixture of 15 g SCB, cut into 10–20 mm strips, and *Saccharomyces cerevisiae* yeast was prepared in two ratios (1:6 and 1:10 by weight) and kneaded to ensure contact for fermentation. These yeast–SCB mixtures were integrated into the sand-lime-hair blend either uniformly (mixed samples) or in alternating layers (layered samples), producing four cylindrical samples (38 mm diameter, 76 mm height) molded in plastic pipes. A filter system of scrubbers and 8–10 mm gravel ensured full saturation and drainage. Samples were saturated with water, with effluent recycled to retain yeast, and allowed to ferment for five days to generate CO_2. From day 6, lime water was added every alternate day to maintain alkaline conditions (pH>8), promoting carbonate ion formation and $CaCO_3$ precipitation. Effluent was reintroduced to maximize CO_2 utilization. After 15 days of curing, samples were demolded, retaining their structural integrity for UCS testing.

The materials used in the preparation of the stabilized sand specimens are illustrated in Figure 13.1. These include lime as a calcium source, human hair fibers as reinforcing elements, yeast as the fermenting microorganism, and sugarcane bagasse as the organic substrate for carbon dioxide generation during the bio-cementation process.

Figure 13.2 presents the sample preparation procedure and specimen configurations used in the study, including layered and mixed arrangements with different yeast-to-substrate ratios designed to promote fermentation and subsequent carbonate precipitation.

FIGURE 13.1 (a) Lime (CaO) used for sample preparation, (b) sand mixed with human hair fibers, (c) yeast used for sample preparation, and (d) sugarcane bagasse used for sample preparation.

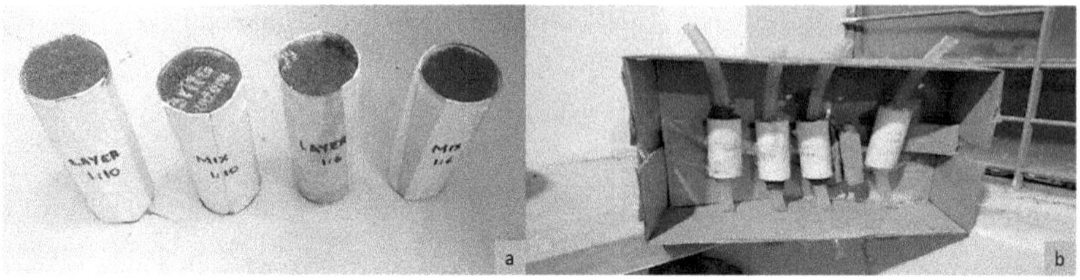

FIGURE 13.2 (a) Preparation of molds containing samples. (b) Configuration of molds containing samples with yeast-to-substrate ratios of 1:6 and 1:10, prepared in layered and mixed arrangements, respectively.

13.3 MECHANISM OF BIO-FERMENTATION-BASED STABILIZATION

The stabilization of sandy soils, particularly to mitigate issues like seismic liquefaction, is a critical challenge in geotechnical engineering. A novel approach utilizing fermentation-based MICP) offers an environmentally sustainable alternative to conventional methods. This process leverages yeast-driven alcoholic fermentation of (SCB to produce carbon dioxide (CO_2), which facilitates the formation of calcium carbonate ($CaCO_2$) as a binding agent within the soil matrix. The incorporation of human hair fibers as micro-reinforcements further enhances the mechanical properties of the stabilized soil. This section provides a detailed explanation of the biochemical and mechanical mechanisms involved, ensuring clarity and originality for professional audiences.

13.3.1 Biochemical Mechanism of Fermentation-Based MICP

The fermentation-based MICP process utilizes *Saccharomyces cerevisiae* (yeast) to ferment the sucrose and glucose present in SCB, a lignocellulosic agricultural by-product, under anaerobic conditions. This process generates CO_2, which is critical for the subsequent formation of $CaCO_2$, a natural cementing agent that binds soil particles to enhance strength and stability.

13.3.1.1 Step 1: Sucrose Hydrolysis

SCB contains residual sucrose (approximately 1–2% by weight) and glucose, which serve as substrates for fermentation. The initial step involves the hydrolysis of sucrose, a disaccharide, into its constituent monosaccharides, glucose and fructose. This reaction occurs in the presence of water and is catalyzed by enzymes such as invertase produced by yeast:

$$C_{\{12\}H_{\{22\}O_{\{11\}}}} + H_{2O} \rightarrow C_{6H_{\{12\}O_{6(\{glucose\})}}} + C_{6H_{\{12\}O_{6(\{fructose\})}}} \qquad (13.1)$$

This hydrolysis ensures that the sugars are in a form readily metabolizable by yeast, providing the energy source for fermentation.

13.3.1.2 Step 2: Alcoholic Fermentation

Under anaerobic conditions, yeast metabolizes glucose and fructose through alcoholic fermentation, converting these sugars into ethanol and CO_2. The reaction for glucose (and similarly for fructose) is as follows:

$$C_{6H_{\{12\}O_6}} \rightarrow 2C_{2H_{5OH}} + 2CO_2 \tag{13.2}$$

For sucrose, the process yields double the amount of ethanol and CO_2 per mole due to its disaccharide nature:

$$C_{\{12\}H_{\{22\}O_{\{11\}}}} + H_{2O} \rightarrow 4C_{2H_{5OH}} + 4CO_2 \tag{13.3}$$

The CO_2 produced dissolves in the pore water within the soil matrix, creating an aqueous solution conducive to further chemical reactions. The ethanol by-product remains in the solution and does not interfere with the stabilization process, though its acidic nature may slightly lower the pH, necessitating alkaline conditions for subsequent steps.

13.3.1.3 Step 3: CO_2 Conversion to Carbonate Ions

The dissolved CO_2 in the pore water undergoes a series of equilibrium reactions influenced by the pH of the environment. In alkaline conditions (pH > 8), typically achieved by the addition of lime (calcium oxide, CaO, or calcium hydroxide, $Ca(OH)_2$), CO_2 reacts with water to form carbonic acid, which dissociates into bicarbonate and carbonate ions:

$$CO_2 + H_{2O} \leftrightharpoons H_{2CO_3} \leftrightharpoons HCO_3^- + H^+ \tag{13.4}$$

$$HCO_3^- + OH^- \rightleftharpoons CO_3^{\{2-\}} + H_2O \tag{13.5}$$

The presence of hydroxide ions (OH^-) from lime shifts the equilibrium toward the formation of carbonate ions (CO_3^{2-}), which are essential for the precipitation of $CaCO_3$. Maintaining a pH above 8 ensures that carbonate ions predominate, facilitating the subsequent reaction.

13.3.1.4 Step 4: Calcite Precipitation

Calcium ions (Ca^{2+}), supplied by lime in the form of calcium hydroxide or limewater, react with carbonate ions in the soil pore water to precipitate calcium carbonate:

$$Ca^{\{2+\}} + CO_3^{\{2-\}} \rightarrow CaCO_3 \downarrow \tag{13.6}$$

This precipitation occurs primarily at particle-particle contacts within the soil matrix, forming a cementitious matrix that binds sand grains together. The $CaCO_2$ crystals enhance interparticle bonding, increasing the shear strength and stiffness of the soil, thereby reducing its susceptibility to liquefaction under dynamic loads.

13.4 RESULTS AND OBSERVATIONS

The stabilization process of sandy soil samples using fermentation-based MICP was monitored over 15 days, with observations recorded for both layered and mixed samples prepared with yeast-to-SCB ratios of 1:6 and 1:10. On day 1, both sample types were fully saturated with water, with mixed samples exhibiting higher permeability than layered samples. By day 6, limewater was introduced to promote carbonate ion formation, and a pungent odor was noted, stronger in mixed samples than in layered ones. On day 8, additional limewater was added, with both sample types showing reduced permeability, indicated by slower flow rates. On tenth day, limewater addition intensified the pungent smell, particularly in mixed samples. No new observations were recorded on twelfth day after limewater addition. On the fourteenth day, the final limewater dose was applied, revealing significantly reduced permeability in layered samples compared to Day 1, alongside visible black decomposition patches near SCB layers; mixed samples, while less permeable than initially, retained higher permeability than layered samples and showed no dark patches. On fifteenth day, samples were gently demolded, with mixed samples exhibiting a notably stronger pungent odor due to greater SCB decomposition compared to layered samples, which maintained structural integrity with reduced permeability and localized decomposition. The chronological observations recorded during anaerobic fermentation-based CO_2 generation are presented in Table 13.2. The observed progression of gas evolution and mineral deposition is consistent with reported fermentation behavior in yeast-mediated sugarcane bagasse systems.

The appearance of the stabilized specimen after 15 days of curing is shown in Figure 13.3. The sample retained its structural integrity under unconfined conditions, indicating the feasibility of the proposed fermentation-based MICP approach involving sugarcane bagasse, yeast, lime, and human hair fibers.

FIGURE 13.3 After 15 days of the Microbially Induced Calcite Precipitation (MICP) process, the resulting stabilized sand sample was obtained.

TABLE 13.2
Chronological Observations During Anaerobic Fermentation-Based CO_2 Generation

Days	Observations
1	No visible activity or physical changes observed.
3	Initial gas bubbles began forming near the base of the reaction vessel.
5	Intense effervescence observed at the vessel bottom, indicating peak fermentation activity.
7	Increased gas production with the onset of surface-level aerobic degradation; upper layers show organic breakdown.
9	Noticeable foul odor developed due to aerobic microbial processes; however, anaerobic CO_2 generation continued actively at lower depths.
11	White deposition appeared along the internal walls and connecting tube, suggesting calcite formation; gas production began to decline.

13.5 DISCUSSION

The development of eco-friendly soil stabilization techniques is increasingly vital in geotechnical engineering, particularly for addressing the susceptibility of sandy soils to seismic liquefaction. The study under consideration introduces a novel fermentation-based approach to MICP, utilizing yeast-driven fermentation of SCB to produce carbon dioxide (CO_2), which facilitates calcium carbonate ($CaCO_3$) precipitation in the presence of lime. This method, which leverages bio-precipitation for structural enhancement, directly parallels the application of bacterial MICP in the structural realm, notably in the field of self-healing concrete. In this domain, specialized bacteria, often *Bacillus* species, are incorporated into the concrete matrix (sometimes as spores) to autonomously precipitate upon crack formation, thereby sealing damage, increasing durability, and extending the material's structural lifespan [18]. This method is complemented by the incorporation of human hair fibers as micro-reinforcements and lime as a dual-purpose stabilizer. This analysis elaborates on the biochemical and mechanical mechanisms, comparative advantages, and structural considerations of this approach, providing a rigorous scientific perspective for professional audiences.

13.5.1 Biochemical Efficacy of Fermentation-Based MICP

The fermentation-based MICP process leverages *Saccharomyces cerevisiae* (yeast) to metabolize the sucrose and glucose present in SCB, an agro-industrial by-product, under anaerobic conditions. This process generates CO_2, which dissolves in the soil's pore water and, in an alkaline environment (pH > 8) provided by lime, forms carbonate ions (CO^{32-}) that react with calcium ions (Ca^{2+}) to precipitate $CaCO_2$. The study confirms the feasibility of this approach, demonstrating consistent CO_2 production over a 15-day curing period, with peak fermentation activity observed between days 5 and 10. The resulting $CaCO_2$ precipitation effectively binds sand particles, enhancing cohesion and reducing the risk of liquefaction under dynamic loading.

The biochemical pathway begins with the hydrolysis of sucrose in SCB into glucose and fructose, catalyzed by yeast enzymes such as invertase:

$$C_{\{12\}H_{\{22\}O_{\{11\}}}} + H_{2O} \rightarrow C_{6H_{\{12\}O_{6(\{glucose\})}}} + C_{6H_{\{12\}O_{6(\{fructose\})}}} \tag{13.7}$$

Subsequent fermentation converts these monosaccharides into ethanol and CO_2:

$$C_{6H_{\{12\}O_6}} \rightarrow 2C_{2H_{5OH}} + 2CO_2 \tag{13.8}$$

The CO_2 dissolves in pore water, forming carbonic acid, which dissociates under alkaline conditions:

$$CO_2 + H_{2O} \leftrightharpoons H_{2CO_3} \leftrightharpoons HCO_3^- + H^+ \tag{13.9}$$

$$HCO_3^- + OH^- \leftrightharpoons CO_3^{\{2-\}} + H_2O \tag{13.10}$$

Calcium ions from lime ($Ca(OH)_2$ or CaO) react with carbonate ions to form $CaCO_2$:

$$Ca^{\{2+\}} + CO_3^{\{2-\}} \rightarrow CaCO_3 \downarrow \tag{13.11}$$

This precipitation occurs preferentially at particle-particle contacts, creating a cementitious matrix that enhances the soil's shear strength and stiffness. The study's feasibility experiments, using both laboratory-grade and commercial yeast, confirmed CO_2 production, evidenced by partial milkiness in limewater and $CaCO_3$ deposition in experimental setups, indicating a successful carbonate formation.

13.5.2 Environmental and Economic Advantages

Compared to conventional ureolytic MICP, which relies on urease-producing bacteria such as *Sporosarcina pasteurii* to hydrolyze urea, the fermentation-based approach offers significant environmental benefits. Ureolytic MICP generates ammonia (NH_3) as a by-product, which can contaminate groundwater and pose health risks in confined spaces. The fermentation method eliminates ammonia production, as CO_2 and ethanol are the primary by-products, with ethanol being benign in this context. This enhances the environmental safety of the process, making it suitable for applications in sensitive ecosystems or urban settings.

Economically, the method leverages readily available and low-cost materials. SCB, a by-product of the sugar industry, is abundant, with approximately 270–280 kg generated per ton of sugarcane processed. Its use repurposes waste that would otherwise require disposal, aligning with circular economy principles. Yeast, particularly commercial baker's yeast, is inexpensive and widely accessible, unlike specialized bacterial cultures used in ureolytic MICP, which require controlled cultivation and nutrient-rich media. Lime, a traditional stabilizer, is also cost-effective and serves a dual role as a calcium source and chemical stabilizer, further reducing material costs.

13.5.3 Mechanical Enhancement with Human Hair Fibers

The incorporation of human hair fibers at 1.5% by weight of dry sand significantly enhances the mechanical properties of the stabilized soil. Human hair, a biodegradable waste material, possesses a tensile strength of 150–270 MPa and high elasticity, making it an effective micro-reinforcement. The fibers are randomly distributed within the soil matrix, providing several mechanical benefits:

- **Tensile Reinforcement**: The fibers resist tensile stresses, reducing crack initiation and propagation under external loads.
- **Deformation Control**: By bridging soil particles, human hair fibers limit relative movement, enhancing the soil's ability to resist deformation during shear or compressive loading.
- **Shear Strength Improvement**: The fibers increase frictional resistance at particle contacts, elevating the peak shear strength and angle of shearing resistance, which is critical for resisting liquefaction.
- **Energy Absorption**: The ductile nature of human hair enhances the soil's capacity to absorb energy, improving its performance under cyclic loading conditions typical of seismic events.

The study observed that sand samples reinforced with human hair fibers exhibited a quick tactile response under saturated conditions, indicating improved cohesion and stiffness. The optimal fiber content of 1.5% balances mechanical enhancement with workability, as higher contents may lead to mixing challenges without proportional strength gains.

13.5.4 Role of Lime in Dual Stabilization

Lime plays a pivotal role in the stabilization process, serving both as a source of calcium ions for $CaCO_2$ precipitation and as a chemical stabilizer to enhance initial soil stiffness. As calcium hydroxide ($Ca(OH)_2$), lime dissolves in water to provide Ca^{2+} ions, which react with carbonate ions to form $CaCO_3$. Additionally, lime's alkaline nature (pH > 12) creates an optimal environment for carbonate ion formation by shifting the CO_2 dissociation equilibrium toward CO_3^{2-}. As a chemical stabilizer, lime reacts with soil particles to form cementitious compounds, such as calcium silicate hydrates, which contribute to early strength development. The study incorporated 5% powdered lime to achieve initial stiffness, with limewater added periodically to sustain $CaCO_2$ precipitation during the 15-day curing period.

The reduced permeability in layered samples suggests a better retention of CO_2 and carbonate ions, potentially enhancing carbonation efficiency. However, the uneven distribution in layered samples led to localized decomposition, which could compromise long-term stability. These findings emphasize the need for thorough mixing to ensure consistent microbial activity and $CaCO_3$ precipitation across the soil matrix.

However, the fermentation-based approach has a slower rate of CO_2 production compared to the rapid urea hydrolysis in ureolytic MICP, which may extend curing times. The study mitigated this by optimizing the yeast-to-SCB ratio and ensuring anaerobic conditions within saturated sand pores, but further research is needed to enhance fermentation kinetics.

13.6 CONCLUSION

This investigation successfully validated a novel, fermentation-based MICP method for stabilizing sandy soils, which offers an environmentally sustainable alternative by eliminating the ammonia generation associated with conventional ureolytic techniques.

The feasibility of the core mechanism was confirmed: *Saccharomyces cerevisiae* successfully ferments SCB – an agro-industrial waste – -under anaerobic conditions to produce CO_2. This CO_2, maintained by lime-induced alkalinity (pH>8), precipitates calcium carbonate ($CaCO_3$) bio-cement, which enhances interparticle bonding and soil cohesion. Consistent CO_2 production was observed over the 15-day curing period, with peak fermentation activity recorded between days 5 and 10. All stabilized sand samples maintained structural integrity under unconfined conditions, validating the effectiveness of the integrated stabilization system. This system relies on the synergy of $CaCO_3$ cementation with 1.5% human hair fibers, which act as micro-reinforcement to increase tensile strength, limit deformation, and enhance energy absorption capabilities necessary for resisting dynamic loads. Lime played a critical dual role, supplying the necessary Ca^{2+} ions for precipitation and providing initial chemical stiffness required during the relatively slow 15-day biological curing process.

The uniform mixing configuration proved strategically superior, resulting in homogenous $CaCO_3$ precipitation due to consistent reactant distribution. While layered samples showed lower end-state permeability, this arrangement introduced structural heterogeneity and risks of localized decomposition, making it less suitable for reliable geotechnical application.

This method eliminates environmental hazards by producing ethanol and CO_2 (instead of ammonia) and leverages low-cost, abundant agro-industrial waste (SCB), significantly enhancing the process's cost-effectiveness and environmental safety. While the slower kinetic rate of CO_2 production is a limitation compared to ureolytic processes, the nontoxic and waste-valorizing nature provides substantial overall advantages. This method shows significant promise for applications in infrastructure stabilization and liquefaction mitigation, advancing eco-conscious engineering solutions.

Future efforts must focus on quantitative mechanical testing (UCS and cyclic triaxial tests) and advanced characterization (SEM/XRD) to confirm performance metrics for field deployment.

REFERENCES

1. Pereira RS, Emmert F, Miguel EP, Gatto A (2018) Soil stabilization with lime for the construction of forest roads. *Floresta e Ambiente* 25. https://doi.org/10.1590/2179-8087.007715
2. Wayal AS, Ameta NK, Purohit DG (2012) Dune sand stabilization using bentonite and lime. *Journal of Engineering Research and Studies* 3:58–60
3. Hejazi SM, Sheikhzadeh M, Abtahi SM, Zadhoush A (2012) A simple review of soil reinforcement by using natural and synthetic fibers. *Construction and Building Materials* 30:100–116. https://doi.org/10.1016/j.conbuildmat.2011.11.045
4. Andrew RM (2018) Global CO_2 emissions from cement production. *Earth System Science Data* 10:195–217. https://doi.org/10.5194/essd-10-195-2018
5. Benhelal E, Zahedi G, Shamsaei E, Bahadori A (2013) Global strategies and potentials to curb CO_2 emissions in cement industry. *Journal of Cleaner Production* 51:142–161. https://doi.org/10.1016/j.jclepro.2012.10.049
6. Achal V, Mukherjee A, Kumari D, Zhang Q (2015) Biomineralization for sustainable construction - a review of processes and applications. *Earth-Science Reviews* 148:1–17. https://doi.org/10.1016/j.earscirev.2015.05.008
7. Sinha S, Tripathi AK, Avchar A, et al (2026) Influence of angle of internal friction and slope face angle on kinematic failures in marble mines: A predictive approach. *Indian Geotechnical Journal* 56:1581–1589. https://doi.org/10.1007/s40098-025-01255-x
8. Hammes F, Boon N, de Villiers J, et al (2003) Strain-Specific ureolytic microbial calcium carbonate precipitation. *Applied and Environmental Microbiology* 69:4901–4909. https://doi.org/10.1128/AEM.69.8.4901-4909.2003
9. Rong H, Qian CX, Li L (2012) Study on microstructure and properties of sandstone cemented by microbe cement. *Construction and Building Materials* 36:687–694. https://doi.org/10.1016/j.conbuildmat.2012.06.063
10. Harkes MP, van Paassen LA, Booster JL, et al (2010) Fixation and distribution of bacterial activity in sand to induce carbonate precipitation for ground reinforcement. *Ecological Engineering* 36:112–117. https://doi.org/10.1016/j.ecoleng.2009.01.004
11. Ivanov V, Chu J (2008) Applications of microorganisms to geotechnical engineering for bioclogging and biocementation of soil in situ. *Reviews in Environmental Science and Biotechnology* 7:139–153. https://doi.org/10.1007/s11157-007-9126-3
12. Soon NW, Lee LM, Khun TC, Ling HS (2013) Improvements in engineering properties of soils through microbial-induced calcite precipitation. *KSCE Journal of Civil Engineering* 17:718–728. https://doi.org/10.1007/s12205-013-0149-8
13. Karatas I (2014) Denitrification-based biocementation for soil stabilization. *Geotechnical and Geological Engineering* 32:1037–1045. https://doi.org/10.1007/s10706-014-9777-7
14. Castanier S, Le Metayer-Levrel G, Perthuisot JP (1999) Ca-carbonates precipitation and limestone genesis-the microbiologist point of view. *Sedimentary Geology* 126:9–23. https://doi.org/10.1016/S0037-0738(99)00028-7
15. Wong YC, Sanggari V (2014) Bioethanol production from sugarcane bagasse using fermentation process. *Oriental Journal of Chemistry* 30:507–513. https://doi.org/10.13005/ojc/300214
16. Parmar S, Marjadi D (2017) Biocementation: A novel technique and approach towards sustainable material. *World Journal of Research and Review (WJRR)* 4:34–41
17. Frankel RB, Bazylinski DA (2003) Biologically induced mineralization by bacteria. *Reviews in Mineralogy and Geochemistry* 54:95–114. https://doi.org/10.2113/0540095
18. Danish P, Mushtaq J, Salem IB, Gilani TA (2025) Bacteria-Induced Self-Healing Concrete: A Critical Review for the Construction Industries. In: Danish P, Rao KSN, Tantary MA, et al (eds) *Innovative Technologies for Smart and Sustainable Infrastructure.* Springer Nature Switzerland, Cham, pp 135–155

19. Almohana AI, Abdulwahid MY, Galobardes I, et al (2022) Producing sustainable concrete with plastic waste: A review. *Environmental Challenges* 9:100626. https://doi.org/10.1016/j.envc.2022.100626
20. Amores I, Ballesteros I, Manzanares P, et al (2013) Ethanol production from sugarcane bagasse pretreated by steam explosion. *Electronic Journal of Energy and Environment* 1:25–36. https://doi.org/10.7770/ejee-V1N1-art486
21. Canilha L, Chandel AK, Milessi TSDS, et al (2012) Bioconversion of sugarcane biomass into ethanol: An overview about composition, pretreatment methods, detoxification of hydrolysates, enzymatic saccharification, and ethanol fermentation. *Journal of Biomedicine and Biotechnology*. https://doi.org/10.1155/2012/989572
22. Alemu D, Lemma E (2023) Stabilization of expansive soil with agricultural waste additives: A review. *International Journal of Geo-Engineering* 14:1–15. https://doi.org/10.1186/s40703-023-00194-x
23. Golovkina K, Zhurishkina EV, Filippova AD, et al (2024) Integration of organic waste for soil stabilization through MICP. *Applied Sciences* 14:62. https://doi.org/10.3390/app14010062
24. Butt WA, Gupta K, Naik H, Bhat SM (2014) Soil sub-grade improvement using human hair fiber. *International Journal of Scientific and Engineering Research* 5:977–981
25. Sahu R, Ayothiraman R, Ramana GV (2018) Shear Behavior of Sand Reinforced with Human Hair Fibers. In: *Proceedings of 19th Southeast Asian Geotechnical Conference and 2nd AGSSEA Conference (19 SEAGC and 2 AGSSEA)*, pp 3–5
26. Narayanan KS, Sharmila SMR (2017) Stabilization of clay with human hair fiber. *International Journal of Civil Engineering and Technology (IJCIET)* 8:662–667
27. Elias T, George S, Zachariah A, et al (2016) Comparative study of soil stabilization using human hair and lime. *International Journal of Scientific & Engineering Research (IJSER)* 7:1323–1326
28. Alavéz-Ramírez R, Montes-García P, Martínez-Reyes J, et al (2012) The use of sugarcane bagasse ash and lime to improve the durability and mechanical properties of compacted soil blocks. *Construction and Building Materials* 34:296–305. https://doi.org/10.1016/j.conbuildmat.2012.02.072
29. Iram M, Asghar U, Irfan M, et al (2018) Production of bioethanol from sugarcane bagasse using yeast strains: A kinetic study. *Energy Sources, Part A: Recovery, Utilization and Environmental Effects* 40:364–372. https://doi.org/10.1080/15567036.2017.1422056
30. Pepin C, Marzzacco C (2015) The fermentation of sugars using yeast: A discovery experiment. *Chem 13 News Magazine*
31. Qabany AA, Mortensen B, Martinez B, et al (2011) Microbial Carbonate Precipitation: Correlation of S-Wave Velocity with Calcite Precipitation. In: *Geo-Frontiers 2011: Advances in Geotechnical Engineering*, pp 3993–4001
32. Li M, Cheng X (2023) State-of-the-art review of soil erosion control by MICP and EICP techniques: Problems, applications, and prospects. *Science of the Total Environment* 858:169016. https://doi.org/10.1016/j.scitotenv.2023.169016
33. Chaparro-Acuna SP, Becerra-Jimenez ML, Martinez-Zambrano JJ, Rojas-Sarmiento HA (2017) Soil bacteria that precipitate calcium carbonate: Mechanism and applications of the process. Acta Agronomica 67:277–288. https://doi.org/10.15446/acag.v67n2.66109
34. Abdel-Ghany MA, El-Sayed AM (2023) A critical review of microbially induced carbonate precipitation for soil stabilization: The global experiences and future prospective. *Pedosphere* 33:717–730. https://doi.org/10.1016/j.pedsph.2023.01.011
35. Bureau of Indian Standards (BIS) (1985) *IS 2720–4: Methods of Test for Soils, Part 4: Grain Size Analysis*. Bureau of Indian Standards
36. Bureau of Indian Standards (1980) *Methods of Test for Soils*, Part 3: Determination of Specific Gravity, Section 2: Fine, Medium and Coarse Grained Soils. Bureau of Indian Standards (BIS), New Delhi
37. Bureau of Indian Standards (1991) *IS 2720–10: Methods of Test for Soils, Part 10: Determination of Unconfined Compressive Strength*. Bureau of Indian Standards

14 Rheological and Topological Optimization Behaviour of Concrete Using Lightweight Green Aggregate (LiGra)

Suvash Chandra Paul, Yi Wei Daniel Tay and Ming Jen Tan

14.1 INTRODUCTION

The global demand for buildings and infrastructure is continually increasing, with concrete serving as the backbone of modern society.[1] Concrete is primarily composed of aggregates and Portland cement, along with other materials like steel, aluminium, and wood. However, concrete production poses significant environmental challenges, particularly due to its composition. Non-renewable natural resources, such as rocks, are used as aggregates,[2] and the production of Portland cement is energy-intensive, releasing substantial amounts of carbon dioxide (CO_2) into the atmosphere. A study estimates that the production of Portland cement accounts for approximately 5 to 9% of global greenhouse gas emissions.[1] To put this into perspective, it is estimated that producing 1 kg of cement generates roughly 0.6 to 1.0 kg of (CO_2).[3,4] In 2021, a total of 4.4 billion tons of cement was produced.[5]

Additionally, aggregates, which include sand, gravel, and crushed stone, constitute a significant volume of concrete and are the most extracted materials by weight on the planet.[6] In many countries, inadequate governance of aggregate extraction has led to severe social and environmental repercussions such as ecosystem destruction and illegal mining.[7] These negative impacts, combined with a demand growth that exceeds the capacity of natural resources, have raised concerns about a global crisis in natural aggregate scarcity.[8] To address these challenges, it is essential to find alternative materials with minimal environmental impact for concrete production. This research focuses on developing Lightweight Self-Compacting Concrete (LWSCC) using lightweight green aggregate (LiGra), which is produced from recycled waste materials. This innovative approach is being spearheaded by JoeGreen Pte Ltd, a Singapore-based company dedicated to creating novel green materials for commercial applications.[9]

DOI: 10.1201/9781003664284-14

Lightweight concrete has a lower mass and density compared to conventional concrete. According to ACI 213R (2014), its density ranges from 1,350 to 1,950 kg/m³, while conventional concrete typically has a density of 2,200 to 2,400 kg/m³. This reduced weight allows for lighter structures that require fewer supports and foundations, simplifying construction efforts and minimizing complications. Similarly, fresh Self-Compacting Concrete (SCC) can settle and consolidate into formwork solely under its own weight, eliminating the need for external vibration typically required with conventional concrete.[10] SCC offers numerous advantages depending on its application. It is characterized by its workability, which includes filling ability, passing ability, and segregation resistance, as well as its engineering properties.[11]

Topology optimization has evolved rapidly in recent decades as a method for economic structural design, finding widespread application in additive manufacturing (AM) technology, including the concrete construction sector.[12] However, traditional construction methods, which rely heavily on formwork, can complicate the optimization of size and shape. Utilizing 3D concrete printing (3DCP) technology in conjunction with topology optimization allows designers greater freedom to create structures that are both manufacturable and aesthetically pleasing, without the need for formwork.[13] This approach results in lighter, more cost-effective, and environmentally sustainable structures, requiring fewer materials and less manpower without compromising structural integrity.[12,13] In this research, we utilized both SCC and topology optimization methods to develop LWSCC using LiGra. We created 3D-printed formworks from polylactic acid (PLA) for three different concrete volume reductions from solid LWSCC samples. Finite element modelling (FEM) software, Abaqus, was employed to analyse the shapes of the optimized samples, as detailed in the methodology section. We conducted three-point loading tests to determine the ultimate load-carrying capacity of the LWSCC samples prior to failure.

Topology optimization in 3DCP offers advantages for self-supporting designs, particularly for overhanging elements.[13] One study found that a 3D-printed, post-tensioned concrete girder with an optimized shape could reduce material consumption by approximately 20% regarding mid-span deflection under self-weight and live load.[14] Oliveira et al. investigated CO_2 emissions from topologically optimized 3DCP structures across various designs.[15] They discovered that, under a fixed distributed compression load of 60 kN on three differently designed structures, properly optimized topology could significantly reduce mass and associated CO_2 emissions compared to monolithic structures.

However, a challenge remains in manufacturing pure tensile or pure compression structures from topology optimization.[16] Nonetheless, optimized structures combined with fibre-reinforced concrete (FRC) can serve as an alternative to traditional reinforcement.[17] This approach allows for the creation of complex geometries with improved environmental performance, indicating a promising direction for sustainable concrete construction.[18] Further research is necessary to fully realize the benefits of this technology, and this study may help define new research avenues.

14.2 MATERIALS FOR METHODS

14.2.1 Materials and Testing

In this study, LiGras (see Figure 14.1a) were utilized as a complete replacement for traditional aggregates like sand and gravel to produce LWSCC, which was cast into topologically optimized fused deposition modelling (FDM)-printed formwork made from PLA. Conventional cube and prism samples were also prepared to assess the properties of LWSCC. Table 14.1 details the material compositions used for the various LWSCC mixes. A total of nine different mixes were created, incorporating three distinct water-to-binder (W/B) ratios of 0.20, 0.25, and 0.30, along with three superplasticizer-to-binder (SP/B) ratios of 0.3%, 0.4%, and 0.5%. The specific compositions of each mix are outlined in Table 14.1. CEM I 42.5 Ordinary Portland Cement (OPC) and silica fume served as the primary binders for LWSCC. All LiGras were prepared in a saturated surface dry (SSD) condition by soaking them in water for a minimum of 24 hours and then drying their surfaces with a towel before incorporating them into the mix.

A combination of three different sizes of LiGra was used as an aggregate for LWSCC. The gradation curves for these LiGra are presented in Figure 14.1b. A Camsizer3D particle size analyser was employed to determine the particle sizes of the aggregates. As shown in Figure 14.1b, the sizes of the small, medium, and large LiGra ranged from 0.3 to 2.0 mm, 2.0 to 4.0 mm, and 4.0 to 8.0 mm, respectively, with their proportions in the mix being 30%, 30%, and 40%. These proportions were selected on the basis of the results of the particle packing factor (PF) test conducted on the LiGra.[19] The test revealed that for different ratios of LiGra, the PF ranged from 40 to 60. Consequently, a balanced mix of 30% small and 30% medium LiGra, combined with 40% large LiGra, was used for all mix designs, as detailed in Table 14.1. The water absorption capacities and bulk densities of small, medium, and large LiGra were approximately 26%, 21%, and 2%, and 341 kg/m^3, 248 kg/m^3, and 882 kg/m^3, respectively.

A 10L Hobart mixer was used to ensure uniform mixing of the compositions, with an average mixing time of approximately 8 minutes. Following mixing, slump flow test using slump cone (see Figure 14.2a) was conducted on the fresh LWSCC. To determine the mechanical characteristics of the hardened LWSCC samples, all samples were cast in moulds and removed after 48 to 60 hours. No external vibration was applied, as the concrete is self-compacting. It is important to note that the samples were lightweight, necessitating additional time for them to harden sufficiently before removal from the moulds. After demoulding, the samples were stored in the laboratory at an ambient temperature of approximately 21 ± 2°C and a relative humidity of 55 ± 5% until testing. For the compression test, 50 mm cube samples were prepared, and prism samples measuring 40 mm × 40 mm × 160 mm were prepared for three-point bending testing, as shown in Figure 14.2b. A universal compression testing machine was used for the compression tests of the cube samples, while an Instron tensile testing machine was employed for the three-point bending tests of the standard prism and optimized beam samples. The loading rates for the cube and prism samples were set at 0.9 MPa/min and 320 N/min, respectively. For each test, a

TABLE 14.1
Lwscc Mix Design for Various Compositions in Kg/M^3 and W/B and Sp/B Ratios

Compositions	20W03SP	20W04SP	20W05SP	25W03SP	25W04SP	25W05SP	30W03SP	30W04SP	30W05SP
Cement	500.2	500.2	500.2	500.2	500.2	500.2	500.2	500.2	500.2
Silica fume	39	39	39	39	39	39	39	39	39
Small LiGra	124.8	124.8	124.8	124.8	124.8	124.8	124.8	124.8	124.8
Medium LiGra	124.8	124.8	124.8	124.8	124.8	124.8	124.8	124.8	124.8
Large LiGra	280.2	280.2	280.2	280.2	280.2	280.2	280.2	280.2	280.2
Water	107.84	107.84	107.84	134.8	134.8	134.8	161.76	161.76	161.76
SP	1.6176	2.1568	2.696	1.6176	2.1568	2.696	1.6176	2.1568	2.696
W/B ratio	0.2	0.2	0.2	0.25	0.25	0.25	0.3	0.3	0.3
SP/B ratio	0.003	0.004	0.005	0.003	0.004	0.005	0.003	0.004	0.005

Note: SP refers to the superplasticizer. Mix 20W03SP indicates the LWSCC mix with a water-to-binder (W/B) ratio of 0.20 and a superplasticizer-to-binder (SP/B) ratio of 0.003. Other mixes are presented in the same sequence, featuring different W/B and SP/B ratios.

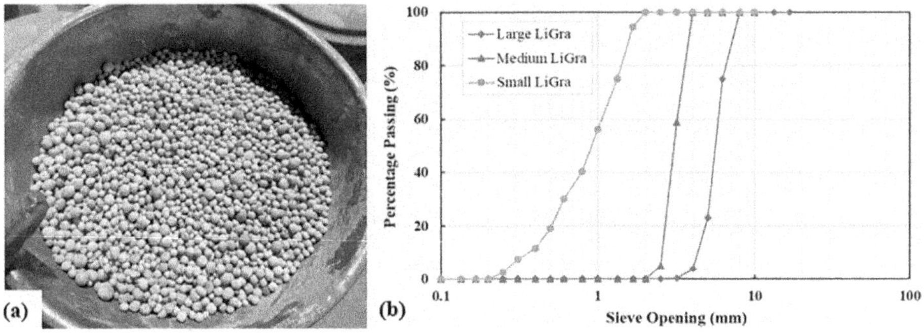

FIGURE 14.1 (a) LiGra used as aggregate and (b) grading curve for different sizes of LiGra used as aggregates.

minimum of three samples were evaluated for both compression and bending after 28 days of curing.

The flow behaviour of LWSCC was evaluated through rheology testing. A modular compact rheometer (MCR) along with Anton Paar (RheoCompass) software, as shown in Figure 14.2c, was used to determine the yield stress of the fresh concrete. Concrete was poured into a metal cup and placed onto the rheometer. The protocol from Saak et al. was adapted, setting the vane to rotate at a rate of 1 rad/s for 7 minutes.[20] The peak stress value recorded from the chart was taken as the yield stress, representing the amount of stress required to initiate flow in the matrix. Among the nine mixes listed in Table 14.1, the 25W03SP mix was selected for creating topology-optimized flexural beam samples, as illustrated in Figure 14.2d.

14.2.2 Testing Protocols for Topology Optimization of the Samples

To investigate the effect of topology optimization on the beam samples, the LWSCC mix of 25W03SP was selected for its excellent flowability and visually observed segregation resistance, while also maintaining acceptable compressive and flexural performance. The optimized formworks were produced using a FDM printer with PLA, as illustrated in Figure 14.3. Each beam measures 250 mm (L) × 80 mm (W) × 80 mm (H) with a thickness of 1 mm. Figure 14.3a to 14.3d panels depict the control beam and beams with 70%, 50%, and 30% concrete volume fractions, denoted as VF70, VF50, and VF30, respectively. The design of each optimized beam was created using ABAQUS FEM software, which strategically places materials in stress-intensive areas while removing them from less critical regions. The goal of the simulation was to minimize strain energy while adhering to volume fraction limits and geometric constraints. The optimized structure was then exported to SolidWorks for further refinement, including filleting sharp corners before printing the formworks, as shown in the figures. A minimum of three samples from each case was tested to determine the flexural strength of the optimized beams. For all sample types, the loading rate of the Instron machine was set to 1228.8 N/min, and their performance was evaluated for different concrete volume fractions.

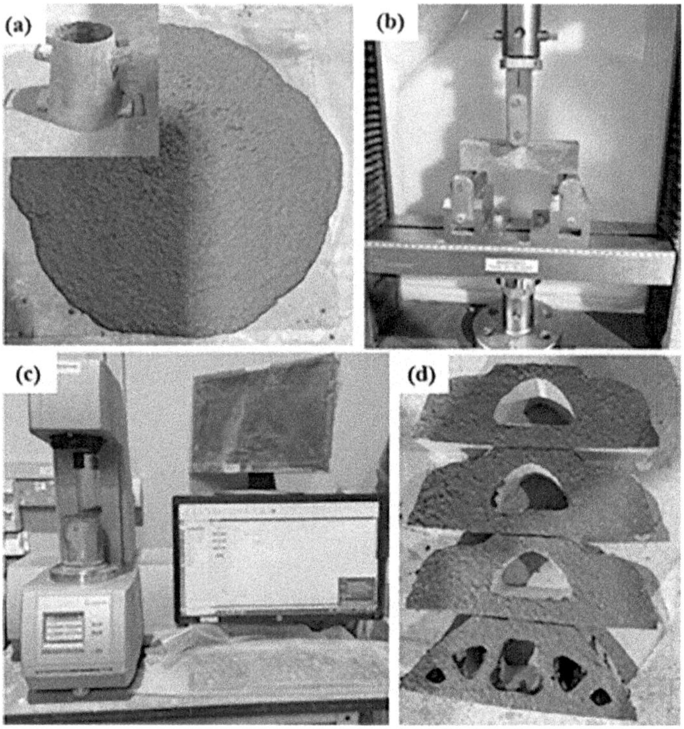

FIGURE 14.2 (a) Slump cone and flow of concrete, (b) three-point bending test, (c) rheology test using Anton Paar equipment, and (d) concrete cast in 3D-printed formworks.

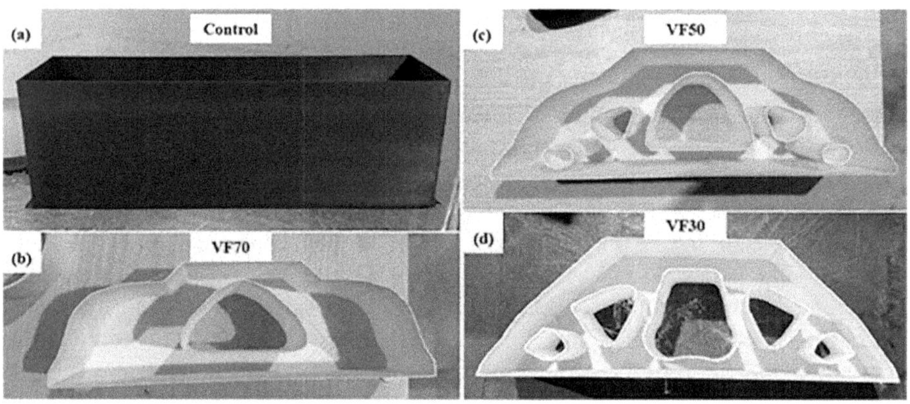

FIGURE 14.3 Formwork of (a) control beam and beam with concrete volume of (b) 70%, (c) 50%, and (d) 30% with an internal dimension of 250 mm × 80 mm × 80 mm.

14.2.3 Scanning Electron Microscopy (SEM)

An SEM was used to investigate the microstructure of hardened concrete samples, focusing specifically on bonding behaviour and porosity after 28 days of curing. The research utilized a JEOL JSM 5600 LV model SEM, capable of magnifications between 60× and 80×, to examine the bonding interface between the cement paste and LiGra. Each sample was cut into a small, thin piece measuring approximately 30 mm (length) × 30 mm (width) × 3 mm (thickness) using a diamond cutter. The cut samples were then soaked in ethanol for seven days to halt further hydration of the binder. Subsequently, the samples were oven-dried at 60°C overnight prior to the SEM analysis. To enhance image resolution and contrast during the SEM process, the samples were coated with a sputter before testing.

14.3 RESULTS AND DISCUSSION

14.3.1 Slump Flow of LWSCC

As shown in Figure 14.4, the flowing ability of fresh concrete improves with an increased content of superplasticizer (SP) in the mix, as expected. Additionally, higher water content also enhances flowability. A comparison of the mixes 20W03SP, 25W03SP, and 30W03SP demonstrates that the amount of water used significantly affects the flowability, particularly at lower W/B ratios. When comparing 20W03SP, the slump flow of the LWSCC for 20W04SP and 20W05SP increased by approximately 29% and 47%, respectively. In comparing with 25W03SP, the differences for

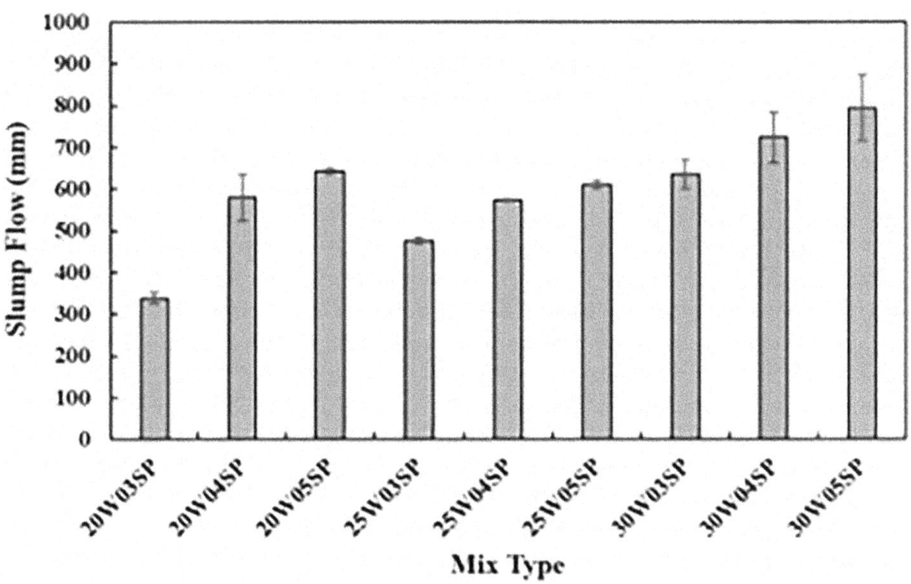

FIGURE 14.4 Slump flow behaviour of different LWSCC mixes.

25W04SP and 25W05SP were 8% and 27%, while for 30W04SP and 30W05SP, the changes were −8% and 18%, respectively, when comparing with 30W03SP.

It is important to note that additional slump flow tests were conducted on various combinations of W/B and SP/B ratios to identify optimal mix compositions that minimize severe segregation and bleeding effects. Visual observations of different mix compositions revealed that in very unworkable mixes, aggregates (i.e. LiGra) tend to clump together in the centre rather than flow. Thus, finding a balance between workability and segregation resistance was crucial.

14.3.2 Yield Stress of Fresh LWSCC

Figure 14.5 illustrates the development of yield stress over time in various LWSCC mixes. Due to its lightweight nature, LiGra tends to float above the cement matrix in certain mix designs, particularly those with higher W/B and SP/B ratios. This phenomenon occurs because the cement paste exhibits less viscous behaviour at elevated W/B and SP/B ratios, making it unable to retain the LiGra in place.

Rheology tests, as shown in Figure 14.5, confirmed a significant reduction in yield stress with increased SP content across all W/B ratios. The highest yield stress, exceeding 3,500 Pa, was observed for the 20W03SP mix. Although the initial yield stress was relatively high, it decreased rapidly within the 7-minute testing period, unlike other mixes with similar W/B ratios. The experiment also indicated that a minimum yield stress of 100 Pa is necessary for the cement paste to effectively prevent the upward movement of LiGra. Furthermore, variations in yield stress development among different mixes can be attributed to the separation of LiGra from the cement matrix. The cement matrix must exert sufficient downward restoring force, as upward hydrostatic forces often dominate the weight of LiGra. Additionally, the density and shape of LiGra are crucial factors in yield stress development within the cement matrix; generally, low-density, spherical LiGra has a greater tendency to float when mixed with a denser liquid like cement paste.

14.3.3 Compressive and Flexural Strength of LWSCC

The compressive strength of all mix designs was evaluated after 28 days of casting, as shown in Figure 14.6. The highest compressive strength recorded was 17.47 MPa for the 20W03SP mix, while the lowest was 13.37 MPa for the 30W03SP mix. Overall, compressive strength decreased with increasing water content. Specifically, the 25W03SP and 30W03SP mixes exhibited strengths that were approximately 13% and 31% lower, respectively, compared to the 20W03SP mix. Additionally, it is evident that compressive strength increases with higher SP content at elevated W/B ratios, such as 0.3. Furthermore, there is a general inverse relationship between workability and compressive strength among the mix designs, as illustrated in Figures 14.4 and 14.6.

In the flexural strength tests of standard prism samples, the highest strength recorded was 3.06 MPa for the 30W05SP mix design, while the lowest strength, 1.50 MPa, was observed in the 25W03SP mix, as shown in Figure 14.7a. Notably, the strength of the mix designs increased with higher SP content. This trend may be

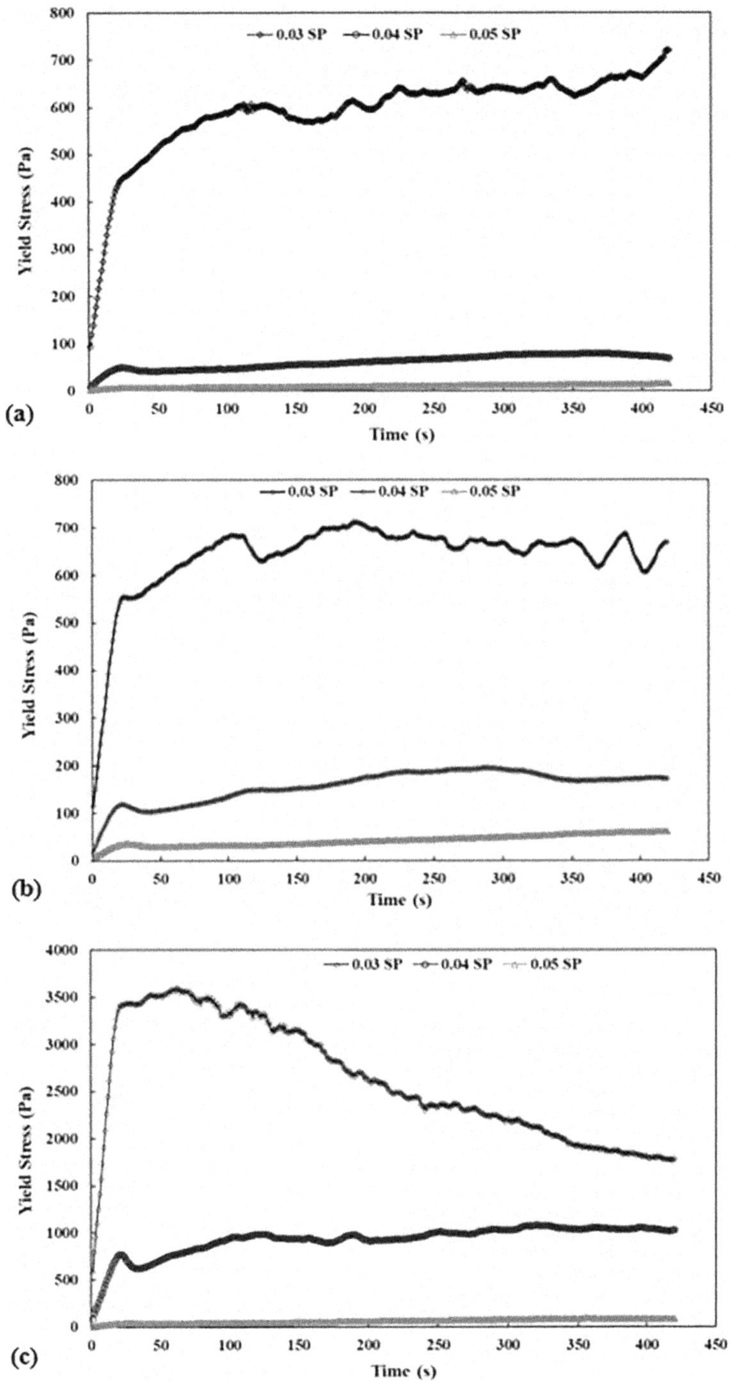

FIGURE 14.5 Plots of yield stress with time determined by Anton Paar rheometer for (a) W/B = 0.30, (b) W/B = 0.25, and (c) W/B = 0.20.

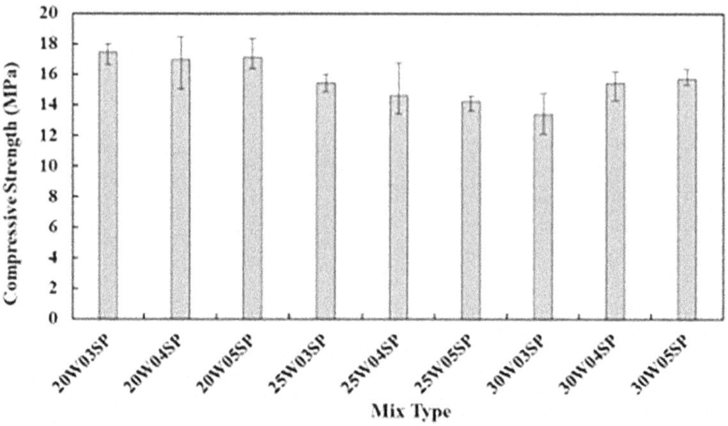

FIGURE 14.6 Compressive strength of different LWSCC mixes after 28 days.

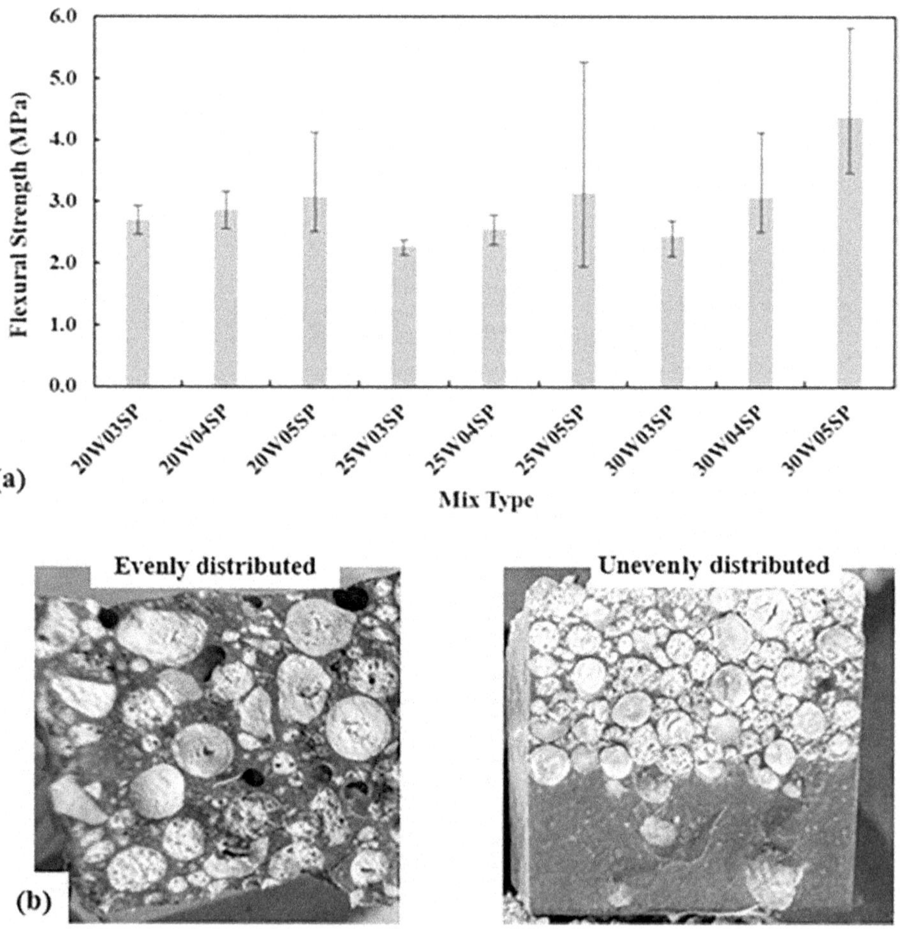

FIGURE 14.7 (a) Flexural strength of different LWSCC mixes after 28 days and (b) distribution of LiGra in low and high SP content.

attributed to the separation of LiGra from the cement paste; during the destruction of the prism samples, it was observed that the bottom half contained less LiGra than the top portion as illustrated in Figure 14.7b. Consequently, samples with higher SP content had more available cement paste than those with lower content. A greater volume of cement pastes and reduced LiGra content may better withstand flexural loads compared to mix designs where LiGra is evenly distributed throughout the prism. Additionally, LiGra exhibits weak interface bonding with cement paste due to its spherical shape, which can lead to lower tensile stress resistance in mix designs with evenly distributed LiGra. This weak interface is prone to failure, resulting in a crack development that can propagate through the prism.

14.3.4 Flexural Strength of Topology Optimized Samples

The 28-day flexural strength and normalized strength-to-weight ratio of topology-optimized samples are presented in Figure 14.8. As illustrated in Figure 14.8a, the flexural strength decreases as the concrete volume factor reduces, indicating a

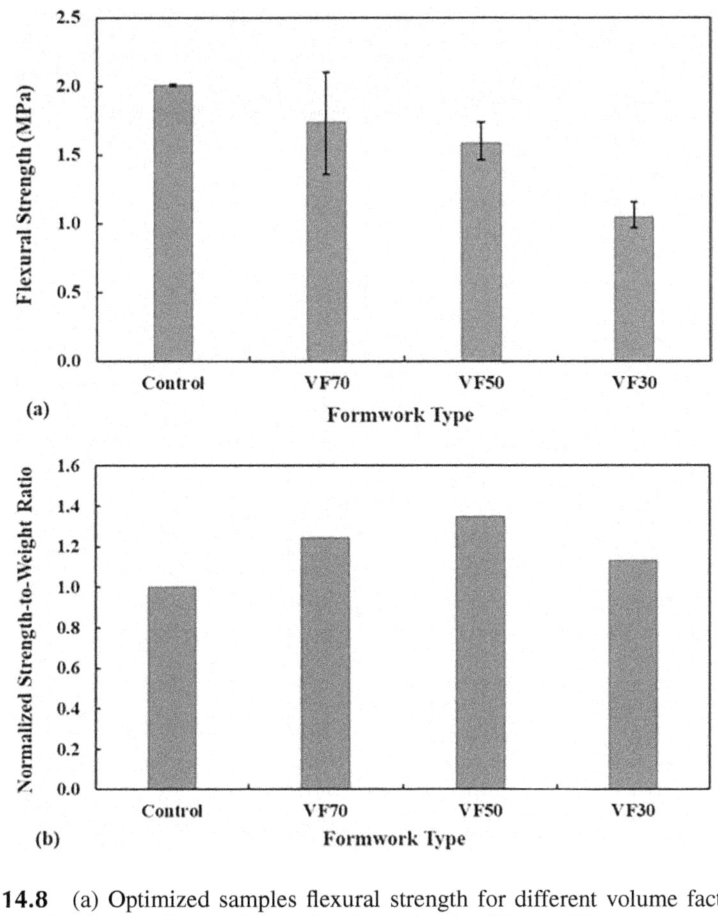

FIGURE 14.8 (a) Optimized samples flexural strength for different volume factors (VFs) and (b) normalized strength-to-weight ratio of samples for different VFs.

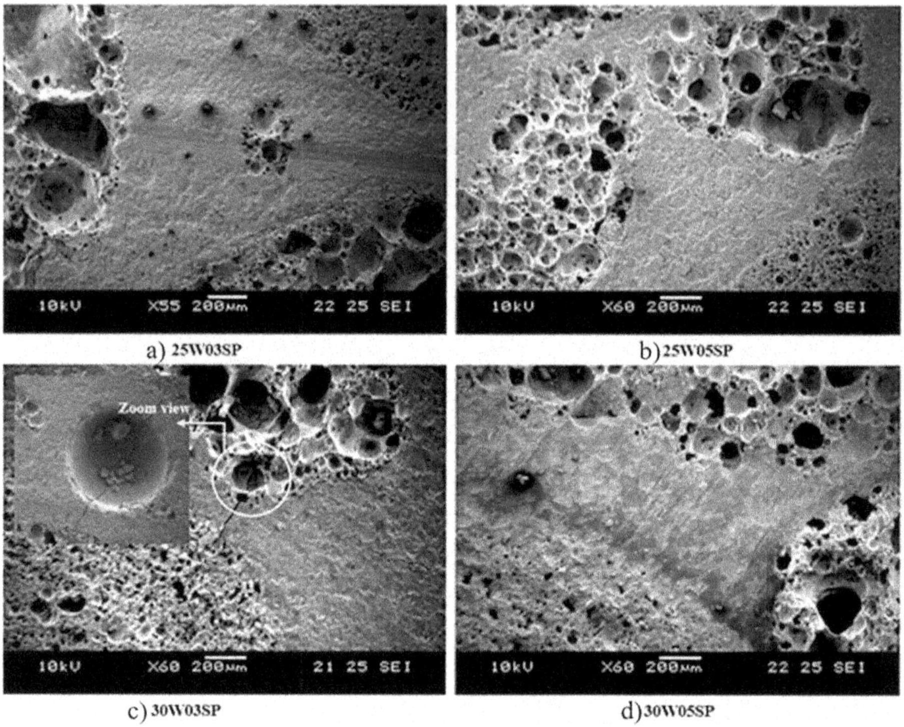

FIGURE 14.9 Scanning Electron Microscopic (SEM) images of different LWSCC samples: a) 25W03SP, b) 25W05SP, c) 30W03SP, and d) 30W05SP after 28 days.

reduction in the weight of the concrete in the samples. The control sample exhibited the highest strength at 2 MPa, while the lowest strength of 1.04 MPa was observed in the VF30 sample. In contrast, the comparison of the strength-to-weight ratio demonstrates that strategic placement of concrete can enhance the overall efficiency of the beam through topology optimization. Figure 14.8b shows that the normalized strength-to-weight ratios for VF70 and VF50 increased to 1.24 and 1.34, respectively, representing a 24% and 34% improvement compared to the control. However, this ratio decreases to 1.13 for the VF30 sample, which is still 13% higher than the control sample. Therefore, it can be concluded that an optimally shaped concrete structure not only reduces the weight of the structure but also lowers material costs.

14.3.5 Microstructures of LWSCC

SEM was used to examine the porosity and interfacial bonding between aggregates and cement paste for various mixes, as illustrated in Figure 14.9. It was observed that large-sized LiGra exhibited lower porosity compared to medium- and small-sized LiGra. This difference in porosity may explain the higher water absorption observed in medium- and small-sized LiGra, as discussed earlier. Furthermore, the lower strength of the LWSCC may be partially attributed to the increased porosity of these two sizes of LiGra, which are expected to be weaker than solid spheres. The samples demonstrated higher porosity at elevated W/B ratios and SP content. Visually, the LiGra aggregates showed "wet" spots, particularly at higher SP content levels (e.g. 0.05%) for all W/B ratios. This may be due to pre-soaked water in the LiGra before incorporation into the mix, which could aid in internally curing the concrete.

Additionally, it was noted that the large LiGra tended to break off at cracks rather than splitting along the crack direction, as shown in Figure 14.9 for the 30W03SP mix. This behaviour is attributed to the spherical shape of the LiGra, which results in weaker interfacial bonding with the cement matrix. In contrast, angular and sharp aggregates may exhibit better interfacial bonding due to their larger contact area with the matrix, leading to higher shear stress. This phenomenon was not observed with medium and small LiGra, likely because these aggregates are significantly weaker and fail before reaching the interfacial bonding shear stress.

14.4 CONCLUSIONS

This study effectively investigates the integration of LiGras and topologically optimized 3D-printed formwork to enhance sustainability in the construction industry. Through material characterization, mix optimization, and structural testing, the research shows that a fully recycled aggregate-based lightweight self-compacting concrete (LWSCC) can be designed to achieve favourable fresh and hardened properties. The workability and yield stress of LWSCC were found to be influenced by the water-to-binder (W/B) and superplasticizer-to-binder (SP/B) ratios in the mix. Additionally, it was determined that the yield stress of the cement paste should exceed 100 Pa to prevent the LiGra from floating. Utilizing topology optimization through finite element simulations allowed for a significant reduction in concrete usage while maintaining structural efficiency, achieving up to a 70% reduction in concrete volume without compromising the strength-to-weight ratio. These successful optimization techniques offer solutions for reducing both the dead load and material costs of concrete structures. However, further studies are needed to explore different concrete volume factors for various optimized shapes, which will assist designers in creating economically beneficial structural designs based on optimization.

REFERENCES

1. Watari T, Cao Z, Serrenho AC, Cullen J. Growing role of concrete in sand and climate crises. *Iscience*. 2023;19:26(5).
2. Abdulrasool AT, Mohammed SS, Kadhim NR, Hussain WA. Sustainable materials used as lightweight aggregate: (An overview). *IOP Conference Series: Earth and Environmental Science*. 2022;961(1):012027.
3. Fayomi GU, Mini SE, Fayomi OS, Ayoola AA. Perspectives on environmental CO_2 emission and energy factor in cement industry. *IOP Conference Series: Earth and Environmental Science*. 2019;331(1):012035.
4. Ellis LD, Badel AF, Chiang ML, Park RJ, Chiang YM. Concrete needs to lose its colossal carbon footprint. *Proc Natl Acad Sci USA*. 2020;117:12584–91.

5. Belaïd F. How does concrete and cement industry transformation contribute to mitigating climate change challenges? *Resources, Conservation & Recycling Advances.* 2022;15:200084.
6. Torres A, Simoni MU, Keiding JK, Müller DB, zu Ermgassen SO, Liu J, Jaeger JA, Winter M, Lambin EF. Sustainability of the global sand system in the Anthropocene. *One Earth.* 2021 May 21;4(5):639–50. https://doi.org/10.1016/j.oneear.2021. 04.011.
7. Marschke M, Rousseau JF. Sand ecologies, livelihoods and governance in Asia: A systematic scoping review. *Resources Policy.* 2022;77:102671.
8. Bendixen M, Best J, Hackney C, Iversen LL. Time is running out for sand. *Nature.* 2019;571(7763):29–31. https://doi.org/10.1038/d41586-019-02042-4.
9. *Lightweight Green Aggregates.* https://www.joegreenpanel.com/en/ligra, accessed on 23 June 2025.
10. Sonebi M, Grünewald S, Walraven J. Filling ability and passing ability of self-consolidating concrete. *ACI Materials Journal.* 2007;104(2):162.
11. Khayat KH, Hu C, Monty H. Stability of self-consolidating concrete, advantages, and potential applications. *PRO 7: 1st International RILEM Symposium on Self-Compacting Concrete.* 1999:7:143.
12. Liu J, Gaynor AT, Chen S, Kang Z, Suresh K, Takezawa A, Li L, Kato J, Tang J, Wang CC, Cheng L. Current and future trends in topology optimization for additive manufacturing. *Structural and Multidisciplinary Optimization.* 2018;57(6):2457–83.
13. Bi M, Tran P, Xia L, Ma G, Xie YM. Topology optimization for 3D concrete printing with various manufacturing constraints. *Additive Manufacturing.* 2022;57:102982.
14. Vantyghem G, De Corte W, Shakour E, Amir O. 3D printing of a post-tensioned concrete girder designed by topology optimization. *Automation in Construction.* 2020; 112:103084.
15. Oliveira FH, Picelli R, Silva EC, Barari A, Romano RC, Pileggi RG, Tsuzuki MS. Topology optimization in 3D concrete printing to reduce greenhouse gas emissions. *IFAC-PapersOnLine.* 2024;58(19):634–9.
16. van de Werfhorst-Bouw I, Hofmeyer H, Boonstra S, Wolfs RJ. Stress-constrained topology optimization of concrete structures: A preliminary study for combining topology optimization and 3D printing. *Heron.* 2022;66(2/3):199–237.
17. Silva, GCAD. *Topological optimization of structures produced through 3D printing of fiber reinforced cementitious materials.* Doctoral dissertation, University of Minho, Portugal, 2021.
18. Lin Y, Bayramvand A, Aghaei Meibodi M. Towards lightweight structure: Coupling topology optimization with non-planar 3D concrete printing. *The International Conference on Computational Design and Robotic Fabrication*, Singapore, 6 July, 2024: pp. 3–12.
19. Hoornahad H, Koenders EA, van Breugel K. Towards the development of self-compacting no-slump concrete mixtures. *Építöanyag (Online).* 2015;1(4):135. http://dx.doi.org/10.14382/epitoanyag-jsbcm.2015.22.
20. Saak AW, Jennings HM, Shah SP. New methodology for designing self-compacting concrete. *Materials Journal.* 2001;98(6):429–39.

15 Behaviour of Standard and High-Strength Concrete Under Cyclic Heating

Vidya Balagam and K. Srinivasa Rao

15.1 INTRODUCTION

Structural concrete is reactive to environmental conditions, especially under high temperatures, and undergoes significant changes during service life. Exposure to elevated temperatures such as in accidental fires causes deterioration primarily due to microstructural transformations in the cement matrix. The breakdown of hydration products increases porosity, weakening concrete mechanical strength. Therefore, assessing concrete condition after fire exposure is crucial for effective repair planning.

Concrete deterioration is primarily associated with temperature-induced physicochemical changes in the cement matrix, which progressively alter its microstructure and mechanical properties.[1, 2] As temperature increases, the decomposition of hydrated cement compounds and calcite leads to increased porosity within the cementitious matrix and the interfacial transition zone (ITZ) between the aggregate and binder. This deterioration weakens the aggregate–paste bond and promotes the initiation and propagation of cracks. Physicochemical processes start with evaporation of absorbed water near 65°C, slightly increasing strength. However, loss of capillary, interlayer, and chemically bound water causes shrinkage and cracking.[3–5] Calcium hydroxide releases chemically bound water around 500°C, leading to notable strength loss, with full dehydration near 600°C. Additional factors include internal vapour pressure at 100°C, thermal incompatibility effects at 200°C, and C-S-H gel decomposition around 700°C.[6–22]

This study focuses on examining how standard concrete (SC) and high-strength concrete (HSC) respond to thermal cycling when exposed to elevated temperatures. Specimens aged 28 days are subjected to temperatures ranging from 100 to 400°C for 8 hours, followed by air cooling for 16 hours – defining one complete thermal cycle. This thermal regime simulates conditions encountered in structural elements such as chimneys, nuclear reactor vessels, and components near industrial furnaces, which typically operate for up to 8 hours daily.

DOI: 10.1201/9781003664284-15

15.2 INVESTIGATIONS ON CONCRETE

15.2.1 Concrete Constituents

A 53-grade Ordinary Portland Cement (OPC), complying with IS 12269 (1989),[23] was utilized in the study. Its chemical and physical characteristics comply with IS 4031 (1988).[24] The mineral admixture used was micro silica, with a specific gravity of 2.2 and a specific surface area of 19,000 m^2/kg, as provided by the manufacturer. The fine aggregate (sand) conformed to grading Zone II as per IS 383 (1970),[25] with a fineness modulus of 2.85 and a specific gravity of 2.66. The coarse aggregate consists of hard blue granite chips. Two size fractions were used: two-thirds of the total coarse aggregate consisted of particles passing through a 20 mm sieve and retained on a 10 mm sieve, while the remaining one-third consisted of particles passing through a 10 mm sieve and retained on a 4.75mm sieve. In the case of M 30 grade concrete, both coarse aggregate size fractions were included, while for M 90 grade concrete, only the fraction smaller than 10 mm was employed. The coarse aggregate had a fineness modulus of 6.30 and a specific gravity of 2.78.

Mixing was carried out using locally sourced potable water with a pH of 7, in accordance with IS 3025(1986).[26] A new-generation chemical admixture based on second-generation polycarboxylic ether polymer was also employed, chosen for its ability to enhance early strength. The admixture possessed a specific gravity of 1.09 and contained a minimum of 34% solids by weight.

15.2.2 Methodology for Specimen Preparation and Testing

Several trial mixes were conducted to establish the optimal mix proportions for M 30 and M 90 concrete grades. The M 30 mix was designed as per IS:10262 (2009),[27] while the M 90 mix followed the guidelines of ACI 211–4R (2008).[28] Table 15.1 presents the detailed proportions of constituent materials used in the M 30 and M 90 concrete mixes.

For mechanical testing, concrete was poured onto a saturated casting surface and moulded into 100 mm cubes, 150 mm × 300 mm cylinders, and 100 mm ×100 mm × 500 mm prisms to determine compressive, split tensile, and flexural strengths,

TABLE 15.1
Constituents of M 30 and M 90 Concrete Mix (Per M³)

Material	M 30	M 90
Cement (kg/m^3)	370	594
Silica fume (kg/m^3)	0	66
Fine aggregate (kg/m^3)	740	650
Coarse aggregate (kg/m^3)	1,214	1,105
Water (L/m^3)	165	145
Superplasticizer (L/m^3)	0	5.28

respectively. After 24 hours in moist air, specimens are demoulded, marked, and cured in water for 28 days. Specimens are then oven-heated following the ISO 834 (1975)[29] time–temperature curve until thermal equilibrium is reached and allowed to cool. Compressive and tensile tests are conducted as per IS: 516 (1959)[30] and IS: 5816 (1970)[31] to assess the residual mechanical properties of concrete after exposure to fire.

15.2.3 TGA Testing Methodology for Concrete Specimens

After curing process, concrete cubes are air-dried at a temperature of 27 ± 2°C, then subjected to elevated temperatures up to 400°C in 100°C intervals using a hot-air oven, with each temperature level sustained for 8 hours. Once cooled to ambient temperature, the cubes are tested for compressive strength. Thermogravimetric analysis (TGA) is conducted on mortar samples extracted from both the exposed surface and the core, in accordance with ASTM 1131 (2008),[32] using a thermogravimetric analyser. This research investigates the impact of thermal exposure on standard and HSCs, focusing on their thermal stability and reporting the physicochemical changes before and after heat exposure.

15.3 OBSERVATIONS AND EVALUATION

15.3.1 TGA Evaluation of M 30 and M 90 Concretes

During cement hydration, the dominant phases that develop are C-S-H (calcium silicate hydrate) and C-H (calcium hydroxide), also known as portlandite. Various studies[15, 33–42] have reported the physico-chemical transformations which occur in cement paste (mortar) across different temperature ranges. These transformations provide insight into the thermal behaviour and decomposition of hydration products in both M 30 and M 90 grade concretes under elevated temperatures, as observed through TGA.

Thermal analysis results for M 30 and M 90 concrete specimens, taken from the surface and core, are illustrated in Figures 15.1 to 15.10. A gradual increase in mass loss is observed for both grades. Up to 400°C, the loss is mainly due to the evaporation of capillary and interlayer water. Between 400 and 450°C, a notable mass reduction occurs due to the release of chemically bound water from portlandite (C-H). A slight mass loss between 600 and 700°C is attributed to the decomposition of calcium carbonate (calcite). Figures 15.11 to 15.14 indicate that mortar specimens extracted from the surface experience higher mass loss than those extracted from the core when exposed to elevated temperatures. This variation is likely attributed to a slower dehydration rate within the core, resulting from the temperature gradient between the surface and inner regions. The thermograms in Figures 15.11 to 15.14, corresponding to surface and core samples of M 30 and M 90 concrete, show that specimens preheated to 400°C exhibit less mass loss of hydrated products than those preheated to 100, 200, and 300°C. This reduction may result from partial or complete physico-chemical transformations occurring during prior heating.

M 30 and M 90 concretes exhibit progressive mass loss with increasing temperature, mainly due to water evaporation, portlandite, and calcite decomposition.

Surface samples lose more mass than core samples because of slower internal dehydration. Preheating to 400°C reduces further mass loss, indicating partial or complete physico-chemical transformations, highlighting the effect of temperature and sample location on thermal stability.

From Figures 15.11 to 15.14, HC@100 indicates concrete exposed to a heat level of 100°C, HC@200 indicates concrete exposed to a heat level of 200°C, HC@300 indicates concrete exposed to a heat level of 300°C, and HC@400 indicates concrete exposed to a heat level of 400°C.

15.3.2 Investigation of Mechanical Behaviour of Concrete

Repeated heating leads to a gradual decline in concrete mechanical properties, resulting in the breakdown of its internal structure at high temperatures. When concrete is exposed to heat, the temperature increase progresses slowly into the deeper layers, creating significant temperature gradients between the surface and core. These gradients induce additional damage to the concrete element. The reduction in concrete strength after heating mainly results from the evaporation of water within the material. This loss of strength is also influenced by physical and chemical changes occurring in the cement matrix, along with the movement of gases and liquids through the concrete pore structure. Prolonged exposure to high temperatures intensifies the dehydration of moisture, which increases the porosity of the concrete, thereby weakening its structural integrity.

This study involved testing M 30 and M 90 concrete specimens for their compressive, split tensile, and flexural strengths, as depicted in Figures 15.15 and 15.16. These results clearly demonstrate the impact of thermal exposure on the mechanical performance of both concrete grades, emphasizing the importance of understanding thermal effects for fire safety and structural durability assessments.

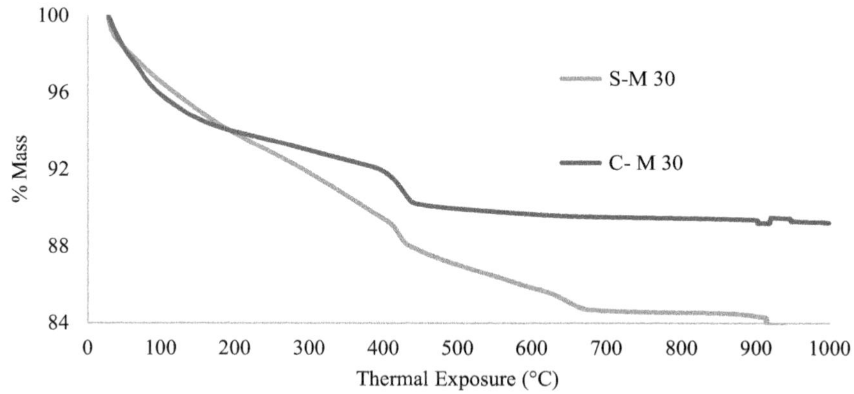

FIGURE 15.1 Thermal profile of surface and core mortar samples extracted from M 30 control specimen.

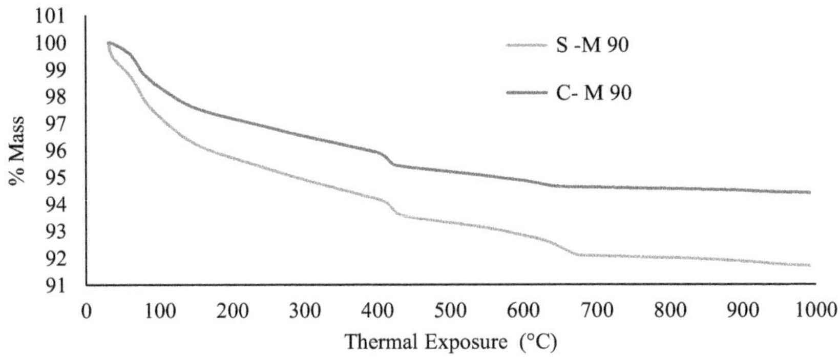

FIGURE 15.2 Thermal profile of surface and core mortar samples extracted from M 90 control specimen.

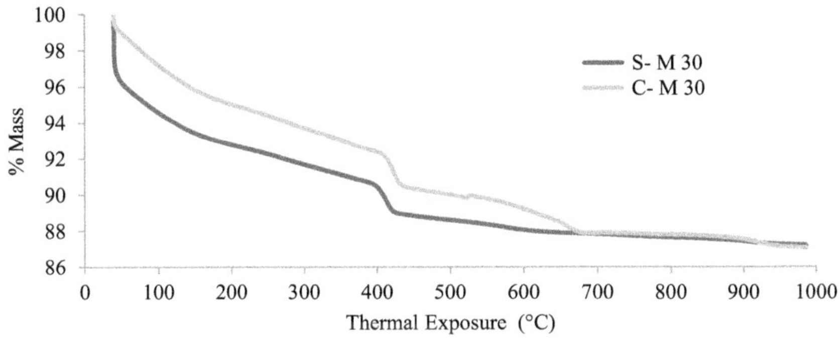

FIGURE 15.3 Surface and core thermal profile of M 30 mortar exposed to 100°C.

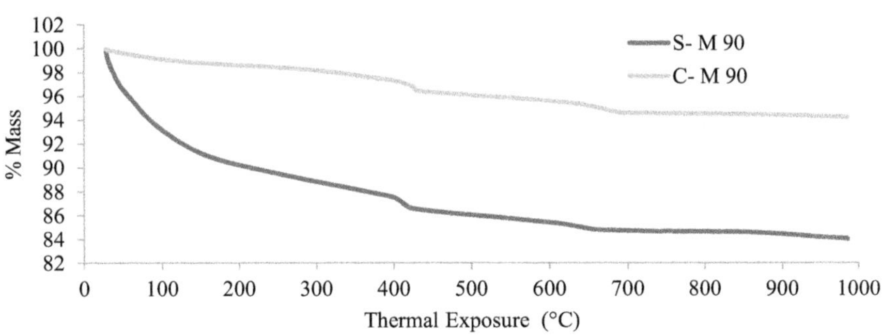

FIGURE 15.4 Surface and core thermal profile of M 90 mortar exposed to 100°C.

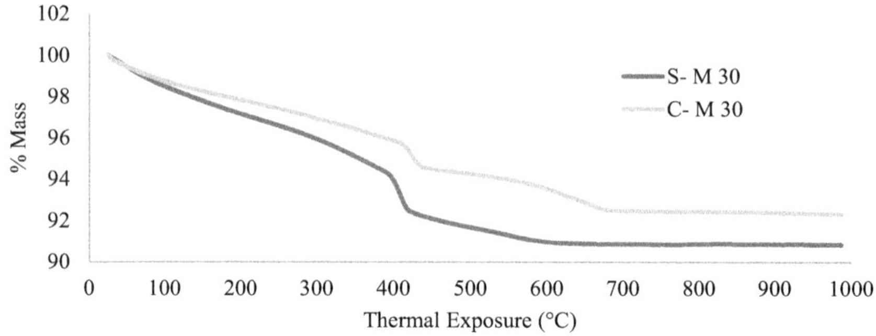

FIGURE 15.5 Surface and core thermal profile of M 30 mortar exposed to 200°C.

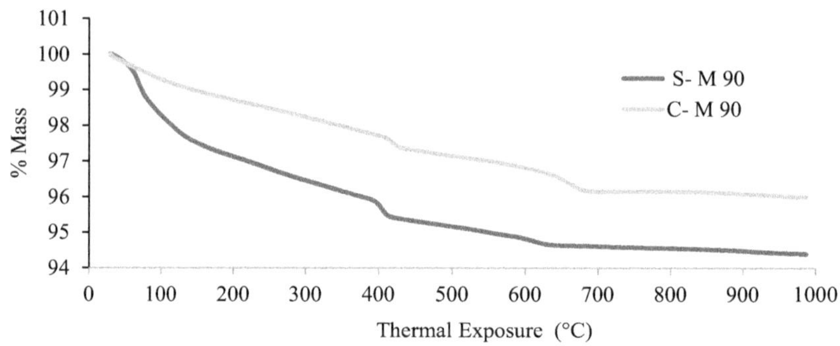

FIGURE 15.6 Surface and core thermal profile of M 90 mortar exposed to 200°C.

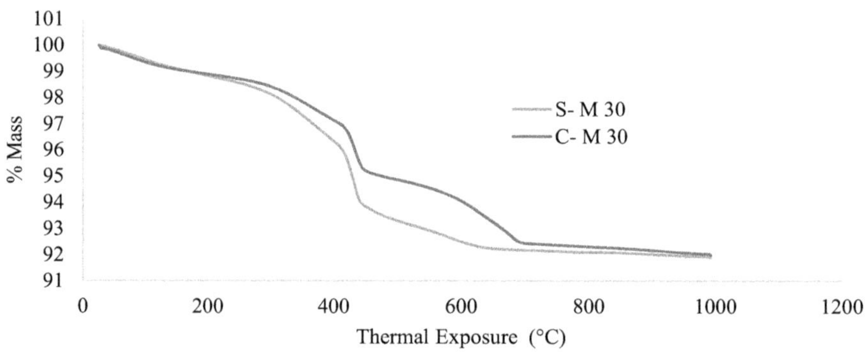

FIGURE 15.7 Surface and core thermal profile of M 30 mortar exposed to 300°C.

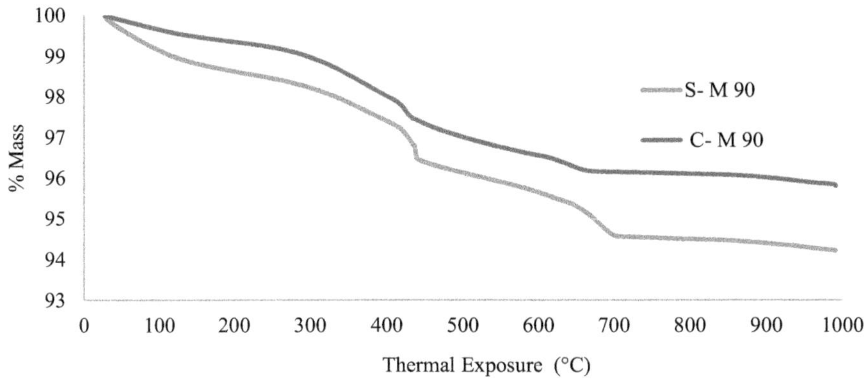

FIGURE 15.8 Surface and core thermal profile of M 90 mortar exposed to 300°C.

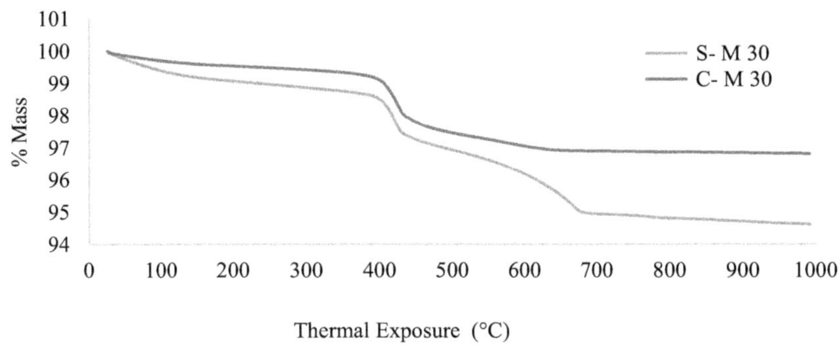

FIGURE 15.9 Surface and core thermal profile of M 30 mortar exposed to 400°C.

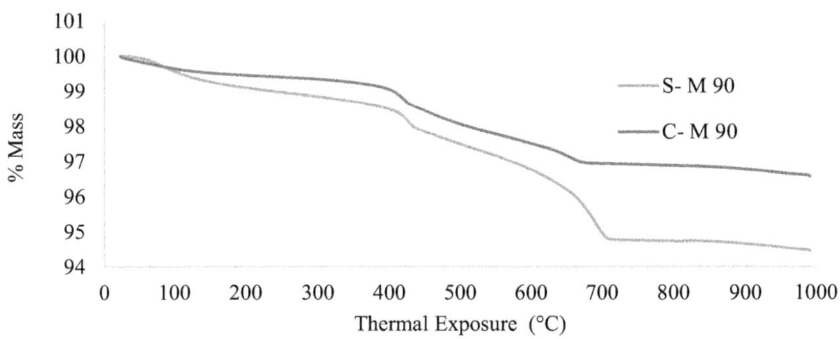

FIGURE 15.10 Surface and core thermal profile of M 90 mortar exposed to 400°C.

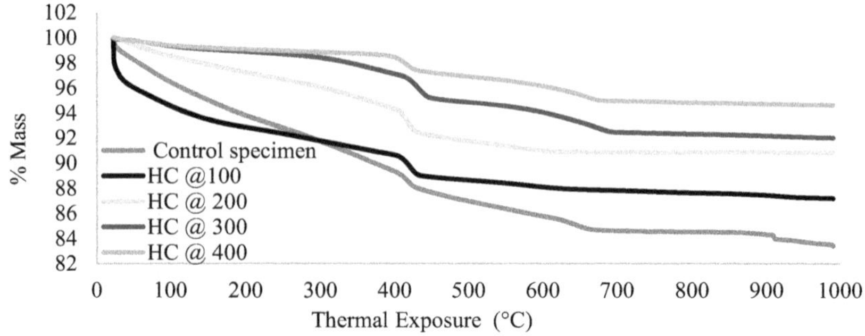

FIGURE 15.11 Surface thermal response of M 30 mortar captured under varying temperature exposures.

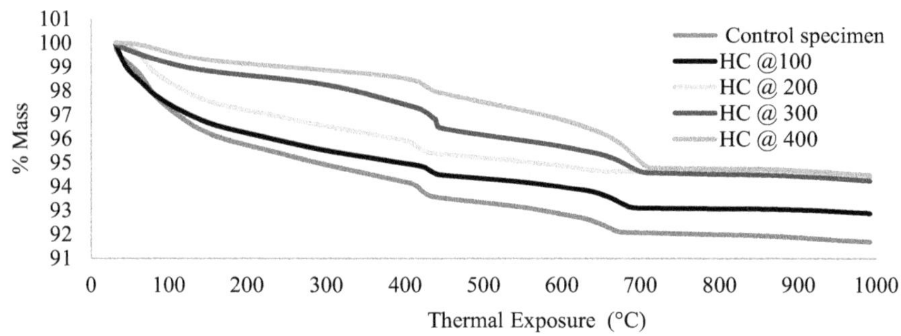

FIGURE 15.12 Surface thermal response of M 90 mortar captured under varying temperature exposures.

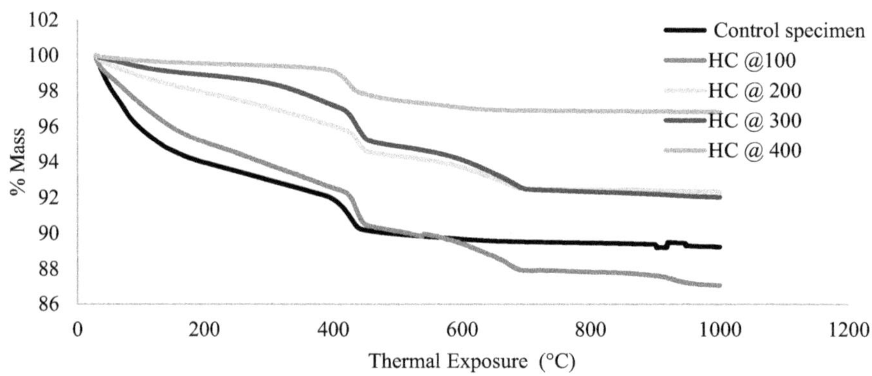

FIGURE 15.13 Core-level thermal analysis of M 30 mortar exposed to different temperature conditions.

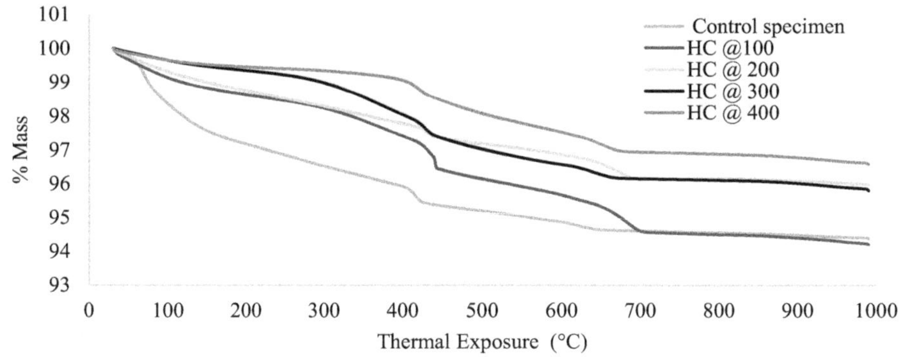

FIGURE 15.14 Core-level thermal analysis of M 90 mortar exposed to different temperature conditions.

FIGURE 15.15 Specimens of M 30 grade of concrete under compression, split tension, and flexure tests.

15.3.2.1 Impact of Thermal Exposure on Compressive Strength

The residual compressive strength percentage of M 30 concrete changes with temperature depending on the number of heating cycles applied as shown in Figure 15.17. It is observed that residual compressive strength gradually decreases as temperature rises across all heating cycles. However, an improvement in compressive strength was noted at 100°C after the first and fifth heating cycles, likely due to accelerated cement hydration during heating. After 50 thermal cycles, the reduction in compressive strength was only 15.56% and 17.78% at 100°C and 200°C, respectively, indicating that M 30 maintained nearly 80% of its initial strength, even after being subjected to prolonged thermal conditions.

A noticeable reduction in compressive strength occurred during the initial heating cycle at elevated temperatures of 200, 300, and 400°C. Despite repeated exposure, M 30 retained approximately 55% of its original strength after 50 heating cycles at 400°C.

Figure 15.18 illustrates the variation in residual compressive strength of M 90 grade concrete subjected to multiple heating cycles. HSC showed a notable reduction

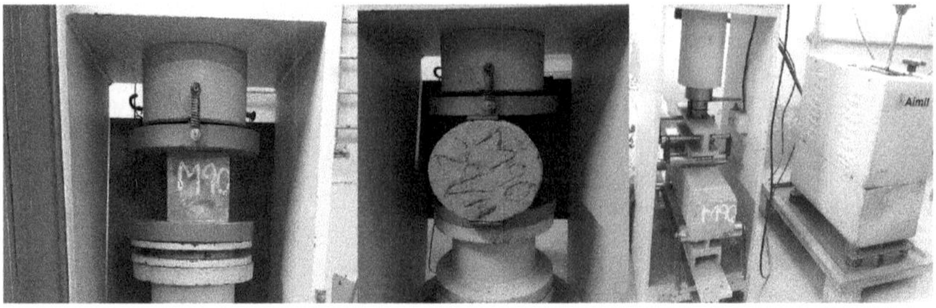

FIGURE 15.16 Specimens of M 90 grade of concrete under compression, split tension, and flexure tests.

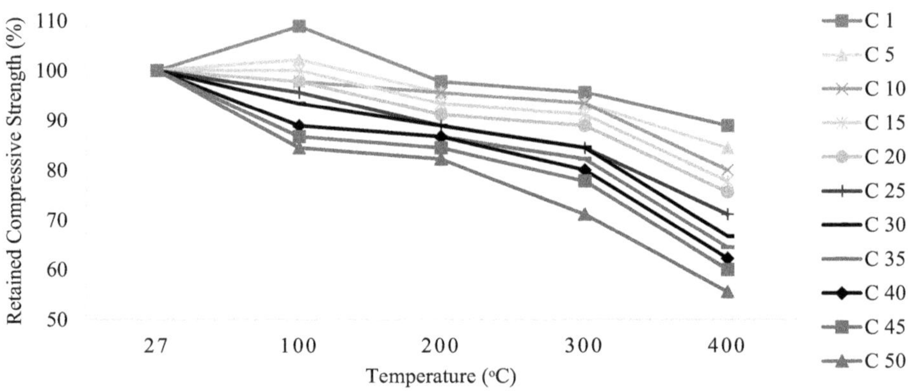

FIGURE 15.17 Temperature-wise variation in retained compressive strength (%) for M 30 concrete under multiple heating cycles.

in compressive strength after the initial heating cycle at 200, 300, and 400°C, with the exception of 100°C. At 100°C, the samples maintained 60% of their initial strength after undergoing 50 heating cycles. When exposed to 400°C, the HSC preserved 50% of its original strength for up to 35 cycles, but after this threshold, the remaining strength dropped further to 44.12%.

Although HSC possesses superior compressive strength compared to SC, it can deteriorate more quickly. This is mainly due to its dense and brittle structure, which restricts the release of internal stresses and increases its vulnerability to cracking. The low water–cement ratio in HSC promotes self-desiccation and the development of fine internal cracks, while the significant heat produced during hydration – especially when additives like silica fume are used – creates thermal stresses that further compromise the concrete integrity.

15.3.2.2 Response of Tensile Strength to Multiple Heating Cycles

Concrete resistance to tension is limited, being roughly one-tenth of its compressive strength in SC and comparatively lower in HSC. Therefore, tensile strength is

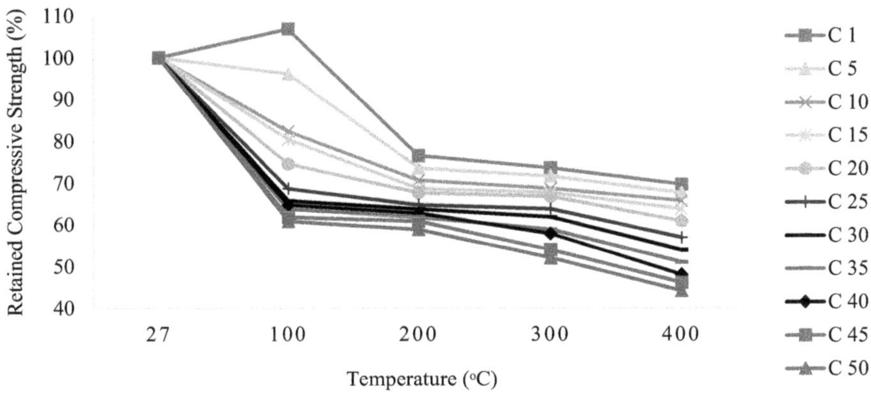

FIGURE 15.18 Temperature-wise variation in retained compressive strength (%) for M 90 concrete under multiple heating cycles.

frequently overlooked in assessments conducted at both ambient and elevated temperatures. However, it is a crucial property as cracking usually starts due to tensile stresses, and structural damage in tension results from micro-crack growth. Under fire conditions, tensile strength becomes especially important, as fire-induced spalling can cause significant damage to concrete members. This makes understanding tensile behaviour essential for assessing the durability and safety of concrete structures exposed to high temperatures.

15.3.2.2.1 Split Tensile Strength Behaviour Under Thermal Conditions

The variation in retained split tensile strength of M 30 grade concrete as temperature increases over multiple heating cycles is shown in Figure 15.19. The results indicate a gradual decrease in residual strength with increases in temperature and cycle count. At 100°C, an initial increase in split tensile strength is noted after the first heating cycle, which is attributed to the accelerated hydration of cement due to heating. However, at higher temperatures – 200, 300, and 400°C – a noticeable reduction in tensile strength occurs right after the initial heating cycle. Standard concrete (SC) maintains 50% of its initial split tensile strength up to 35 heating cycles at 100°C, but this retention declines sharply at higher temperatures, with 50% strength maintained only up to 25, 20, and 15 cycles at 200, 300, and 400°C, respectively. After 50 cycles at 400°C, SC retains just 30.44% of its initial split tensile strength, indicating considerable degradation due to repeated heating.

Figure 15.20 shows the residual split tensile strength behaviour of M 90 grade, or HSC, under similar heating cycle. HSC exhibits substantial strength loss after the first heating cycle at elevated temperatures of 200, 300, and 400°C, except at 100°C where the strength initially increases. After 50 thermal cycles at 100°C, HSC retains 37.18% of its initial strength and is able to maintain 50% strength up to 30 cycles at this temperature. These findings highlight the vulnerability of tensile strength in both standard and HSCs to elevated temperatures and repeated heating, emphasizing the need to consider thermal effects in structural design and assessment.

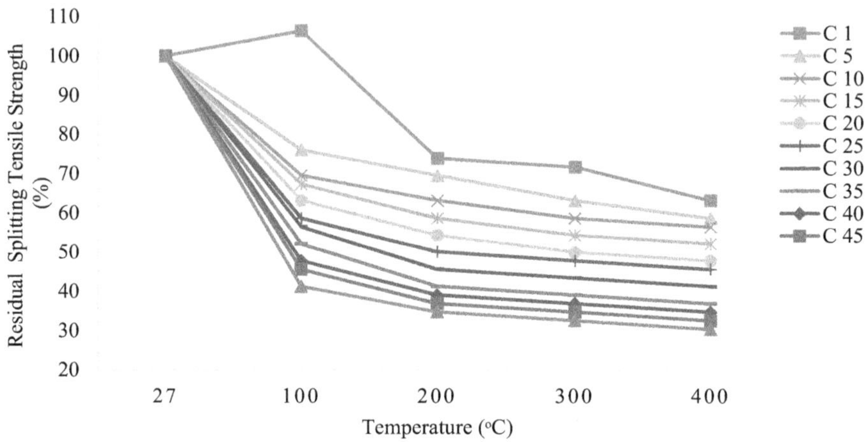

FIGURE 15.19 Temperature-dependent variation in residual split tensile strength (%) of M 30-grade concrete exposed to multiple heating cycles.

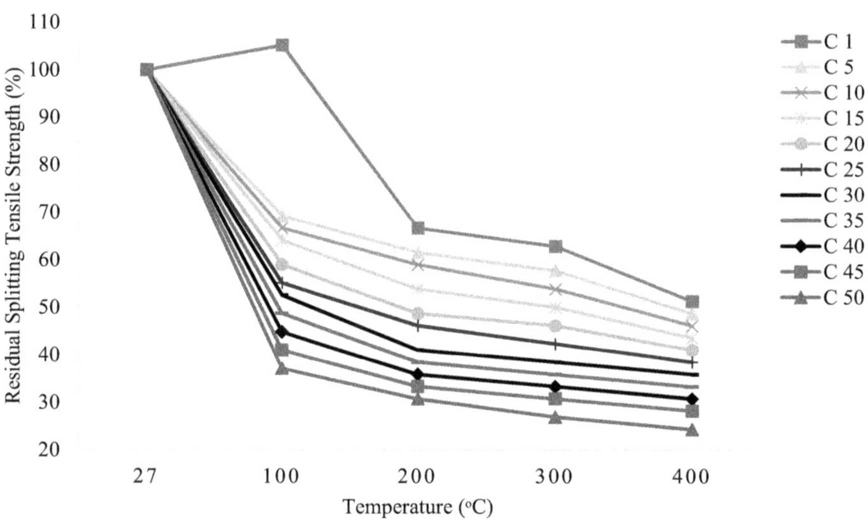

FIGURE 15.20 Temperature-dependent variation in residual split tensile strength (%) of M 90-grade concrete exposed to multiple heating cycles.

Beyond 50 heating cycles, the strength of HSC dropped to 37.18%. HSC was able to retain half of its initial strength at 200 and 300°C for a maximum of 15 cycles. When heated to 400°C, while HSC initially maintained half of its initial split tensile strength after one heating cycle, repeated cycles resulted in a decline to 24.36%.

15.3.2.2.2 Response of Flexural Strength to Thermal Exposure

The variation in percentage residual flexural strength of M 30 grade concrete with increasing temperature and number of heating cycles shown in Figure 15.21.

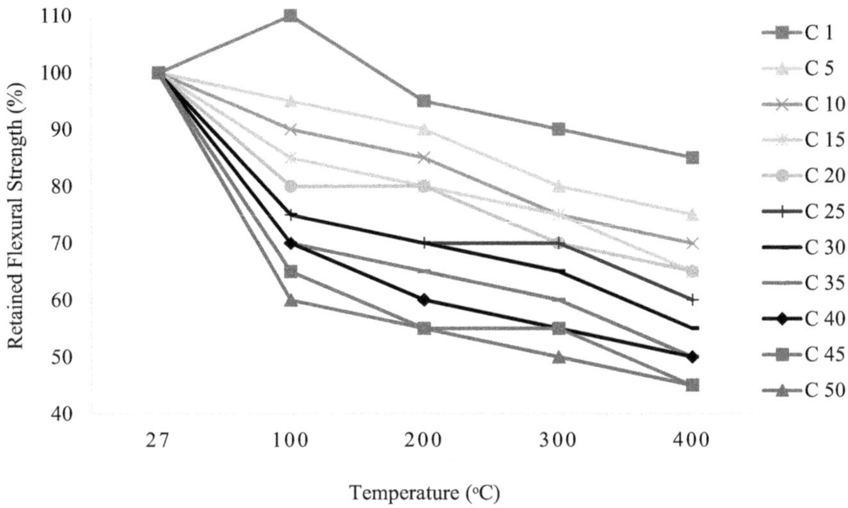

FIGURE 15.21 Temperature-wise variation in retained flexural strength (%) of M 30 concrete for multiple heating cycles.

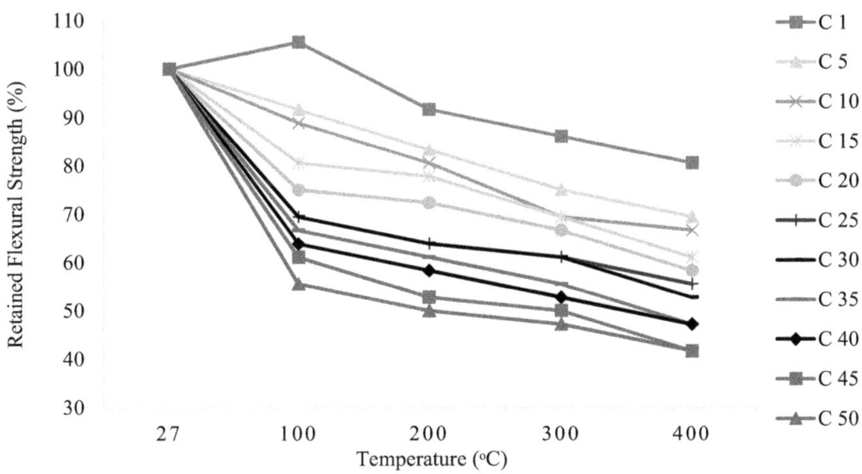

FIGURE 15.22 Temperature-wise variation in retained flexural strength (%) of M 90 concrete for multiple heating cycles.

The data indicate that residual flexural strength decreases as temperature and heating cycles increase. An increase in flexural strength is seen after the first cycle at 100°C, attributed to the rapid hydration process of cement under heat. After 50 heating cycles, the flexural strength reduction was 40% at 100°C and 45% at 200°C, that is, SC retained about 50% of its initial strength up to 200°C. However, at higher temperatures of 300 and 400°C, significant losses in flexural strength were noted. SC preserved half of its initial strength despite undergoing 40 heating cycles at 400°C.

Under similar thermal conditions, the variation in residual flexural strength for M 90-grade concrete is depicted in Figure 15.22. HSC shows a gradual decrease in flexural strength with increasing temperature and thermal cycles. A substantial strength loss occurs at elevated temperatures of 200, 300, and 400°C, except at 100°C where strength initially increases. After 50 thermal cycles at 100°C, HSC retains 55% of its original strength. At 400°C, HSC maintains 50% strength up to 30 thermal cycles, but beyond 35 cycles, residual strength drops further to 41.67%. These results highlight the impact of thermal exposure and cycling on the flexural performance of both standard and HSCs.

15.4 OVERALL OUTCOMES

The study has led to several key conclusions concerning the thermal response and mechanical characteristics of M 30 and M 90 concrete grades when exposed to high temperatures and repeated thermal cycling.

1. TGA findings confirm that both M 30 and M 90 concretes remain stable under thermal exposure up to 400°C, as decomposition of hydration products is observed beyond this point. Additionally, mortar samples taken from the surface of specimens exhibit greater mass loss compared to those taken from the core, likely due to more direct exposure to heat and faster dehydration.
2. Residual compressive and tensile strengths for both standard and HSCs increase after the first heating cycle at 100°C. This strength gain may be attributed to the loss of free (non-chemically bound) water, which accelerates the hydration process of unhydrated cement particles within the concrete core. Specifically, after one heating cycle at 100°C, M 30 showed strength increases of 8.9, 6.5, and 10% in compressive, splitting tensile, and flexural strength, respectively. M 90 concrete exhibited increase of 6.9, 5.2, and 5.6% for the same properties.
3. Thermal cycling at elevated temperatures (200, 300, and 400°C) has a more detrimental effect on HSC compared to standard concrete. A significant reduction in compressive strength is observed during the initial heating cycle, with standard concrete showing strength loss of 2.2%, 4.4%, and 11.1% at temperatures of 200, 300, and 400°C, respectively, while HSC suffers much higher losses of 23.5, 26.5, and 30.4% at the same temperatures. Similarly, split tensile strength losses after the first cycle are 26.1, 28.3, and 36.8% for standard concrete, compared to 33.3, 37.2, and 48.7% for HSC. Flexural strength reductions follow the same trend, with standard concrete losing 5, 10, and 15%, and HSC experiencing 8.3, 13.9, and 19.4% losses.
4. When subjected to 50 cycles at 400°C, HSC undergoes severe strength reductions – 65% in compressive, 75% in split tensile, and 58% in flexural strength – exceeding the corresponding losses in standard concrete, which are 45, 69, and 55%.

5. The rate of deterioration varies among different mechanical properties. At 400°C, standard concrete shows a 50% reduction in compressive strength after 50 heating cycles, while the same level of loss is reached after 15 cycles in split tensile strength and 30 cycles in flexural strength. In contrast, HSC requires only 35 cycles to lose 50% of compressive strength, just one cycle for split tensile strength, and 30 cycles for flexural strength.

Overall, the study demonstrates that the HSC degrades faster than standard concrete at elevated temperatures due to its low permeability, higher cement content, thermal incompatibility, and brittle nature. While it performs better under mechanical loads at ambient temperature, these same features make it less resilient under thermal stress. Consequently, standard concrete performs adequately under thermal cycles up to 300°C, making it suitable for applications involving moderate or extreme temperature exposure resulting from routine operation or fire incidents in industrial facilities such as chimneys, furnaces, kilns, and nuclear reactors. The insights gained from this study will support engineers in developing concrete mixtures capable of preserving their strength and structural integrity under these challenging thermal conditions.

REFERENCES

1. Liu, Yixiang, Hu, Jianguo, and Hou, Shuang. 2005. "SEM analysis of concrete heated on varied temperature", *Fire Safety Science*, 15, pp. 80–83.
2. Liu, Xian, Yuan, Yong, Ye, Guang, and Geert, De Schutter. 2008. "Study on pore structure evolution of high-performance concrete with elevated temperatures", *Journal of Tongji University: Natural Science*, 36, pp. 1473–1478.
3. Crozier, D. A., and Sanjayan, J. G. 1999. "Chemical and physical degradation of concrete at elevated temperatures", *Journal of Concrete Institute of Australia*, 25(1), pp. 18–20.
4. Harmanthy, T. Z. 1970. "Thermal properties of materials", *ASTM Journal of Materials*, 5(1), pp. 47–74.
5. Heo, Y. S., and Kim, K. M. 2011. "Effect of elevated temperatures on chemical properties, microstructure and carbonation of cement paste", *Journal of Ceramic Processing Research*, 16(S.1), pp. 28–32.
6. Schneider, U. 1988. "Concrete at high temperatures – a general review", *Fire Safety Journal*, 13(1), pp. 55–68.
7. Ribeiro, M. C. S., Novoa, P. R., Ferreira, A. J. M., and Marques, A. T. 2004. "Flexural performance of polyester and epoxy polymer mortars under severe thermal conditions", *Cement and Concrete Composites*, 26(7), pp. 803–809.
8. Haddad, H. Rami, and Abendeh, M. Raed. 2004. "Effect of thermal cycling on bond between reinforcement and fiber reinforced concrete", *Cement and Concrete Composites*, 26, pp. 743–752.
9. Sravana, P., Srinivasa, Rao P., and Seshagiri, Rao M. V. 2006. "Effect of thermal cycles on compressive strength of high-volume fly ash concrete", *31st Conference of our World in Concrete and Structures*: 16th–17th August, Singapore.
10. Srinivasa, Rao P., and Seshadri, Sekhar T. 2015. "Experimental investigation on the effect of thermal cycles on the strength properties of glass fibre SCC", *The Indian Concrete Journal*, September, pp. 69–74.
11. Mirdul, Garg, and Manjit, Singh. 2006. "Strength and durability of cementitious binder produced from fly ash- lime sludge- portlant cement", *Indian Journal of Engineering & Materials Sciences*, 13, February, pp. 75–79.

12. Farhat, F. G., and Karihaloo, B. L. 2006. "The behaviour of RC Beams retrofitted with CARDIFRC after thermal cycling", *Concrete Repair, Rehabilitation and Retrofitting*, Alexander (ed.), Taylor and Francis Group, London.
13. Farhat, F., and Assadi, S. 2007. "The effect of hot-dry environment on reinforced concrete beams", *32nd Conference on Our World in Concrete & Structures*, 28–29th August, Singapore.
14. Kanellopoulos, A., Farhat, A. F., Nicolaides, D., and Karihaloo, B. 2009. "Mechanical and fracture properties of cement-based bi-materials after thermal cycling", *Cement and Concrete Research*, 39, pp. 1087–1094.
15. Naus, D. J. 2010. "A compilation of elevated temperature concrete material property data and information for use in assessments of nuclear power plant reinforced concrete structures", Report NUREG/CR-7031, ORNL/TM-2009/175, United States Nuclear Regulatory Commission, Office of Nuclear Regulatory Research, USA.
16. Shokrieh, M. M., Heidari-Rarani, Shakouri. M., and Kashizadeh, E. 2011. "Effect of thermal cycles on mechanical properties of optimized polymer concrete", *Construction and Building Materials*, 25, pp. 3540–3549.
17. Chandramouli, K., Srinivasa Rao, P., Seshadri Sekhar, T., Pannir Selvam, N., and Sravana, P. 2012. "Effect of thermal cycles on compressive strength of different grades of concrete", *International Journal of Engineering Science and Technology*, 3(1), pp. 198–201.
18. Srinivasa Reddy, V., Seshagiri, Rao M. V., Sreenivasa, Rao D., and Rahul, K. 2013. "Studies on bacterial concrete exposed to elevated temperatures and thermal cycles", *International Journal of Engineering Science and Technology*, 3(1), pp. 126–135.
19. Khan. 2014. "Flexural strength of concrete subjected to thermal cyclic loads", *KSCE Journal of Civil Engineering*, 18(1), pp. 249–252.
20. Anders, S. 2016. "Comparison of residual strength of concrete with quarzitic, limestone and slag sand constituents after cyclic high-temperature exposure", *Concrete Repair, Rehabilitation and Retrofitting IV*, Dehn et al. (eds.), Taylor and Francis Group, London.
21. Vidya, B., and Srinivasa, Rao K. 2019. "Effect of thermal cycles on compressive strength of high strength concrete (M 90) compared to normal strength concrete (M 30)", *The Indian Concrete Journal*, 93(6), June, pp. 8–17.
22. Reis, J. M. L., and Ferreira, A. J. M. 2006. "Influence of thermal cycles on the fracture properties of fiber reinforced polymer concrete", *Journal of Polymer Engineering*, 26(6), pp. 605–616.
23. IS: 12269. 1989. "Specifications for 53 grade ordinary Portland cement", Bureau of Indian Standards, New Delhi, India.
24. IS: 4031. 1988. "Methods of physical tests for hydraulic cement", Bureau of Indian Standards, New Delhi, India.
25. IS: 383. 1970. "Specifications for coarse and fine aggregates from natural sources for concrete", Bureau of Indian Standards, New Delhi, India.
26. IS: 3025. 1986. (parts 22 and 24). "Methods of sampling and tests (chemical and physical) for water and waste water", Bureau of Indian Standards, New Delhi, India.
27. IS: 10262. 2009. "Recommended guidelines for concrete mix design", Bureau of Indian Standards", New Delhi, India.
28. ACI 211.4R. 2008. "Guide for selecting proportions for high strength concrete using Portland cement and other cementitious materials", ACI Committee Report, American Concrete Institute, USA.
29. ISO: 834. 1975. "Fire resistant tests – Elements of building", International Standards Organization, Geneva, Switzerland.
30. IS: 516. 1959. "Methods of tests for strength of concrete", Bureau of Indian Standards, New Delhi, India.

31. IS: 5816. 1970. "Method of test splitting tensile strength of concrete", Bureau of Indian Standards, New Delhi, India.
32. ASTM E 1131. 2008. "Standard test method for compositional analysis by thermogravimetry", American Society for Testing and Materials, ASTM International, U.S.
33. Watcharapong, Wongkeo, Pailyn, Thongsanitgarn, Prinya, Chindaprasirt, and Arnon Chaipanich. 2013. "Thermogravimetric of ternary cement blends – effect of different curing methods", *Journal of Thermal Analysis and Calorimetry*, 113(3), pp. 1079–1090.
34. Alma, D. R. M., Martha, L. D. P., Rosa, M. M. A., and Gerardo, A. R. T. 2014. "Characterization by thermogravimetric analysis of polymeric concrete with high density polyethylene mechanically recycled", *Journal of Minerals and Materials Characterization and Engineering*, 2, pp. 259–263.
35. Chatterjee, M. K., and Lahiri, D. 2014. "Estimation of free calcium hydroxide present hydrated cements by differential thermos gravimetric analysis", *Transactions – Indian Ceramic Society*, 23(1), pp. 198–202.
36. Monteagudo, S. M., Moragues, A., Gálvez, J. C., Casati, M. J., and Reyes, E. 2014. "The degree of hydration assessment of blended cement pastes by differential thermal and thermogravimetric analysis- morphological evolution of solid phases", *Thermochimica Acta*, 592, pp. 37–51.
37. Santos Silva, A., Gameiro, A., Grilo, J., Veiga, R., and Velosa, A. 2014. "Long term behaviour of lime metakaolin pastes at ambient temperature and humid curing condition", *Applied Clay Science*, 88–89, pp. 49–55.
38. Adithya Saran, A. S., and Magudeaswaran, P. 2016. "Concrete micro structure – a review", *Imperial Journal of Interdisciplinary Research*, 2(12), pp. 1670–1673.
39. Collier, N. C. 2016. "Transition and decomposition temperatures of cement phases- a collection of thermal analysis data", *Ceramics-Silikaty*, 60(4), pp. 338–343.
40. Youngsun, Heo, Guncheol, Lee, and Gunyoung, Lee. 2016. "Effect of elevated temperatures on chemical properties, microstructure and carbonation of cement paste", *Journal of Ceramic Processing Research*, 17(6), pp. 648–652.
41. Tantawy, M. A. 2017. "Effect of high temperatures on the microstructure of cement paste", *Journal of Material Science and Chemical Engineering*, 5, pp. 33–48.
42. Mathew, George, and Joseph, Benny. 2017. "Microstructural analysis of fly ash based geopolymer exposed to elevated temperatures", *ACTA Technica corviniensis – Bulletin of engineering*, Tome X, Fascicule 2, pp. 117–122.

16 Green Mortar Innovation
Integrating Titanium Dioxide for Structural and Environmental Benefits

Chandramouli Viswanathan, Aryamanki Pala, Iaemlangi Ymbon, Rajib Ahmed, Babitdor Marwein, Rafiq Gymbad Rymbai, Halton Dan, Girija T.R. and Mrinal Kanti Sen

16.1 INTRODUCTION

Rapid urbanization and increasing air pollution demand innovative construction materials which serve structural and environmental functions (Liang et al., 2019). Mortar remains a foundational material in modern construction, extensively employed in developing residential buildings, transportation infrastructure, airports, and high-rise structures. The word "mortar," from the Latin root meaning "to bind together," refers to the harmonious blending of its parts – cement, sand, and water – into a homogenous mixture. However, air pollution poses serious risks to human health and ecological balance and is becoming a bigger problem in modern urban environments. Key pollutants like carbon monoxide, sulfur dioxide, nitrogen oxides (NO_x), volatile organic compounds (VOCs), lead, and particulate matter are rising at alarming rates, especially in densely populated cities, according to the U.S. Environmental Protection Agency (Lippmann et al., 2003). Numerous health issues, such as respiratory and cardiovascular disorders, cognitive decline, and even cancer, are exacerbated by these pollutants. Vehicle emissions are a significant source of this pollution. As a result, there is increasing interest in creating "smart" mortars that contain functional additives that can break down airborne pollutants. One such material is titanium dioxide (TiO_2), which is generally considered safe by the U.S. Food and Drug Administration (Weir et al., 2012). About four million tonnes of TiO_2 are used yearly in various products such as paints, plastics, food items, medications, and cosmetics.

The anatase phase has the highest photocatalytic activity of its three naturally occurring forms: rutile, brookite, and anatase. Incorporating anatase TiO_2 into cementitious matrices and exposing them to heat and light cause harmful pollutants like NO_x and VOCs to break down into less dangerous compounds, potentially providing a solution for pollution mitigation in urban infrastructure.

DOI: 10.1201/9781003664284-16

Because of its capacity to break down airborne contaminants like nitrogen oxides (NO_x) when exposed to ultraviolet (UV) light, TiO_2 has become a promising photocatalyst. Its potential as a cementitious additive can simultaneously affect the mechanical qualities of mortar and concrete. This study investigates how TiO_2 can strengthen mortar while offering self-cleaning and air-purifying qualities.

16.2 LITERATURE REVIEW

To enhance the mechanical and functional qualities of building materials, incorporating nanomaterials into cementitious composites has recently attracted much attention. Titanium dioxide (TiO_2) has shown promise as an additive, providing both strength enhancement and photocatalytic properties.

16.2.1 Photocatalytic Properties of TiO_2

TiO_2 shows a strong photocatalytic activity under ultraviolet (UV) light, especially in its anatase form. This allows it to break down airborne pollutants like formaldehyde (HCHO), nitrogen oxides (NO_x), and volatile organic compounds (VOCs). The first work in this field was done by Fujishima et al. (2000), who showed how TiO_2 can start redox reactions when exposed to UV light, generating reactive oxygen species (ROS) that can oxidize dangerous pollutants into innocuous by-products like CO_2 and H_2O. To passively lower urban air pollution, this property has been used in tunnel linings, paving blocks, and building facades (Chen and Poon, 2009; Russell et al., 2021).

16.2.2 Mechanical Impact of *TiO₂* Addition

Incorporating TiO_2 nanoparticles into cement-based materials can enhance compressive and flexural strengths, particularly at low dosages (Bunea et al., 2023). Orakzai (2021) evaluated that up to 1% cement replacement by weight with TiO_2 can increase strength due to improved packing density and filler effects. However, excessive TiO_2 content may lead to agglomeration and pore disruption, adversely impacting strength (Zhou et al., 2013). Thus, identifying the optimal dosage is critical to balancing structural and functional gains.

16.2.3 Air-Purifying Mortars and Concretes

TiO_2-functionalized mortars have been used successfully for air purification. For instance, Poon and Cheung (2007) explained that mortar samples coated with TiO_2 were able to remove significant portions of NOx gases in a controlled environment. Ballari et al. (2010) evaluate similar results in field studies, showing that pavements treated with TiO_2 reduced local NOx concentrations. Recent works have extended these findings to VOCs, suggesting that TiO_2-modified concrete can target a broad spectrum of indoor and outdoor pollutants (Yu et al., 2002).

16.2.4 Self-Cleaning and Durability Enhancement

TiO_2-based mortars not only purify the air but also can clean themselves due to their superhydrophilicity and ability to degrade organic deposits through photocatalysis. Surfaces with TiO_2 remained cleaner for more extended periods of time, requiring less maintenance, according to Chen and Poon (2009). Additionally, the photocatalytic activity added value in high-humidity and healthcare environments by promoting antifungal and antibacterial effects (Sunada et al., 2003).

16.2.5 Research Gaps and Need for Optimization

Most existing studies focus on specific pollutants (e.g., NOx), with limited attention to HCHO and TVOCs, which are prevalent in indoor environments. Moreover, few studies have addressed the synergistic relationship between TiO_2 content, curing conditions, and pollutant degradation efficiency. Long-term durability, economic feasibility, and real-world performance under varying environmental conditions remain underexplored.

Collectively, prior research substantiates the potential of TiO_2-modified mortars as innovative construction materials offering both mechanical strength and environmental remediation. However, the optimal dosage, method of incorporation, and long-term behavior under field conditions warrant further investigation. The current study addresses some of these gaps by evaluating the compressive strength and VOC degradation potential of TiO_2-modified mortar under natural light exposure, with specific focus on formaldehyde and TVOC reduction. Based on the research gap, the objectives of the study are considered as follows:

- To evaluate the influence of different titanium dioxide (TiO_2) dosages on the compressive strength characteristics of cement mortar,
- To assess the preliminary air purification capability of TiO_2-modified mortar cubes exposed to urban outdoor conditions.

16.3 MATERIALS AND METHODS

Figure 16.1 shows the framework of the work on TiO_2-modified green mortar, encompassing structural and environmental performance enhancement. The work begins by recognizing the pressing issue of urban air pollution and the need for sustainable construction materials. Cementitious composites offer an ideal platform for embedding smart functionalities like pollutant degradation. The primary aim is to investigate how TiO_2 affects the compressive strength and air purification capabilities of cement mortar. Standard construction materials – OPC (43 grade), river sand, and anatase-phase TiO_2 – were selected to ensure replicability and relevance in practical construction scenarios. TiO_2 was added in varying proportions to identify an optimal dosage. Here, two key tests were conducted: compressive strength tests at 28 days, and a smog chamber test to assess pollutant absorption (HCHO and TVOC) under natural sunlight exposure. Then the results are analyzed, and based on the results, the study is concluded.

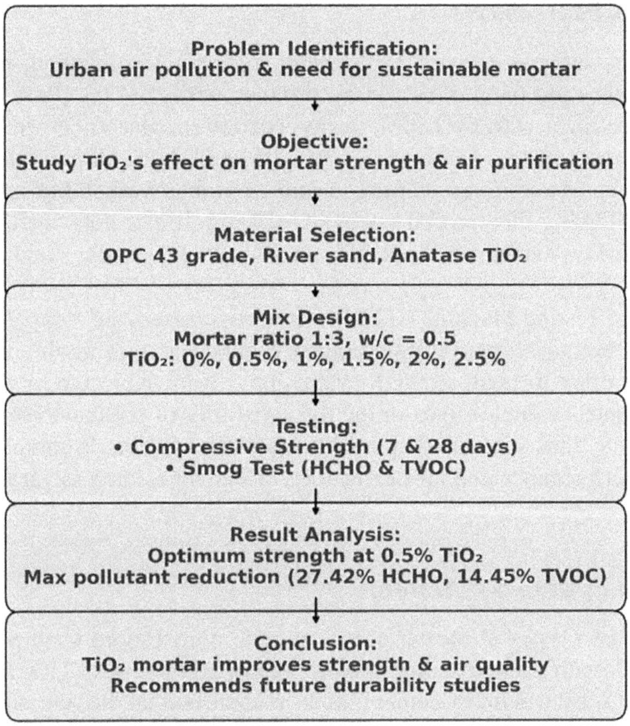

FIGURE 16.1 Framework of the study.

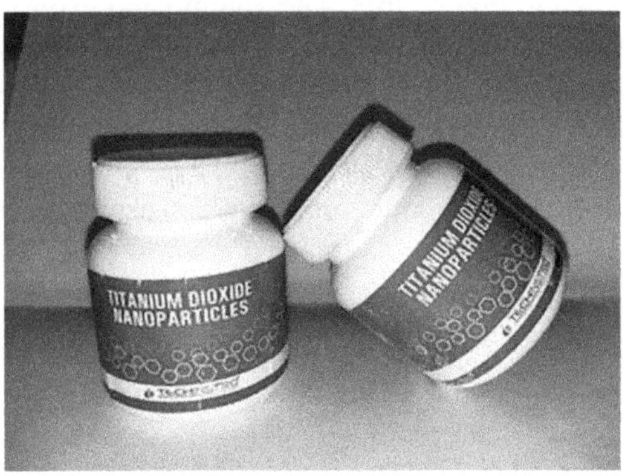

FIGURE 16.2 Titanium dioxide used for the study.

16.3.1 Materials Used

Ordinary Portland Cement (43 Grade), river sand conforming to Zone II, potable water, and commercial anatase form of TiO_2 powder (>99% purity) were used as shown in Figure 16.2.

16.3.2 Mix Proportions

A standard mortar mix ratio of 1:3 (cement to sand) was adopted, maintaining a water-to-cement ratio (w/c) of 0.5, in accordance with IS 4031 (Part 6):1988. TiO_2 , in its anatase form with over 99% purity, partially replaced cement by weight at incremental levels of 0%, 0.5%, 1%, 1.5%, 2%, and 2.5%. A consistency test was conducted to ensure uniform dispersion and desirable workability across all mortar mixes. Following this, mortar cubes of 70.7 mm dimensions were cast for each TiO_2 concentration and cured in potable water for 28 days under standard laboratory conditions.

A Universal Testing Machine (UTM) was used to assess the compressive strength at the end of the curing period. Subsequently, cubes containing the optimum TiO_2 content – identified through strength evaluation – were subjected to smog chamber testing and control samples to examine the capability of pollutant absorption under natural sunlight. This structured approach enabled a systematic comparison of both mechanical performance and air purification effectiveness across various TiO_2 dosage levels.

16.3.4 Air Purification Testing

In this study, two types of mortar cubes (sample preparation) were prepared using 43-grade OPC, such as control mortar (without any additive) and TiO_2 mortar (mixed with 0.5% TiO_2 by weight of cement, as it is identified as the optimum from prior strength tests).

All samples were cured for 28 days under standard laboratory conditions. Each mortar cube (control and TiO_2-modified) was placed in a separate, sealed transparent glass case, as shown in Figure 16.3. The volume and exposure conditions of the chambers were kept identical to ensure uniformity. A controlled amount of indoor air pollutants, specifically HCHO (formaldehyde) and TVOC, was introduced or allowed to stabilize naturally within each chamber. The test was conducted in natural sunlight to activate the photocatalytic properties of TiO_2. The exposure was maintained for a fixed duration (e.g., several hours or a day), consistent across both setups. This

FIGURE 16.3 Set up for air pollution absorption test.

setup simulates a micro-environment, directly comparing pollutant removal between plain and TiO_2-enhanced mortars. Sealed glass cases ensure that external variables (like wind or additional pollutants) do not interfere. Natural light exposure mimics real-world conditions under which the photocatalytic reaction occurs.

16.4 RESULTS AND DISCUSSION

16.4.1 Compressive Strength

Figure 16.4 shows the variation in crushing load with increasing percentages of TiO_2 in cement mortar. The crushing load increases sharply from 18.5 kN at 0% TiO_2 (control) to a peak of 35.7 kN at 0.5% TiO_2, representing a 92.97% improvement in compressive strength. This enhancement is attributed to the filler effect of TiO_2 nanoparticles, which densify the matrix and improve particle packing. However, beyond 0.5%, the strength progressively declines, dropping to 30.9 kN at 1%, 27.9 kN at 1.5%, 22.3 kN at 2%, and 19.1 kN at 2.5%. This reduction may be due to nanoparticle agglomeration and interference with cement hydration, leading to microstructural discontinuities. The trend indicates an optimal TiO_2 dosage at 0.5%, beyond which adverse effects outweigh the benefits. Thus, appropriate TiO_2 content is critical for maximizing mechanical performance while avoiding adverse impacts on mortar integrity.

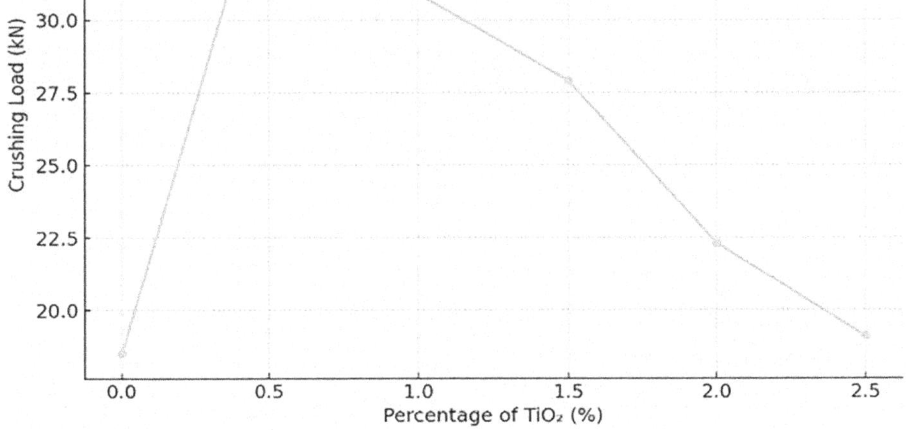

FIGURE 16.4 Variation of crushing load with percentage of titanium dioxide (TiO_2) in cement mortar.

16.4.2 Air Purification Observations

This study measured pollutant concentrations inside each sample/chamber using appropriate air quality sensors (e.g., for HCHO and TVOC) after exposure, as shown in Table 16.1. The TiO_2-modified mortar significantly reduced pollutant concentrations compared to the control. HCHO levels decreased by 27.42%, indicating a strong photocatalytic degradation efficiency. This aligns with TiO_2 's known ability to generate ROS under UV light, which oxidize harmful organic compounds into less toxic by-products like CO_2 and H_2O. TVOC levels showed a 14.45% reduction, suggesting effective breakdown of complex organic compounds in indoor environments. These findings validate the dual-functional capability of TiO_2 in cement mortar – enhancing compressive strength and contributing to passive air purification. Using TiO_2 in construction materials thus presents a sustainable solution to mitigate indoor air pollution, especially in urban environments.

Beyond its structural advantages, titanium-dioxide-based concrete also demonstrates self-cleaning capabilities, which can extend the structure's service life and reduce the need for frequent maintenance. This property is primarily attributed to TiO_2 's photocatalytic activity and super-hydrophilic nature. When exposed to light, TiO_2 facilitates the breakdown of airborne and surface pollutants by generating ROS such as hydroxyl and oxygen radicals from moisture and atmospheric oxygen. These radicals oxidize harmful compounds into less hazardous substances, such as soluble salts, which are easily removed from the surface. Additionally, due to the strong polarity of both TiO_2 and water, water spreads evenly across the treated surface rather than forming droplets. This sheeting effect promotes effective rinsing of contaminants, ensuring a cleaner surface over time and enhancing the aesthetic and functional durability of the concrete.

TiO_2in mortar has a measurable impact on reducing air pollutants, specifically HCHO and TVOC. The optimum amount of TiO_2 could potentially result in pollutant reductions, as TiO_2 is known for its photocatalytic properties that help break down pollutants in the presence of light. These findings underscore the potential of TiO_2 infused "green" cement materials for improving urban air quality. Photocatalytic cements can reduce smog precursors by oxidizing VOCs and even NO_x in ambient air. For example, previous work has shown that cementitious composites containing TiO_2 significantly "plunge the concentration of urban pollutants" such as NO_x and are already used in self-cleaning, air-purifying building surfaces. Our results extend this concept to volatile organics: the observed HCHO and TVOC reductions are consistent

TABLE 16.1
Effect of TiO_2 Addition on Pollution Reduction (Smog Test Readings): Quantitative Comparison

Pollutant	Control Mortar (Mg/M³)	TiO_2 Mortar (Mg/M³)	% Reduction
HCHO	0.124	0.090	27.42%
TVOC	0.693	0.593	14.45%

with reports that TiO_2-bearing mortars and concretes can rapidly degrade airborne organics using the ROS described above. Large-area applications of photocatalytic concrete (e.g., paving blocks, facade panels, tunnel linings) have demonstrated measurable drops in NO_x and VOC levels in polluted environments. Thus, incorporating TiO_2 into cement can create passive air-cleaning infrastructure. Such photocatalytic cements represent an emerging sustainable technology as they add no hazardous by-products, require only ambient light, and turn ubiquitous construction materials into functional pollution sinks. In the broader context of urban sustainability, this experimental data validates that TiO_2-modified mortar can non-mechanically lower key pollutant concentrations. When scaled to real structures, this approach could complement regulatory and emission-control measures, helping to mitigate smog and improve public health in high-density urban areas. The present work adds experimental confirmation that even a low TiO_2 dosage (0.5%) yields appreciable pollutant degradation, illustrating the promise of green-cement technologies for air purification.

16.5 CONCLUSION AND FUTURE SCOPE

This study highlights the dual-functional potential of titanium dioxide (TiO_2) as a sustainable additive in cement mortar, offering significant enhancements in structural and environmental performance. The experimental results revealed that incorporating 0.5% TiO_2 by weight of cement yielded the maximum compressive strength, with an improvement of over 90% compared to the control mix. Beyond this dosage, a gradual decline in strength was observed, likely due to nanoparticle agglomeration and disruption of the hydration process. Additionally, the TiO_2-modified mortar demonstrated notable photocatalytic activity, achieving a 27.42% reduction in formaldehyde (HCHO) and a 14.45% reduction in total volatile organic compounds (TVOCs) under natural sunlight exposure. These findings validate the potential of TiO_2-based “green” mortars as multifunctional construction materials capable of contributing to air quality improvement in urban environments while maintaining structural reliability. The observed self-cleaning properties further support such materials’ longevity and low-maintenance benefits, making them suitable for facade panels, pavements, and indoor applications.

However, the study is subject to certain limitations. The air purification tests were conducted under controlled small-scale conditions, which may not fully replicate real-world variability such as weather conditions, pollutant flux, or long-term durability. Also, only two indoor air pollutants (HCHO and TVOC) were considered, while broader pollutant categories (e.g., NO_x, SO_2, ozone) remain unexamined. To address these gaps, the following recommendations and future research directions are proposed:

- Long-term exposure studies should be conducted to evaluate the aging, wear resistance, and sustained photocatalytic activity of TiO_2-modified mortars.
- Advanced imaging and spectroscopy techniques (e.g., Scanning Electron Microscopy (SEM), X-Ray Diffraction (XRD), Fourier Transform Infrared Spectrometer (FTIR)) are needed to analyze the dispersion and interaction of TiO_2 within the cement matrix.

- Full-scale implementation under real environmental conditions would help validate laboratory findings and assess practical feasibility.
- A techno-economic assessment should be conducted to evaluate the scalability, affordability, and life cycle impacts of using TiO_2 in construction.
- Future experiments should include multiple air pollutants and study their degradation kinetics under varying environmental conditions (e.g., light intensity, humidity).

TiO_2-incorporated mortar presents a promising pathway toward eco-efficient and high-performance construction materials. With further optimization and field validation, such materials could be pivotal in promoting sustainable urban development and public health.

REFERENCES

Ballari, M. M., Hunger, M., & Brouwers, H. J. H. (2010). NOx photocatalytic degradation employing concrete pavement containing titanium dioxide. *Applied Catalysis B: Environmental*, 95(3–4), 245–254.

Bunea, G., Alexa-Stratulat, S. M., Mihai, P., & Toma, I. O. (2023). Use of clay and titanium dioxide nanoparticles in mortar and concrete – a state-of-the-art analysis. *Coatings*, 13(3), 506.

Bureau of Indian Standards. (1988). *IS 4031 (Part 6): Methods of physical tests for hydraulic cement*. New Delhi: BIS.

Chen, J., & Poon, C. S. (2009). Photocatalytic construction and building materials: From fundamentals to applications. *Building and Environment*, 44(9), 1899–1906.

Fujishima, A., Rao, T. N., & Tryk, D. A. (2000). Titanium dioxide photocatalysis. *Journal of Photochemistry and Photobiology C: Photochemistry Reviews*, 1(1), 1–21.

Liang, L., Wang, Z., & Li, J. (2019). The effect of urbanization on environmental pollution in rapidly developing urban agglomerations. *Journal of Cleaner Production*, 237, 117649.

Lippmann, M., Frampton, M., Schwartz, J., Dockery, D., Schlesinger, R., Koutrakis, P., & Zelikoff, J. (2003). The US Environmental Protection Agency particulate matter health effects research centers program: A midcourse report of status, progress, and plans. *Environmental Health Perspectives*, 111(8), 1074–1092.

Orakzai, M. A. (2021). Hybrid effect of nano-alumina and nano-titanium dioxide on mechanical properties of concrete. *Case Studies in Construction Materials*, 14, e00483.

Poon, C. S., & Cheung, E. (2007). NO removal efficiency of photocatalytic paving blocks prepared with recycled materials. *Construction and Building Materials*, 21(8), 1746–1753.

Russell, H. S., Frederickson, L. B., Hertel, O., Ellermann, T., & Jensen, S. S. (2021). A review of photocatalytic materials for urban NOx remediation. *Catalysts*, 11(6), 675.

Sunada, K., Watanabe, T., & Hashimoto, K. (2003). Studies on photokilling of bacteria on TiO_2 thin film. *Journal of Photochemistry and Photobiology A: Chemistry*, 156(1–3), 227–233.

Weir, A., Westerhoff, P., Fabricius, L., Hristovski, K., & Von Goetz, N. (2012). Titanium dioxide nanoparticles in food and personal care products. *Environmental Science & Technology*, 46(4), 2242–2250.

Yu, J. C., Yu, J., Ho, W., Jiang, Z., & Zhang, L. (2002). Effects of F-doping on the photocatalytic activity and microstructures of nanocrystalline TiO_2 powders. *Chemistry of Materials*, 14(9), 3808–3816.

Zhou, D., Ji, Z., Jiang, X., Dunphy, D. R., Brinker, J., & Keller, A. A. (2013). Influence of material properties on TiO_2 nanoparticle agglomeration. *PLoS ONE*, 8(11), e81239.

17 Life Cycle Assessment in Sustainable Waste Management

Tools, Trends, and Impact Pathways

Alice Ogunlolu, Awodele Imoleayo Abraham and Saidu Minin Mohammed

17.1 INTRODUCTION

Construction and Demolition Waste (C&DW) has gained recognition as a critical global sustainability challenge, accounting for roughly one-third of all solid waste generated worldwide, the largest waste stream by mass. According to Bonifazi *et al.* (2025), the wastes are partly from construction, renovation, renewal, dismantling of buildings, roads, bridges, and other structures, due to rapid population growth and urbanisation, amongst others. They also cited other causes to be partly responsible, such as earthquakes, hurricanes, floods, and tsunamis. If not managed properly, this voluminous waste stream can cause serious environmental harm through air, water, and soil pollution (Asdrubali *et al.*, 2013; Victar & Waidyasekara, 2025; Kara De Maeijer, 2025). Ensuring sustainable waste management, particularly for Construction and Demolition (C&D) debris is, therefore, an urgent priority in the push for greener and more circular economies.

Al-Raqeb *et al.* (2023) are advocating for the raising of awareness among construction stakeholders and the public on the benefits of waste management and recycling. The circular economy (CE) principles, according to Bonifazi (2025) citing others, have resulted in such actions as Refuse, Rethink, Repair, Refurbish, Remanufacture, Repurpose, and Recover. Martin *et al.* (2024) had earlier studied the CE implementation strategies and also assessed challenges associated with their adoption. Abubakar *et al.* (2024) call for government regulations, incentives for recycling facilities, and educational initiatives, highlighting the advantages of applying CE also. Iyiola *et al.* (2024) identified Blockchain, Internet of Things (IoT), Artificial Intelligence (AI), Machine Learning (ML), Robotics, Computer Vision (CV), and Building Information Modelling (BIM) playing significant roles in promoting efficient C&DW Management in the Construction Industry. Unegbu and Yawas (2024) emphasised the need for an integrated approach to improving C&DW management

DOI: 10.1201/9781003664284-17

in Nigeria, by calling for enhanced regulatory measures, infrastructure development, and continuous education and training programmes. Shajidha and Mortula (2025) emphasised that source reduction should be mandatory, while recycling and reuse should complement waste reduction efforts.

However, Islam *et al.* (2024) had identified several study gaps, including limitations in operational aspects, the lack of monitoring of legislation and regulations, an inadequate market for recycled products, inefficient stakeholder engagement and management, limitations of effective framework/models for sustainable C&DW management, limited research on social performance, and inadequate integration of information technology in C&D waste. Also, Swarnakar and Khalfan (2024) found from literature that no studies discussed Life Cycle Assessment (LCA) indicators, among other findings in their search. Wu *et al.* (2025a) suggest future research to cover, among others, inclusive studies throughout all life cycle phases of C&DW management. Zakerhosseini *et al.* (2024) found that LCA of C&DW management showed major environmental effects on global warming, fine particulate matter formation, human carcinogenic toxicity, and human non-carcinogenic toxicity. In the light of the above, attention has increasingly turned to rigorous assessment methods which can quantify the environmental implications of waste management choices and guide policy and practice towards lower-impact outcomes.

LCA has rapidly emerged as a pivotal tool for evaluating the environmental performance of waste management strategies, including those for C&D waste (Wu *et al.*, 2025b). This is also alluded to by several researchers, including the few mentioned above. By holistically accounting for impacts across a waste material's life cycle from raw material extraction and construction through use, demolition, and end-of-life processing, LCA enables objective comparison of alternatives such as landfilling, recycling, and on-site reuse. Studies consistently demonstrate that prioritising material reuse and recycling over disposal can substantially reduce overall environmental burdens in line with the waste hierarchy (Cheng & Ma, 2013). Consequently, LCA-based analyses have proliferated in the past decade as researchers and practitioners seek science-based evidence to inform sustainable waste management decisions. Emerging research in this domain has also begun to integrate digital technologies with LCA to improve data accuracy and to explore waste reduction within broader CE frameworks (Gao *et al.*, 2025). These developments underscore a dynamic and evolving research landscape at the nexus of LCA and sustainable waste management (Donthu *et al.*, 2021). Despite the growing body of literature applying LCA to waste management, there remains a notable knowledge gap: no consolidated synthesis currently maps the intellectual structure, thematic evolution, and methodological advancements of this field. The relevant research is dispersed across diverse studies, making it difficult to discern overarching patterns or identify how the field has progressed (Lu & Tam, 2013). In particular, the literature lacks an integrated perspective on which analytical tools and frameworks have been employed across LCA-based waste management studies, what thematic clusters and trends have emerged, and how the knowledge has evolved and generated impact. A bibliometric approach is well suited to address this gap as bibliometric analysis quantitatively examines publication data to reveal the development of a research domain, uncovering key research themes, collaborations, and knowledge structures. By leveraging such an approach, one can systematically map how LCA applications in sustainable waste management have developed over time, identify current research frontiers, and highlight under-explored areas requiring further inquiry (Pomponi & Moncaster, 2016; Yazdani & Lakzian, 2023).

To provide this much-needed synthesis, this study conducts a comprehensive bibliometric analysis of the global research on LCA in sustainable waste management with a particular focus on C&DW complemented by a systematic review of pertinent literature. This combined approach allows to not only quantify publication trends and networks but also qualitatively contextualise the findings. A comprehensive dataset of relevant studies was analysed to identify prevalent analytical tools, evaluate methodological approaches, uncover dominant and emerging research themes, and investigate the evolution of research impact pathways within the field (Farina *et al.*, 2020). Through mapping these elements, the study offers a consolidated insight into LCA-based sustainable waste management research. The findings illuminate the knowledge evolution and intellectual trends in this domain, demonstrating the state of the art and highlighting opportunities for future research (Kasulla *et al.*, 2024; Marinković *et al.*, 2017). This contribution is significant in that it not only fills a gap in understanding the landscape of LCA applications in waste management but also enhances methodological rigour in the field by pinpointing best practices and prevailing gaps, thereby guiding researchers and decision-makers towards more impactful and well-founded studies in the pursuit of sustainable waste management.

17.2 METHODOLOGY

17.2.1 Research Design

This study employs a quantitative bibliometric analysis to examine global research trends in the application of LCA to Sustainable Construction and Demolition Waste (C&DW) Management. Bibliometric analysis offers a rigorous, replicable method for evaluating large volumes of scientific literature, uncovering the development of a research field and identifying emerging themes and collaboration networks (Donthu *et al.*, 2021; Alimi *et al.* 2025). By mapping publication patterns and intellectual structures in this domain, the study provides an evidence-based overview of how LCA has been applied as a decision-support tool in Sustainable C&DW Management. All methodological steps were designed to be transparent and reproducible, as detailed below.

17.2.2 Data Source and Search Strategy

The Scopus database was selected as the sole data source for the bibliometric study due to its broad coverage of high-quality publications in engineering, environmental science, and sustainability fields. While Web of Science offers comparable indexing, both databases have significant content overlap, and Scopus's inclusion of a wider range of journals and recent publications makes it well suited for an exhaustive literature survey in this field (Zhang *et al.*, 2016). A comprehensive search strategy was developed to capture literature at the intersection of LCA, sustainability, and C&DW management. The search was conducted on 10th July 2025 and restricted to English-language documents and focused on peer-reviewed journal articles, books, conference review, conference proceedings, and editorials (this is for their depth and reliability, to include emerging research and scrutinised publications) (He & Cha, 2022). The core search string used in Scopus was:

> TITLE-ABS-KEY ("Life Cycle Assessment" OR "LCA") AND TITLE-ABS-KEY ("construction waste" OR "demolition waste" OR "C&D waste" OR "construction and demolition waste") AND TITLE-ABS-KEY ("sustainable" OR "sustainability" OR "environmental impact")

This query targeted publications with those keywords in the title, abstract, or author keywords. The initial broad search yielded a large pool of documents, which was intended to capture the contemporary evolution of the field while including its foundational studies, for this first search, a total of 1,043 documents was generated. The results were then refined through iterative filtering: first by screening titles and abstracts to ensure a focus on C&DW management in an LCA context and then by removing any clearly out-of-scope records and spanning approximately the last decade (2015–2025). All retrieved bibliographic records (metadata) were exported from Scopus in a standardised format (CSV), including information on authors, titles, publication years, source titles, abstracts, keywords, citations, and author affiliations. A thorough data cleaning and preprocessing to improve the reliability of the analysis were performed. This involved removing duplicate entries and non-relevant items, as well as standardising key fields. Author names were normalised to merge variants of the same individual, and keyword nomenclature was unified. These cleaning steps were carried out using the Bibliometrix package in R (Swarnakar & Khalfan, 2024) together with manual verification. By refining the raw Scopus data in this way, it ensured that the subsequent bibliometric analysis would not be skewed by data quality issues. After this refinement process, a final dataset of 403 relevant documents was obtained for analysis. This set of publications constitutes the basis for all subsequent bibliometric and qualitative analysis.

General publication metrics were first analysed to evaluate the research output and influence within the domain of LCA in sustainable C&DW management. This included examining annual publication trends and identifying the most prolific contributors. Key performance indicators were:

- Publication output per year: The number of publications in the dataset by year (2015–2025) was tallied to illustrate growth trends and significant shifts in research activity.
- Leading authors and regions: Publications and total citations and the region contributing the most work, highlighting which researchers and regions are driving the field, were noted.

- Citation analysis: Highly cited papers within the dataset as a proxy for influential works and computed citation counts per publication and per year to understand the impact and knowledge diffusion of key studies were identified.

This performance analysis provided a quantitative overview of the landscape, including which years saw surges in interest, which countries or regions were most active (by aggregating authors' affiliations), and which publications and venues shaped the scholarly conversation.

Science Mapping: To uncover the intellectual structure and collaborative networks of the field, several science mapping analyses were performed using the bibliometric data:

- Co-authorship network analysis: Collaboration networks were mapped among authors (and, separately, among institutions and countries) to understand how researchers connect and form knowledge-sharing clusters. By analysing co-authorship, we could see the extent of international collaboration and identify prominent research groups or cross-country partnerships in LCA for C&D waste.
- Keyword co-occurrence analysis: The author keywords and index keywords from the publications were analysed to find out which terms frequently appear together, using this as an indication of thematic clusters. High-frequency keywords signify important research topics, and when keywords co-occur in papers, it suggests a conceptual link. We used these co-occurrence patterns to automatically cluster related keywords, revealing the major research themes (e.g. groups of publications focusing on "circular economy," "recycled concrete aggregates," "policy assessment," etc., within the LCA and waste management context).
- Thematic evolution: Using a temporal overlay on the bibliometric mapping research, themes were explored on how they have evolved over the last decade. By partitioning the dataset into earlier versus later years and observing changes in keyword prominence and network structure helped illustrate the field's trajectory and shifting priorities over time.

All network analyses and visualisations were carried out with VOSviewer software (Balasbaneh & Sher, 2024). VOSviewer is particularly well suited for mapping bibliometric networks because of its capacity to handle large datasets and its use of a distance-based visualisation layout, where the proximity of nodes (e.g. keywords or authors) reflects their relatedness. VOSviewer (version 1.X) was used to construct maps for co-authorship and co-occurrence networks, applying its built-in clustering algorithm to identify groups of closely related items. The resulting maps including network diagrams, overlay timelines, and density visualisations were then interpreted to characterise the research structure of LCA in sustainable waste management. Clusters of keywords from the co-occurrence analysis, for instance, were labelled and discussed as distinct thematic areas in the literature.

17.2.3 Limitations

Every research approach has limitations, and the findings of this study must be interpreted with an awareness of the following considerations:

> While Scopus's coverage is broad and largely overlaps with other major databases, there is a possibility that some niche or very recent studies were not captured. Similarly, the restriction to English-language documents may exclude pertinent research reported in other languages.

Also, the accuracy of bibliometric analysis depends on the quality of the metadata provided. Inconsistencies or errors in author names, affiliations, or keyword indexing in the source data can influence the results. This was mitigated through careful data cleaning to ensure that incomplete citation or misclassified documents are eliminated.

Despite these limitations, this study aimed to ensure a transparent and reproducible methodology. The search strategy, selection criteria, and analysis techniques are documented so that other researchers can replicate the study or update it with new data in the future. All interpretations of the bibliometric results are made cautiously, with the recognition of the above constraints, to provide a balanced and credible mapping of LCA research in sustainable C&DW management.

17.3 FINDINGS

17.3.1 Top Contributing Countries by Publications, Citations, and Collaboration Strength in LCA and Sustainable Waste Management Research

The global distribution of scholarly contributions to LCA in the context of sustainable waste management reveals a concentrated research effort among a core group of countries, with significant regional disparities in output, impact, and collaboration intensity. Countries with at least 10 documents publication have been included in Table 17.1.

China leads the global output with 55 publications and the highest citation count of 2,762, suggesting both prolific activity and high impact. This is followed closely by India (54 documents, 2,135 citations), which demonstrates strong research engagement but a comparatively lower total link strength (TLS) (31) than its counterparts. The UK and Australia, despite publishing fewer documents (37 and 51, respectively), exhibit the highest TLS (79), indicating deep international collaboration and integration within global LCA research networks.

The US contributes 24 documents with 862 citations and a notable link strength of 45, positioning it as an active yet somewhat regionally collaborative player. European countries such as Italy, Portugal, and Spain contribute moderately (ranging between 13 and 29 documents), with Portugal showing a relatively high citation-per-document ratio, highlighting impactful research despite modest volume. Canada (17 documents) stands out with 1,058 citations and a solid collaboration profile (link strength 36), reflecting high-quality outputs.

TABLE 17.1
Global Distribution of Research Output on Life Cycle Assessment

Country	Documents	Citations	Total Link Strength
China	55	2,762	68
India	54	2,135	31
Australia	51	1,651	79
The United Kingdom	37	1,584	79
Italy	29	416	13
The United States	24	862	45
Portugal	22	876	18
Canada	17	1,058	36
Brazil	15	760	24
Malaysia	15	1,736	7
Spain	13	774	30
Turkey	13	344	9
Hong Kong	12	462	17
New Zealand	11	19	4

FIGURE 17.1 Global distribution of research output on Life Cycle Assessment in sustainable waste management.

Interestingly, Malaysia, despite producing only 15 documents, has amassed 1,736 citations, yielding the highest citation-per-document ratio (115.7) among all countries analysed. This suggests a niche but highly influential research footprint. In contrast, New Zealand ranks lowest across all metrics, with 11 documents, 19 citations, and a link strength of 4, implying limited reach and engagement. The world map in Figure 17.1 shows regions lagging publications in this context.

The bibliometric landscape suggests a mature and diversified global research field, with high-output nations contributing both volume and influence and smaller or emerging nations offering specialised, high-impact insights. This analysis supports the need for enhanced global collaboration, particularly involving underrepresented regions, to further enrich the collective knowledge base on sustainable waste management practices.

17.3.2 Publication Output Per Year

The distribution of publications between 2015 and 2025, as illustrated in Figure 17.2, reflects the evolving academic interest in LCA within the domain of sustainable waste management. A steady increase in research output is observed over the decade, with publication counts rising from 4 in 2015 to a peak of 101 in 2024, although currently being 67 research outputs by July 2025, there is a high possibility. This trajectory underscores a significant amplification of scholarly engagement, particularly over the last five years.

Between 2015 and 2017, the field experienced relatively limited activity, indicative of its nascent stage. However, from 2018 onwards, a notable acceleration in publication volume is evident, with 2019–2023 recording consistent year-on-year growth and consolidating the field's emergence as a key area of inquiry within sustainability science. The dramatic surge in 2024 may be attributed to increased global policy emphasis on CE transitions, heightened construction waste regulations, and the wider application of LCA as a tool for environmental performance benchmarking.

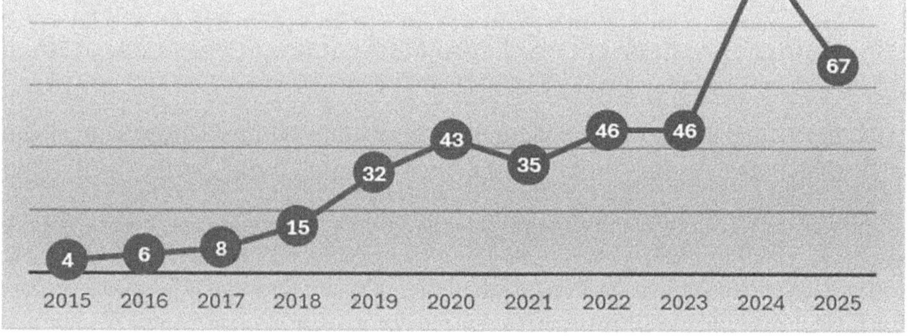

FIGURE 17.2 Publication output per year.

The surge in publication output, particularly in the post-2020 period, signifies a paradigm shift from isolated environmental assessments towards more integrated and systems-based approaches to waste management.

17.3.3 High-Frequency Keywords and Network Connectivity in LCA and Sustainable Waste Management Research

An in-depth bibliometric analysis of author keywords reveals the dominant conceptual foci and thematic structures shaping the scholarly landscape of LCA in Sustainable Waste Management. This analysis is based on keyword occurrence and TLS, a measure of co-occurrence and network connectivity both of which indicate the frequency and relational importance of a given term within the bibliographic network.

The most frequently occurring keyword is "demolition" (158 occurrences; TLS = 964), followed closely by "construction and demolition waste" (154; TLS = 912) and "sustainable development" (131; TLS = 783). These keywords form the conceptual backbone of the field, emphasising the intersection of environmental sustainability and construction sector waste streams. Their high TLSs underscore their central role in linking subfields and co-occurring themes.

"Recycling" (126 occurrences; TLS = 820) and "waste management" (95; TLS = 496) also emerge as critical nodes, highlighting the practical application of LCA in resource recovery and CE initiatives. These are complemented by "sustainability" (75; TLS = 360) and "circular economy" (75; TLS = 353), signalling the normative frameworks guiding the discourse on sustainable waste management practices.

Keywords directly referencing methodological constructs, such as "life cycle" (85; TLS = 598), "life cycle assessment" (52; TLS = 332), and "life cycle assessment (LCA)" (35; TLS = 237), affirm the analytical orientation of the field. These terms demonstrate a widespread usage of LCA tools and frameworks in environmental impact evaluation, particularly in the context of built environment studies.

Material-related terms such as "concrete aggregates" (51; TLS = 388), "concrete" (31; TLS = 232), and "recycled aggregates" (35; TLS = 258) point to a robust subdiscipline focused on material performance and resource optimisation. These keywords are particularly significant within engineering and construction literature where environmental assessments of structural materials are integrated into LCA models.

Other notable keywords include "environmental impact" (89; TLS = 570), "construction industry" (81; TLS = 500), and "literature review" (32; TLS = 205), reflecting both the domain of application and the knowledge synthesis efforts aimed at consolidating methodological developments. Emerging or specialised terms such as "durability" (21; TLS = 117), "recycled aggregate concrete" (25; TLS = 185) and "project management" (20; TLS = 108) indicate areas of growing research attention.

In summary, the frequency and link strength of these keywords demonstrate that the field is anchored by well-established concepts such as sustainable development, recycling, and LCA, while simultaneously expanding into more technical and application-specific areas such as recycled construction materials, circularity, and review-based synthesis. This maturity reflects a multidimensional and interdisciplinary

TABLE 17.2
Keywords and Network Connectivity in Lca

S/no	Keyword	Occurrences	Total Link strength
1	aggregates	25	196
2	circular economy	75	353
3	compressive strength	24	147
4	concrete	31	232
5	concrete aggregates	51	388
6	concretes	26	187
7	construction	37	202
8	construction and demolition waste	154	912
9	construction industry	81	500
10	construction waste	22	97
11	construction wastes	36	173
12	demolition	158	964
13	demolition wastes	24	159
14	durability	21	117
15	environmental impact	89	570
16	environmental management	28	188
17	lca	20	89
18	life cycle	85	598
19	life cycle analysis	27	207
20	life cycle assessment	52	332
21	life cycle assessment (lca)	35	237
22	literature review	32	205
23	project management	20	108
24	recycled aggregate concrete	25	185
25	recycled aggregates	35	258
26	recycled concrete aggregates	20	129
27	recycling	126	820
28	review	37	264
29	sustainability	75	360
30	sustainable construction	24	155
31	sustainable development	131	783
32	waste disposal	30	208
33	waste management	95	496
34	waste treatment	23	147

approach to sustainable waste management in the construction and built environment sectors.

Table 17.2 reveals that demolition, construction and demolition waste, recycling, and sustainable development exhibit the highest occurrence frequencies and network link strengths, indicating that waste-related sustainability themes constitute the dominant and most strongly interconnected research clusters within life cycle assessment studies.

17.3.4 Thematic Landscape of Research Keywords in LCA for Sustainable Waste Management

A co-occurrence analysis of author keywords, generated using VOSviewer, provides insight into the thematic structure and intellectual focus areas within the research domain of LCA in sustainable waste management. The analysis identified several keyword clusters, each representing a distinct research trajectory within the broader field.

The core cluster, anchored around the term "life cycle assessment," emerges as the most dominant, with high frequency and centrality. This keyword serves as the methodological backbone across diverse applications, demonstrating its integrative role in environmental sustainability evaluations.

Another prominent cluster revolves around "municipal solid waste," "waste management," and "recycling," reflecting the practical and policy-driven concerns which shape much of the applied LCA literature. These terms highlight the sectoral focus of the field on urban waste systems and the comparative environmental performance of various waste treatment options, including incineration, composting, and landfilling.

A third major thematic cluster is oriented towards "environmental impact," "sustainability," and "circular economy." These keywords reflect the normative goals underpinning LCA research minimising environmental burdens and promoting sustainable development. The appearance of "carbon footprint," "global warming potential," and "greenhouse gas emissions" indicates a strong alignment with climate change mitigation objectives.

Additionally, technical and methodological terms such as "multi-criteria decision analysis (MCDA)," "input-output analysis," and "SimaPro" appear in isolated but meaningful nodes, suggesting growing methodological diversification. These are often linked to more advanced LCA studies which integrate other decision-support frameworks or rely on software tools to enhance model robustness.

Emerging keywords such as "biochar," "construction and demolition waste," and "food waste" signal evolving areas of interest. These terms suggest that the field is expanding to address specialised waste streams and sector-specific challenges, especially in light of increasing global urbanisation and food insecurity.

The network visualisation also reveals strong interlinkages among environmental indicators (such as, eutrophication, acidification, ecotoxicity), denoting comprehensive impact assessments typical in high-quality LCA research. The density of links and co-occurrence strength across clusters implies an interdisciplinary character, integrating environmental science, engineering, urban studies, and public policy. The keyword co-occurrence analysis illustrates a mature yet dynamically evolving field. While "life cycle assessment" remains the intellectual anchor, new terminologies and interdisciplinary linkages reflect the field's responsiveness to emerging sustainability challenges and methodological innovations.

The keyword co-occurrence diagram in Figure 17.3 displays three clusters of keywords as described below.

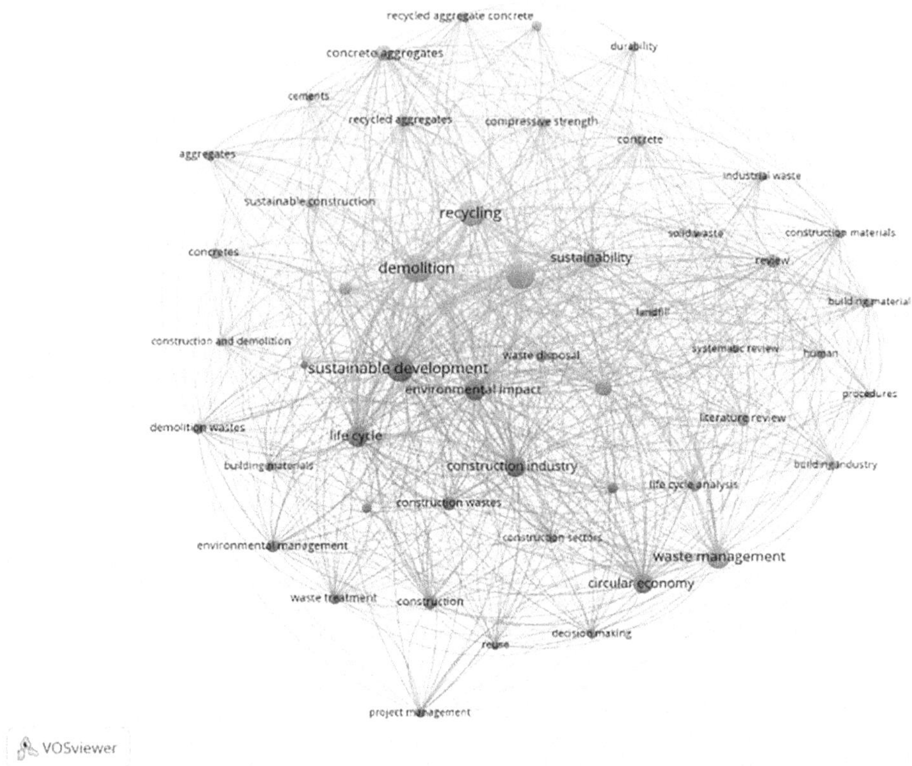

FIGURE 17.3 Keyword co-occurrence clusters in LCA and sustainable waste management research.

Cluster 1: Construction Waste, Circular Economy, and Environmental Governance (18 items)

This cluster is centred around keywords such as *construction waste*, *demolition waste*, *building materials*, *circular economy,* and *environmental impact*, reflecting a dominant thematic stream which connects sustainable construction practices with waste minimisation strategies. Terms like *decision making*, *project management,* and *reuse* suggest the growing role of LCA in informing managerial and policy-oriented decisions in the construction sector. The inclusion of *greenhouse gases*, *intelligent buildings,* and *sustainable development* indicates an integration of LCA with climate action goals and smart city agendas. This cluster reveals a comprehensive application of LCA in assessing and improving the sustainability performance of construction processes and waste treatment practices.

Cluster 2: Concrete-Based Materials, Recycled Aggregates, and Material Performance (16 items)

Keywords in this cluster, such as *aggregates*, *cements*, *concrete*, *compressive strength,* and *durability*, underscore a technically focused body of research on the environmental and structural performance of construction materials. A significant emphasis is placed on *recycled aggregates*, *recycled concrete,* and *construction and demolition waste*, reflecting the growing interest in circular material flows and low-impact material alternatives. The presence of *life cycle assessment* within this cluster suggests that LCA is being employed to assess not only environmental impacts but also the material feasibility and engineering performance of sustainable construction products. This cluster is heavily technical and material science-oriented, indicating a strong empirical foundation for sustainability assessments in construction.

Cluster 3: LCA Methodologies, Review Studies, and Waste Management Practices (16 items)

Cluster 3 encapsulates terms such as *life cycle analysis*, *systematic review*, *literature review*, *procedures,* and *sustainability*, signalling a meta-analytical and methodological focus. The presence of *solid waste*, *waste disposal*, *landfill,* and *waste management* links the cluster directly to end-of-life waste practices and disposal strategies. Keywords like *construction materials*, *building industry,* and *human* suggest the socio-technical context of waste generation and assessment. This cluster reflects a synthesis-driven research trend, where scholars use LCA frameworks to review, critique, and consolidate knowledge on sustainable waste practices across diverse sectors.

The three clusters demonstrate the multidimensional nature of LCA research in sustainable waste management:

- Cluster 1 highlights macro-level integration with CE and policy applications,
- Cluster 2 focuses on material performance and technical assessments,
- Cluster 3 encapsulates methodological and review-based inquiries into waste systems.

Together, these clusters reflect a mature, interdisciplinary research field with ongoing diversification towards sector-specific innovations and evidence-based policy development.

17.3.5 Temporal Trends in Keyword Co-Occurrence (2015–2025)

The VOSviewer overlay visualisation reveals the temporal evolution of key themes within the domain of LCA in sustainable waste management by assigning colour codes based on the average publication year associated with each keyword. This analytical layer enables a nuanced understanding of how research interests have shifted over time, providing insight into both foundational and emerging topics within the field.

From Figure 17.4, early-stage (pre-2020) keywords, highlighted in grey, include foundational concepts such as "life cycle," "sustainable development," "environmental impact," "construction waste," "building materials," and "waste treatment." These terms represent the core environmental and technical pillars of LCA research, where scholars initially focused on assessing environmental performance across construction and demolition processes and material life cycles.

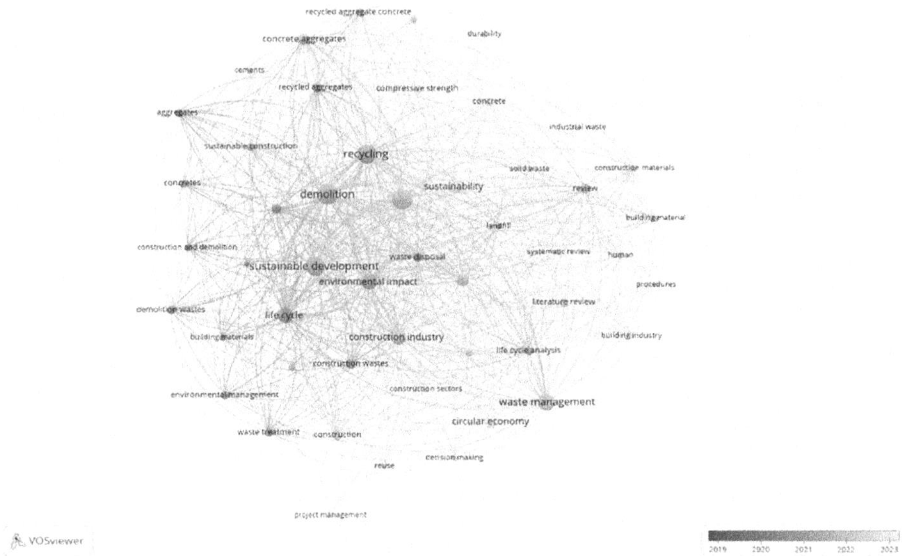

FIGURE 17.4 Overlay visualisation of author keyword co-occurrence in LCA research on sustainable waste management (2019–2023).

As the timeline progresses towards 2021, the brown-coloured keywords such as "construction industry," "environmental management," "sustainability," and "recycling" indicate the rise of integrative and policy-relevant themes. This period marks a transition from purely analytical applications of LCA to a more holistic approach encompassing waste valorisation, resource circularity, and strategic decision-making.

Most notably, recent research themes (2022–2023), depicted in yellow hues, centre around terms like "circular economy," "project management," "reuse," "review," "systematic review," "durability," and "recycled aggregate concrete." These newer terms indicate a strong pivot towards evidence synthesis, material innovation, and operationalising sustainability principles through circular practices. The emergence of systematic review and literature review also suggests a maturing field, characterised by the consolidation of knowledge and critical evaluation of methodological frameworks.

The visualisation also underscores the interconnectedness of topics, with "sustainability" and "waste management" appearing as central hubs which link various sub-themes. Their high co-occurrence frequency and positioning suggest their continued relevance as anchoring themes across evolving LCA applications.

Research trajectory in LCA for sustainable waste management has evolved from foundational impact assessments and technical evaluations towards a more comprehensive, circular, and management-oriented discourse (Uwa & Agboola, 2025; Weihan *et al.*, 2025). This temporal shift reflects broader global priorities on sustainable infrastructure, circular resource use, and knowledge integration through evidence-based reviews.

17.3.6 Core Research Foci in LCA and Sustainable Waste Management

The density visualisation map generated in Figure 17.5 illustrates the concentration and intensity of co-occurrence among keywords used in publications on LCA in sustainable waste management. In this visualisation, grey nodes represent keywords with the highest frequency and centrality, indicating strong research attention and widespread thematic relevance. Green areas denote moderately frequent terms, while blue regions highlight peripheral or emerging topics.

At the core of the map, "sustainability," "recycling," "demolition," "sustainable development," and "life cycle" emerge as the most intensively researched themes. These keywords form the intellectual nucleus of the field, representing the critical environmental principles and processes underpinning the use of LCA in assessing waste management practices. Their high density indicates their central role in framing contemporary research narratives and methodological approaches.

Notably, "recycled aggregate concrete," "compressive strength," "cement," "durability," and "aggregates" form a distinct cluster in the upper region of the map, suggesting an active sub-field centred on the environmental and structural evaluation of recycled construction materials. These technical terms highlight LCA's role not just in policy or impact assessment but also in advancing sustainable material innovation.

Peripheral keywords such as "project management," "decision making," "literature review," "procedures," and "human" indicate emerging or interdisciplinary dimensions. Their lower density does not diminish their importance, as these may represent newer lines of inquiry which integrate managerial, behavioural, and evaluative dimensions into the LCA discourse.

This density map confirms the thematic maturity and coherence of LCA research in sustainable waste management. While foundational concepts dominate the

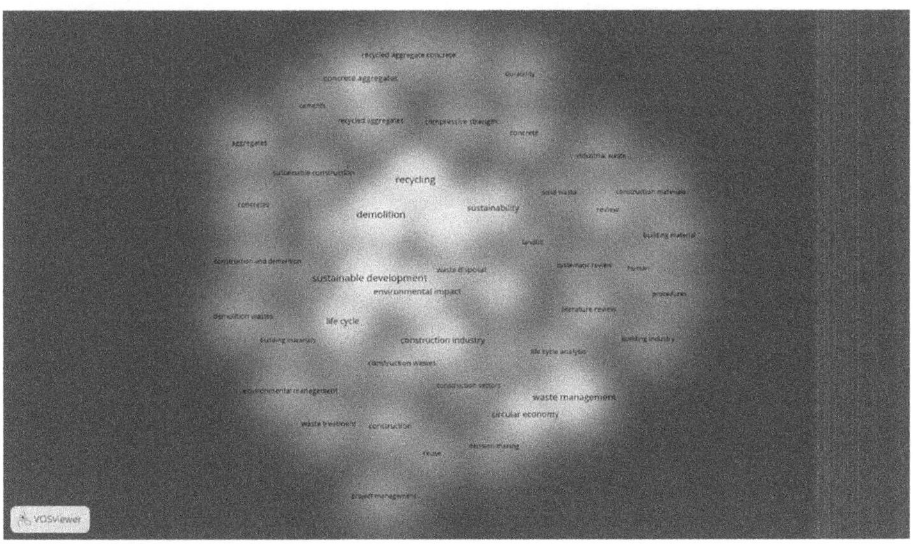

FIGURE 17.5 Keyword density visualisation.

intellectual landscape, the visualisation also reveals an expanding interest in CE applications, recycled material performance, and systematic assessment methods, pointing towards a progressively integrative and practice-oriented research trajectory.

17.4 IMPLICATION OF FINDINGS

This study reveals a rapidly evolving research landscape at the intersection of LCA and sustainable construction and demolition waste management (Prasittisopin *et al.*, 2025). The bibliometric analysis shows exponential growth in global scholarly output, with concentrated contributions from countries like China, India, the UK, and Australia. However, regional disparities highlight the need for more inclusive research collaboration. Three major thematic clusters emerged: (1) policy integration and CE linkages, (2) material and technical innovations, and (3) methodological advancements in LCA. These trends reflect the field's interdisciplinary nature, bridging engineering, environmental science, and policy. Emerging themes such as BIM-enabled LCA, AI-based waste sorting, and IoT-enhanced tracking signal future directions for research focused on digital transformation and real-time data integration. Academically, the findings encourage deeper exploration into underrepresented regions, upstream waste prevention, and the integration of social and economic metrics. Methodological innovations like multi-criteria decision analysis and hybrid LCA approaches also offer fertile ground for future inquiry (Rota *et al.*, 2024; Wang *et al.*, 2024).

Practically, the evidence supports the implementation of waste diversion strategies and the use of recycled materials in construction. LCA findings provide a strong justification for policy shifts, including stricter landfill regulations, recycling incentives, and sustainable procurement practices. Industry professionals and urban planners can leverage these insights, especially when integrated with digital tools to optimise waste strategies and align with climate and CE goals.

17.5 CONCLUSION

This study presents a comprehensive synthesis of the global research landscape on LCA in sustainable construction and demolition waste management. Through bibliometric and thematic analyses, it maps the field's evolution from being early foundational work to a mature, interdisciplinary domain aligned with sustainability goals and global policy imperatives. Key findings highlight a surge in research output, the diversification of contributors, and the emergence of thematic clusters around CE integration, material innovations, and advanced LCA methodologies. The temporal trend analysis reveals a notable shift towards holistic and circularity-focused approaches, with increasing emphasis on evidence synthesis, digital tools, and decision-support techniques. These trends reflect the field's growing responsiveness to pressing global challenges such as climate change, resource efficiency, and sustainable urban development (Neupane *et al.*, 2025; Farshadfar *et al.*, 2024).

The contribution of this study lies in its integrative perspective; it consolidates fragmented literature into a coherent, panoramic view, identifying key influencers, dominant tools, and methodological innovations. Unlike isolated case-based studies, it reveals meaningful connections across disciplines, such as the convergence of material science, policy studies, and digital innovation within LCA research. Importantly, the study flags underexplored areas including socio-economic impact metrics, upstream waste prevention, and regional disparities particularly in developing economies. These gaps point towards a future research agenda centred on broader sustainability assessments, context-sensitive applications, and cross-disciplinary integration. Beyond academia, the findings offer practical value for policymakers, urban planners, and industry stakeholders. The consolidated evidence supports the adoption of CE strategies, data-driven tools, and lifecycle thinking in construction waste management. In summary, this study not only documents the state of the art in LCA for C&D waste but also sets the stage for its future evolution. It bridges research and practice, provides a strategic reference for decision-makers, and contributes to the collective effort towards achieving global sustainability in the built environment.

REFERENCES

Abubakar, A., Mambo, A. D., Mogbo, O. I., Akeem, A., & Kolo, M. U. (2024). Sustainability assessment of construction and demolition waste material in the federal capital. *Nile Journal of Engineering and Applied Science*, 2(1), 320–327.

Al-Raqeba, H., Ghaffar, S. H., Al-Kheetan, M. J., & Chougan, M. (2023). Understanding the challenges of construction demolition waste management towards circular construction: Kuwait stakeholder's perspective. *Cleaner Waste Systems*, 4, 100075. https://doi.org/10.1016/j.clwas.2023.100075

Alimi, K., Jin, R., Nguyen, B. N., Nguyen, Q., Chen, W., & Hosking, L. (2025). Exploring artificial intelligence applications in construction and demolition waste management: A review of existing literature. *Journal of Science and Transport Technology*, 104–136.

Asdrubali, F., Baldassarri, C., & Fthenakis, V. (2013). Life cycle analysis in the construction sector: Guiding the optimization of conventional Italian buildings. *Energy and Buildings*, 64, 73–89.

Balasbaneh, A. T., & Sher, W. (2024). A systematic literature review of life cycle sustainability assessment of mass timber in the construction industry toward circular economy. *Environment, Development and Sustainability*, 1–37.

Bonifazi, G., Grosso, C., Palmieri, R., & Serranti, S. (2025). Current trends and challenges in construction and demolition waste recycling. *Current Opinion in Green and Sustainable Chemistry*, 53, 101032.

Cheng, J. C. P., & Ma, L. Y. H. (2013). A BIM-based system for demolition and renovation waste estimation and planning. *Waste Management*, 33(6), 1539–1551.

Donthu, N., Kumar, S., Mukherjee, D., Pandey, N., & Lim, W. M. (2021). How to conduct a bibliometric analysis: An overview and guidelines. *Journal of Business Research*, 133, 285–296.

Farina, I., Colangelo, F., Petrillo, A., Ferraro, A., Moccia, I., & Cioffi, R. (2020). LCA of concrete with construction and demolition waste. In *Advances in Construction and Demolition Waste Recycling* (pp. 501–513). Woodhead Publishing.

Farshadfar, Z., Mucha, T., & Tanskanen, K. (2024). Leveraging machine learning for advancing circular supply chains: A systematic literature review. *Logistics*, 8(4), 108.

Gao, Y., Yiu, T. W., Shen, X., & Tam, V. W. (2025). Life cycle insights into construction and demolition waste management: Past, present and emerging futures. *Journal of Building Engineering*, 113441.

He, X., & Cha, E. J. (2022). State of the research on disaster risk management of interdependent infrastructure systems for community resilience planning. *Sustainable and Resilient Infrastructure*, 7(5), 391–420.

Islam, N., Sandanayake, M., Muthukumaran, S., & Navaratna, D. (2024). Review on sustainable construction and demolition waste management – challenges and research prospects. *Sustainability*, 16, 3289. https://doi.org/10.3390/su16083289

Iyiola, C. O., Shakantu, W., & Daniel, E. I. (2024). Digital technologies for promoting construction and demolition waste management: A systematic review. *Buildings*, 14, 3234. https://doi.org/10.3390/buildings14103234

Kara De Maeijer, P. (2025). Innovative solutions for concrete applications. *Infrastructures*, 10(3), 59.

Kasulla, S., Malik, S. J., Baxla, S. P., & Zafar, S. (2024). The role of Iot in waste management and sustainability. *Partners Universal International Research Journal*, 3(2), 76–88.

Lu, W., & Tam, V. W. (2013). Construction waste management policies and their effectiveness in Hong Kong: A longitudinal review. *Renewable and Sustainable Energy Reviews*, 23, 214–223.

Marinković, S., Dragaš, J., Ignjatović, I., & Tošić, N. (2017). Environmental assessment of green concretes for structural use. *Journal of Cleaner Production*, 154, 633–649.

Martin, H., Chebrolu, D., Chadee, A., & Brooks, T. (2024). Too good to waste: Examining circular economy opportunities, barriers, and indicators for sustainable construction and demolition waste management. *Sustainable Production and Consumption*, 48, 460–480.

Neupane, R. P., Devi, N. R., Imjai, T., Rajput, A., & Noguchi, T. (2025). Cutting-edge techniques and environmental insights in recycled concrete aggregate production: A comprehensive review. *Resources, Conservation & Recycling Advances*, 25, 200241.

Pomponi, F., & Moncaster, A. M. (2016). Embodied carbon mitigation and reduction in the built environment–What does the evidence say? *Journal of Environmental Management*, 181, 687–700.

Prasittisopin, L., Tuvayanond, W., Kang, T. H. K., & Kaewunruen, S. (2025). Concrete mix design of recycled concrete aggregate (RCA): Analysis of review papers, characteristics, research trends, and underexplored topics. *Resources*, 14(2), 21.

Rota, A., Lamberti, V., & Fiorito, F. (2024). Reuse of facade materials across benefits, challenges and potential opportunities. In *International Conference of Ar. Tec. (Scientific Society of Architectural Engineering)* (pp. 647–664). Cham: Springer Nature Switzerland.

Shajidha, H. & Mortula, M. M. (2025). Sustainable waste management in the construction industry. *Frontiers in Sustainable Cities*, 7, 1582239. https://doi.org/10.3389/frsc.2025.1582239

Swarnakar, V., & Khalfan, M. (2024). Circular economy in construction and demolition waste management: An in-depth review and future perspectives in the construction sector. *Smart and Sustainable Built Environment*. https://doi.org/10.1108/SASBE-02-2024-0056

Unegbu, H. C. O., & Yawas, D. S. (2024). Optimizing construction and demolition waste management in nigeria: Challenges, regulatory frameworks, and policy solutions. *Discover Civil Engineering*, 1(141). https://doi.org/10.1007/s44290-024-00142-3

Uwa, J. N., & Agboola, O. P. (2025). Applications of immersive technology in building construction phase: Fostering sustainability in the global context. In *Applications of Immersive Technology in Architecture, Engineering and Construction* (pp. 93–104). Routledge.

Victar, H. C., & Waidyasekara, A. S. (2025). Optimising construction waste management in Sri Lanka through Circular economy strategies: A focus on construction and renovation and use and operate stages. *Engineering, Construction and Architectural Management*, 32(7), 4275–4309.

Wang, Z., Zhou, Y., Wang, T., & Zhao, N. (2024). Efficiency of construction waste and carbon reduction in the construction industry: Based on improved three stage SBM-DEA model in China. *Engineering, Construction and Architectural Management*. https://doi.org/10.1108/ECAM-10-2023-1088

Weihan, S., Quddus, T., Guomin, Z., Song, A., Lei, H., Jingxuan, Z., & Wang, S. (2025). An analytical review of construction and demolition waste management and quantification methods using a science mapping approach. *Recycling*, 10(3), 115.

Wu, J., Ye, X., & Cui, H. (2025a). Recycled materials in construction: Trends, status, and future of research. *Sustainability*, 17(6), 2636.

Wu, Z., Pei, T., Bao, Z., Thomas NG, S., Lu, G., & Chen, K. (2025b). Utilizing intelligent technologies in construction and demolition waste management: From a systematic review to an implementation framework. *Frontiers of Engineering Management*, 12(1), 1–23.

Yazdani, S., & Lakzian, E. (2023). Conservation; waste reduction/zero waste. In *Pragmatic Engineering and Lifestyle* (pp. 131–152). Emerald Publishing Limited.

Zakerhosseini, A., Abdoli, M. A., Molayzahedi, S. M., & Salmi, F. K. (2024). Life cycle assessment of construction and demolition waste management: A case study of Mashhad, Iran. *Environment, Development and Sustainability*, 26, 25717–25743.

Zhang, L., Rousseau, R., & Glänzel, W. (2016). Diversity of references as an indicator of the interdisciplinarity of journals: Taking similarity between subject fields into account. *Journal of the Association for Information Science and Technology*, 67(5), 1257–1265.

18 Competitive Adsorption of Aqueous Selenium & Phenol Species onto Activated Carbon: RSM-based Modelling

Mehvish Bilal

18.1 INTRODUCTION

In view of the potential health hazards associated with selenium (SE) exposure, the allowable limits in drinking water and wastewater discharge are stringent. The most commonly reported aqueous SE species are $\left(SeO_3^{2-}\right)$ and selenate $\left(SeO_4^{2-}\right)$, which are produced by various industrial sources, including the petroleum refineries [1–3]. However, SE species are not the only harmful components in the refinery wastewater; it also includes phenol, an organic co-pollutant which is dangerous to the environment [4, 7].

The presence of synergistic effects between co-contaminants makes it especially difficult to treat SE in mixed wastewater streams. This complication is a reason to conduct more studies on the effective methods to remove either SE or organic pollutants, for example, phenol, under competitive adsorption conditions. Activated carbon (AC)-based adsorption is one of the most widely used approaches that has been established to eliminate organic and inorganic contaminants [8,11]. The removal of SE in aqueous solution using AC has also been investigation under different conditions [12, 15]. Several low-cost carbons and biosorbents have been explored in the elimination of Se (IV) and Se (VI), and most of them have increased uptake with Se (IV) at acidic pH [16, 17]. Granular activated carbon (GAC) has demonstrated versatility in removing other water contaminants, such as bromate [18], and coconut shell-based GAC has proven effective in improving water quality by reducing excess chlorine and neutralising physico-chemical parameters [19]. Although selenium removal through adsorption has been extensively studied, the treatment of selenium-phenol mixed wastewater streams using AC has not been previously explored. The current study addresses this research gap by focusing on adsorption efficiency and process optimisation. The following sections contain detailed findings.

DOI: 10.1201/9781003664284-18

18.2 MATERIALS AND METHODOLOGY

18.2.1 Materials

The following reagents were employed in this study: phenol, hydrochloric acid, sodium hydroxide (NaOH) to alter pH, and selenium salts (SeO_3^{2-} and SeO_4^{2-}). The commercial activated carbon (FILTRASORB 400) was used as an adsorbent. The pH meter and electrode system were standardised using buffer solutions of pH 4, 7, and 10. The concentrations of SeO_3^{2-} and SeO_4^{2-} were measured using an ion chromatography (IC) system (Metrohm, Switzerland) equipped with a high-sensitivity conductivity detector. Samples were prepared by filtration using 0.2 μm cellulose nitrate membranes (Whatman, Germany).

18.2.2 Phenol Analysis

The concentration of phenol was examined using a UV-Vis spectrophotometer (UV-1601 PC, Shimadzu, Japan). Samples were filtered through 0.23 μm membranes were put in spectrophotometric samples, and a drop or two of potassium dihydrogen phosphate buffer (KH_2PO_4) was added to ensure that pH was maintained. The absorbance at 269 nm was then recorded. Calibration was performed using reagent-grade phenol standards before each batch, ensuring accuracy and consistency. This approach aligns with commonly used UV-vis analytical protocols where phenol quantification in aqueous samples is reliably carried out around 260-270 nm (e.g. 260 nm detection wavelength) for trace phenol monitoring in environmental waters [20].

18.2.3 Response Surface Methodology

Response surface methodology (RSM), which is based on Central Composite Design (CCD), was used to design the experiment [21, 22]. The adsorbent dosage (g/L), SeO_3^{2-}, SeO_4^{2-}, and phenol were taken as independent variables. Accordingly, the design of experiments (DOE) was designed as demonstrated in Table 18.1. The experimental runs were generated according to the CCD matrix, resulting in 25 runs as shown in Table 18.2.

TABLE 18.1
Factor Levels Used in the Experimental Design

Factors	Level		
	−1	0	1
A = Adsorbent dose (g/L)	0.2	0.6	1.0
B = SeO_3^{2-} (mg/L)	2.5	5.0	7.5
C = SeO_4^{2-}(mg/L)	5.0	10	15
D = Phenol (mg/L)	2.5	5.0	7.5

TABLE 18.2
Design of Experiments With Factor Levels for Each Experimental Run

Run	Factor 1 A: Dose g/l	Factor 2 B: SeO_3^{2-} mg/L	Factor 3 C: SeO_4^{2-} mg/L	Factor 4 D: Phenol mg/L
1	1	2.5	15	2.5
2	1	5	10	5
3	0.2	5	10	5
4	0.6	5	10	7.5
5	1	7.5	5	2.5
6	1	2.5	5	7.5
7	1	2.5	15	7.5
8	0.2	2.5	5	2.5
9	0.2	7.5	5	7.5
10	1	2.5	5	2.5
11	0.6	5	10	5
12	1	7.5	15	2.5
13	0.2	7.5	5	2.5
14	1	7.5	5	7.5
15	0.2	2.5	5	7.5
16	0.6	5	5	5
17	0.2	7.5	15	2.5
18	0.6	5	10	2.5
19	0.2	7.5	15	7.5
20	0.2	2.5	15	2.5
21	0.6	2.5	10	5
22	1	7.5	15	7.5
23	0.6	5	15	5
24	0.2	2.5	15	7.5
25	0.6	7.5	10	5

18.3 RESULTS AND DISCUSSION

18.3.1 Competitive Removal of Phenol and Selenium Species

This work uses Filtrasorb granular activated carbon (FSGAC) as an adsorbent to investigate the competitive adsorption of phenol and selenium species from aqueous solutions. The adsorption experiments were systematically designed using RSM coupled with a CCD to evaluate the interaction and significance of multiple variables. Four independent variables were selected: adsorbent dosage (g/L), initial concentration of

SeO_3^{2-}, SeO_4^{2-}, and phenol (Table 18.1). The corresponding experimental matrix and response values are provided in Table 18.2. The experimental responses were defined as the percentage removal of SeO_3^{2-}, SeO_4^{2-}, and phenol. The observed removal efficiencies under various operating conditions are summarised in Table 18.3.

Among the three compounds tested, phenol showed the strongest attraction to FSGAC, followed by SeO_3^{2-} and SeO_4^{2-}. This means that even when all three substances were present together, phenol adsorption was only minimally affected. In contrast, SeO_3^{2-} and then SeO_4^{2-} are more polar and structurally different from phenol,

TABLE 18.3
Experimental Responses for Selenite, Selenate, and Phenol Removal Efficiency

Run	Factor A A: Dose g/l	Factor B B: SeO_3^{2-} ppm	Factor C C: SeO_4^{2-} ppm	Factor D D: Phenol ppm	Experimental Response for B Removal SeO_3^{2-} %	Experimental Response for C Removal SeO_4^{2-} %	Experimental Response for D Removal Phenol %
1	1	2.5	15	2.5	0	0	96
2	1	5	10	5	48	46	98
3	0.2	5	10	5	3	11	78
4	0.6	5	10	7.5	13	33	94
5	1	7.5	5	2.5	24	28	92
6	1	2.5	5	7.5	46	34	94
7	1	2.5	15	7.5	38	55	93
8	0.2	2.5	5	2.5	50	23	80
9	0.2	7.5	5	7.5	32	30	76
10	1	2.5	5	2.5	41	18	94
11	0.6	5	10	5	71	15	89
12	1	7.5	15	2.5	61	40	97
13	0.2	7.5	5	2.5	44	28	77
14	1	7.5	5	7.5	39	48	97
15	0.2	2.5	5	7.5	48	60	77
16	0.6	5	5	5	58	60	87
17	0.2	7.5	15	2.5	60	41	90
18	0.6	5	10	2.5	31	45	87
19	0.2	7.5	15	7.5	41	17	72
20	0.2	2.5	15	2.5	6	31	78
21	0.6	2.5	10	5	41	20	97
22	1	7.5	15	7.5	50	37	95
23	0.6	5	15	5	32	34	0
24	0.2	2.5	15	7.5	50	27	0
25	0.6	7.5	10	5	49	42	0

making them less competitive in the adsorption process when all three compounds are present. To model these observations, RSM was employed to generate predictive equations for each contaminant. The resulting RSM-based models for SeO_3^{2-}, SeO_4^{2-}, and phenol removal are provided in Equations (18.1), (18.2), and (18.3), respectively. The models achieved high predictive accuracy and were statistically validated using analysis of variance (ANOVA), like other adsorption studies utilising RSM for process optimisation [23].

Removal efficiency of SeO_3^{2-}:

$$\text{Removal}_{\text{SeO3}}{}^{2-} = 54.53 + 18.21A - 1.85B - 1.31D - 2.17AB - 5.51B^2 - 7.29D^2 \quad (18.1)$$

Removal efficiency of selenate $\left(SeO_4^{2-}\right)$:

$$\text{RemovalSeO}_4^{2-} = 47.53 + 17.044 \quad (18.2)$$

Removal efficiency of phenol:

$$\text{Removal}_{\text{Phenol}} = 93.94 + 8.844 - 0.88D - 4.28A^2 - 3.60\,D^2 \quad (18.3)$$

Where A represents adsorbent dosage (g/L), B is SeO_3^{2-}, C is SeO_4^{2-}, D is phenol, and the remaining terms denote respective interaction and quadratic factors.

The statistical fitting parameters corresponding to the RSM models are given in Equations (18.1), (18.2), and (18.3). Similarly, for phenol removal, ANOVA results showed that adsorbent dosage was the most important factor ($p < 0.0001$), followed by phenol concentration and their quadratic terms, as summarised in Table 18.4. Another critical layer of model validation involves examining the p-values of individual model terms. A p-value less than 0.05 indicates that the corresponding factor or interaction significantly influences the response. Detailed p-values for each term are provided in Tables 18.5, 18.6, and 18.7. Furthermore, in Figure 18.1a-c, the normal probability plots of residuals for SeO_3^{2-}, SeO_4^{2-}, and phenol display residuals aligned closely along a straight line, confirming that the errors follow a normal distribution, an important assumption of linear regression models. In Figure 18.2a-c, the predicted versus actual response plots show data points clustered around the 45° reference line, demonstrating an good predictive accuracy and model fit.

TABLE 18.4
Fit Statistics of Generated Models

Response	R^2	Adjusted R^2	Predicted R^2	Adequate Precision
RE. SeO_3^{2-}	0.9707	0.9609	0.9382	25.2559
RE. SeO_4^{2}	0.9106	0.9067	0.8963	25.5142
RE. Phenol	0.9708	0.965	0.9467	31.2681

TABLE 18.5
RSM Details of the % Selenite Removal Model

Source	Sum of Squares	df	Mean Square	F-value	p-value	
Model	6316.25	6	1052.71	99.23	< 0.0001	Significant
A-Dose	5971.98	1	5971.98	562.92	< 0.0001	
B-SeO_3^{2-}	61.51	1	61.51	5.8	0.027	
D-Phenol	30.67	1	30.67	2.89	0.1063	
AB	75.55	1	75.55	7.12	0.0157	
B^2	97.48	1	97.48	9.19	0.0072	
D^2	170.22	1	170.22	16.05	0.0008	
Residual	190.96	18	10.61			
Cor Total	6507.21	24				

TABLE 18.6
RSM Details of the % Selenate Removal Model

Source	Sum of Squares	df	Mean Square	F-value	p-value	
Model	5225.74	1	5225.74	234.35	< 0.0001	Significant
A-Dose	5225.74	1	5225.74	234.35	< 0.0001	
Residual	512.87	23	22.3			
Cor Total	5738.61	24				

TABLE 18.7
RSM Details of the % Phenol Removal Model

Source	Sum of Squares	df	Mean Square	F-value	p-Value	
Model	1672.24	4	418.06	166.37	< 0.0001	Significant
A-Dose	1406.9	1	1406.9	559.9	< 0.0001	
D-Phenol	14	1	14	5.57	0.0285	
A^2	58.67	1	58.67	23.35	0.0001	
D^2	41.62	1	41.62	16.56	0.0006	
Residual	50.26	20	2.51			
Cor Total	1722.5	24				

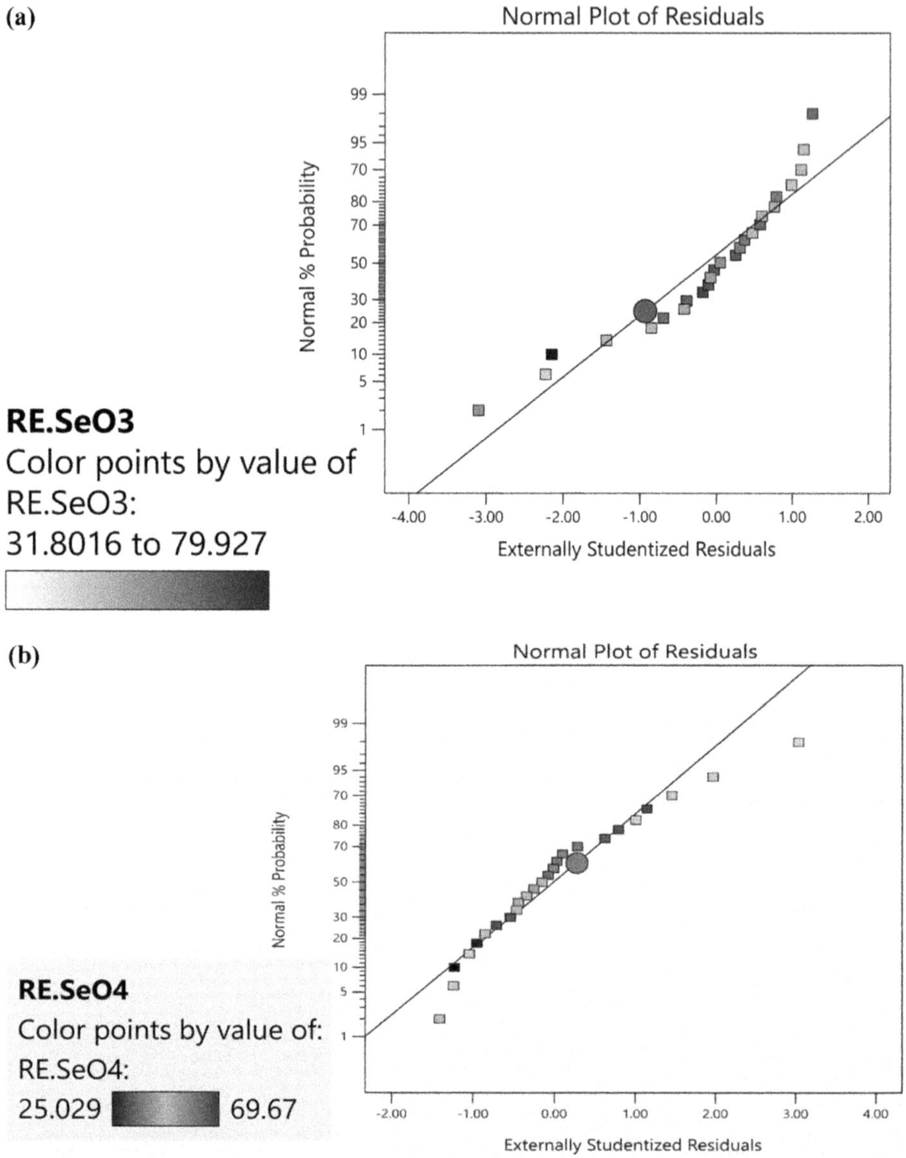

FIGURE 18.1 RSM-based "Normal Probability Plot of Residuals" for (a) selenite, (b) selenate, and (c) phenol.

(c)

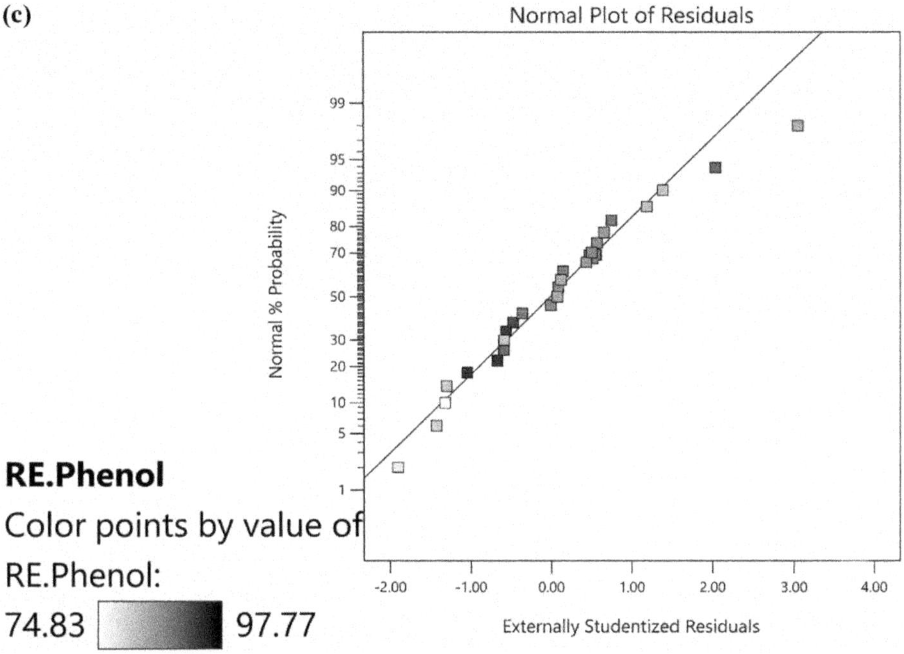

FIGURE 18.1 (*Continued*)

(a)

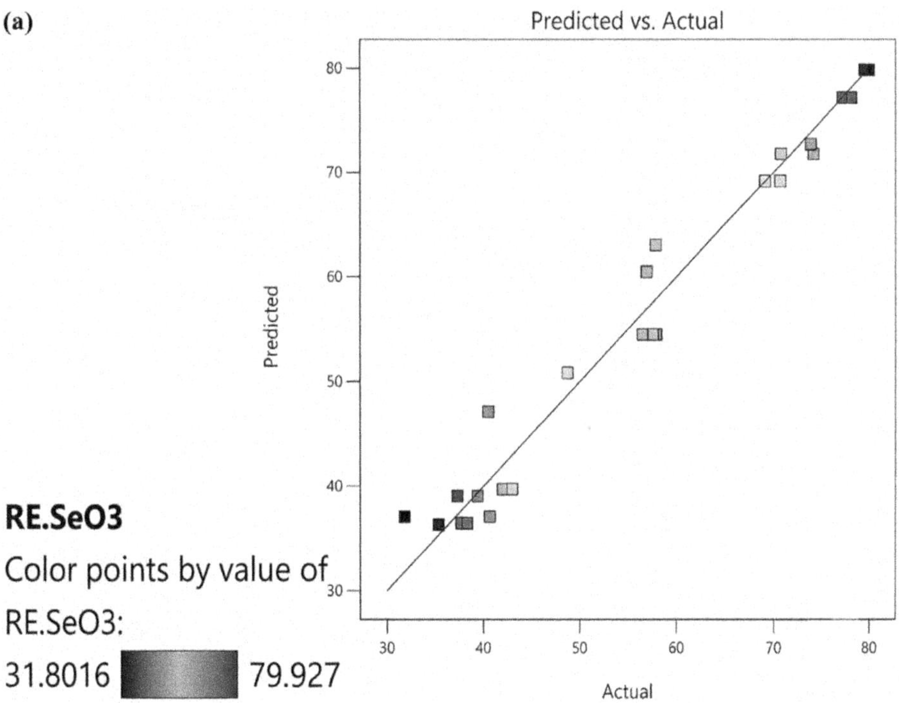

FIGURE 18.2 RSM-based "Predicted vs. Actual Plots" for (a) selenite, (b) selenate, and (c) phenol.

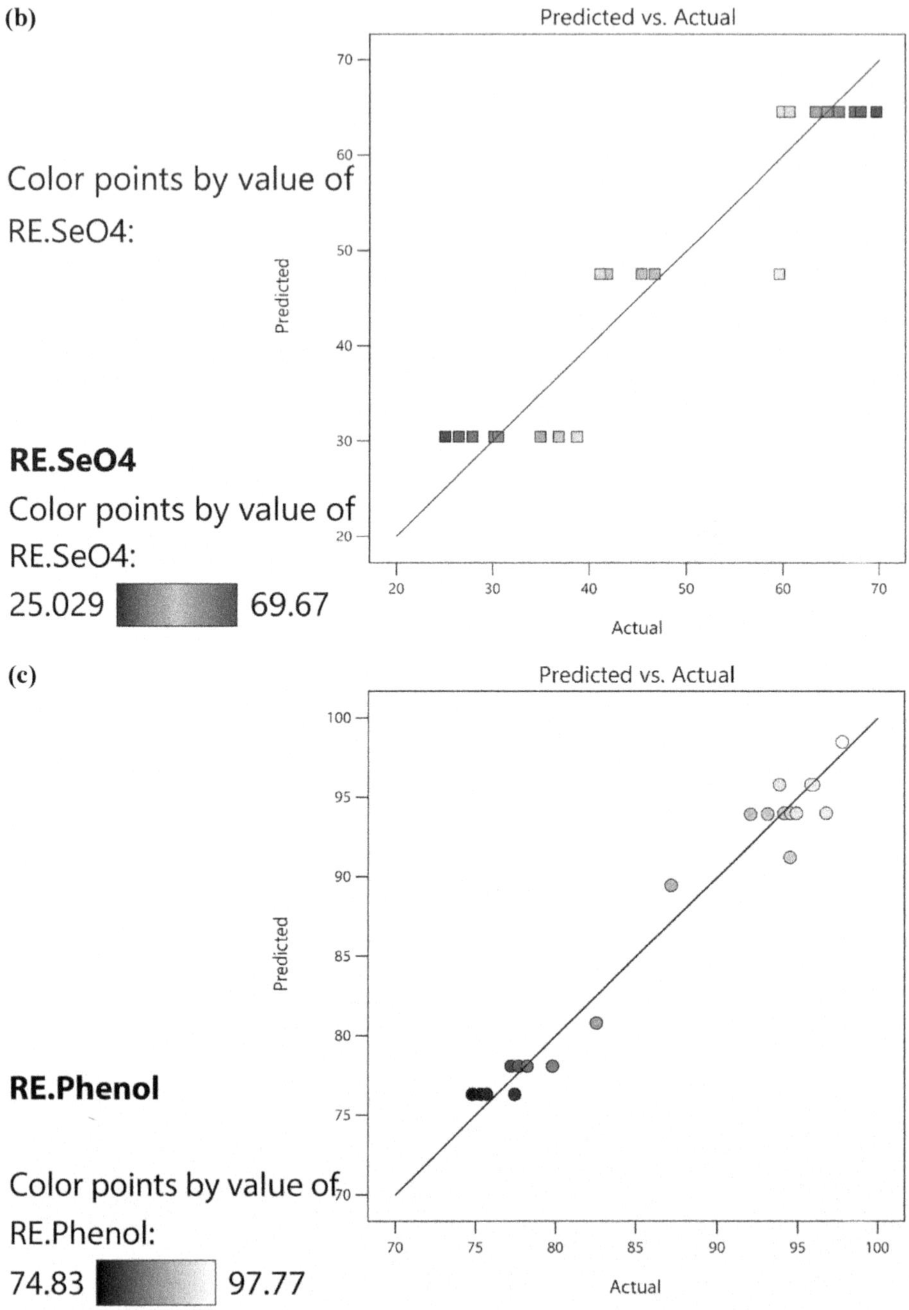

FIGURE 18.2 *(Continued)*

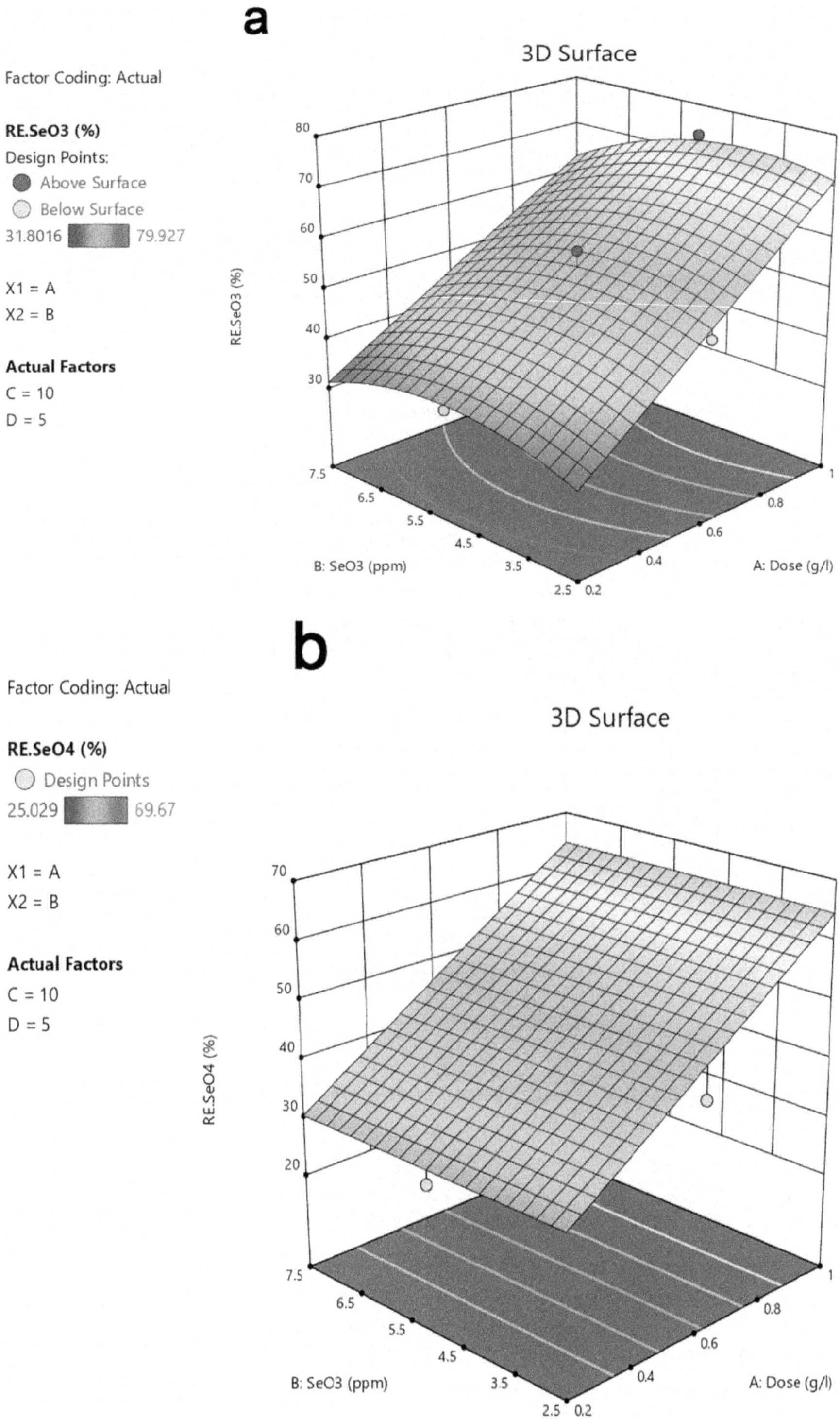

FIGURE 18.3 The RSM plots for (a) % selenite removal, (b) % selenate removal, and (c) % phenol removal, as a function of selenite concentration and adsorbent dosage.

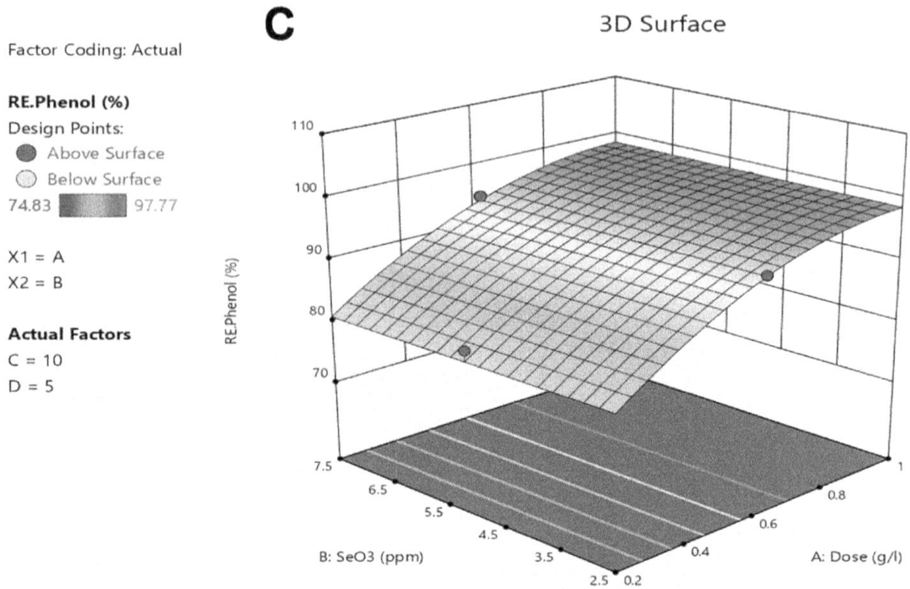

FIGURE 18.3 (*Continued*)

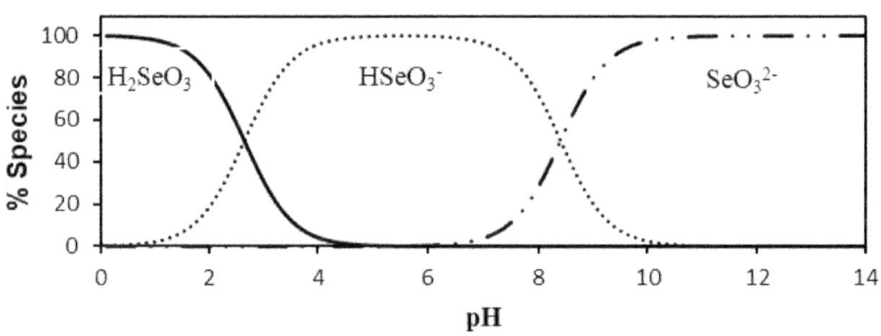

FIGURE 18.4 Qualitative selenite speciation diagram as a function of pH [pK_{a1} 2.63 and pK_{a2} 8.4].

18.3.2 Effect of Main Parameters on the Removal of Selenium Species

The adsorption behaviour of SeO_3^{2-}, SeO_4^{2-}, and phenol was assessed using 3D response surface plots (Figures 18.3a-c). In the case of SeO_4^{2-}, removal efficiency was observed to increase significantly with higher adsorbent dosage, as shown in Figure 18.3b. At higher solution pH, the surface of the GAC becomes increasingly negatively charged, which reduces the electrostatic attraction between the GAC surfaces and negatively charged (anionic) species in solution. To better understand this phenomenon, qualitative aqueous speciation diagrams for SeO_3^{2-} and SeO_4^{2-} are provided in Figures 18.4 and 18.5, respectively. In contrast, the phenol removal

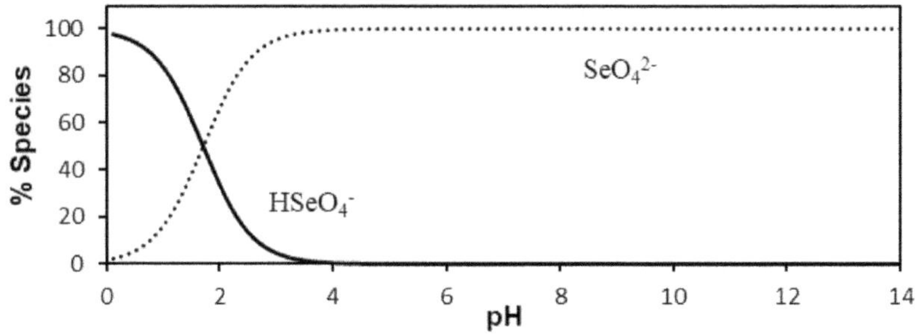

FIGURE 18.5 Qualitative selenate speciation diagram as a function of pH [pK_{a2} 1.7].

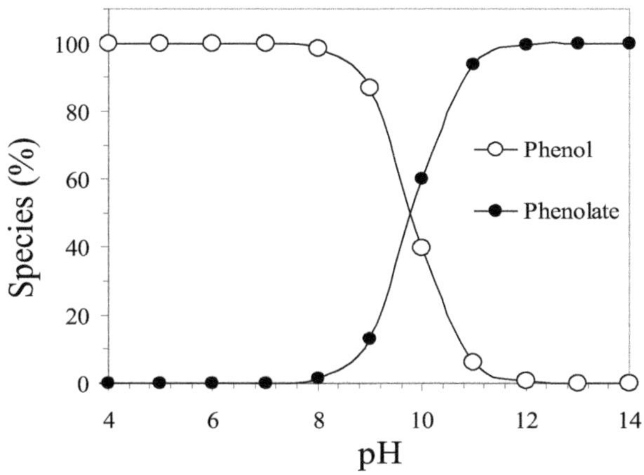

FIGURE 18.6 Qualitative phenol speciation diagram as a function of pH [using pK_a 9.82].

results (Figure 18.3c) demonstrate nearly complete removal at higher adsorbent dosages. The same adsorption behaviour was observed in earlier works, whereby selenite showed significantly greater uptake on GAC-based adsorbents in relation to other oxyanions [13]. The aqueous speciation of phenol as a function of pH can account for this phenomenon, illustrated in Figure 18.6.

18.4 CONCLUSION

This study examined the feasibility of employing FSGAC for the treatment of mixed aqueous streams contaminated with SeO_3^{2-}, SeO_4^{2-}, and phenol. The competitive adsorption behaviour of these pollutants followed a consistent trend, with phenol exhibiting the highest removal efficiency, followed by SeO_3^{2-} and then SeO_4^{2-}. Phenol removal was found to be particularly efficient at the experimental pH of 4, where phenol exists predominantly in its neutral molecular form. RSM modelling yielded non-linear models for both phenol and SeO_3^{2-} removal, while SeO_4^{2-} removal followed a linear trend. The statistical parameters indicated strong model fits and good predictive capability. With the increase in adsorbent dosage, the removal efficiency of all the contaminants generally increased. The removal of SeO_3^{2-} had a hump-shaped response with respect to the initial concentration: the removal of efficiency was maximum at a specific concentration, beyond which it decreased. This behaviour may be attributed to mass transfer limitations during adsorption, particularly from the bulk liquid phase to the solid adsorbent surface. SeO_4^{2-} removal also improved with higher adsorbent dosages and appeared largely unaffected by the presence of SeO_3^{2-}. However, SeO_4^{2-} removal remained lower than that of SeO_3^{2-}, which may be due to its larger molecular size and its divalent anionic form at pH 4, both of which reduce its affinity for the negatively charged GAC surface. In summary, the results demonstrate that FSGAC can effectively remove phenol, SeO_3^{2-}, and SeO_4^{2-} from competitive aqueous systems, with phenol showing the most favourable adsorption. These findings can inform the design and optimisation of GAC-based treatment systems for the remediation of contaminated water containing multiple organic and inorganic pollutants.

REFERENCES

1. N. Miekeley, R. C. Pereira, E. A. Casartelli, A. C. Almeida, and M. De F.b. Carvalho, “Inorganic speciation analysis of selenium by ion chromatography-inductively coupled plasma-mass spectrometry and its application to effluents from a petroleum refinery,” *Spectrochim. Acta – Part B At. Spectrosc.*, vol. 60, no. 5, pp. 633–641, 2005, https://doi.org/10.1016/j.sab.2005.02.024.
2. G. B. Tonietto, J. M. Godoy, A. C. Oliveira, and M. V. De Souza, “Simultaneous speciation of arsenic (As(III), MMA, DMA, and As(V)) and selenium (Se(IV), Se(VI), and SeCN-) in petroleum refinery aqueous streams,” *Anal. Bioanal. Chem.*, vol. 397, no. 5, pp. 1755–1761, 2010, https://doi.org/10.1007/s00216-010-3764-y.
3. D. Wallschläger and N. S. Bloom, “Determination of selenite, selenate and selenocyanate in waters by ion chromatography-hydride generation-atomic fluorescence spectrometry (IC-HG-AFS),” *J. Anal. At. Spectrom.*, vol. 16, no. 11, pp. 1322–1328, 2001, https://doi.org/10.1039/b103108m.
4. H. K. Hansen, S. F. Peña, C. Gutiérrez, A. Lazo, P. Lazo, and L. M. Ottosen, “Selenium removal from petroleum refinery wastewater using an electrocoagulation technique,” *J. Hazard. Mater.*, vol. 364, no. October 2018, pp. 78–81, 2019, https://doi.org/10.1016/j.jhazmat.2018.09.090.
7. Y. Yavuz, A. S. Koparal, and Ü. B. Öğütveren, “Treatment of petroleum refinery wastewater by electrochemical methods,” *Desalination*, vol. 258, no. 1–3, pp. 201–205, 2010, https://doi.org/10.1016/j.desal.2010.03.013.
8. R. R. Gil, B. Ruiz, M. S. Lozano, M. J. Martín, and E. Fuente, “VOCs removal by adsorption onto activated carbons from biocollagenic wastes of vegetable tanning,” *Chem. Eng. J.*, vol. 245, pp. 80–88, 2014, https://doi.org/10.1016/j.cej.2014.02.012.
11. J. Guo, W. S. Xu, Y. L. Chen, and A. C. Lua, “Adsorption of NH 3 onto activated carbon prepared from palm shells impregnated with H_2SO_4,” *J. Colloid Interface Sci.*, 2005, https://doi.org/10.1016/j.jcis.2004.08.101.
12. C. Namasivayam, D. Sangeetha, and R. Gunasekaran, “Removal of anions, heavy metals, organics and dyes from water by adsorption onto a new activated carbon from Jatropha husk, an agro-industrial solid waste,” *Process Saf. Environ. Prot.*, vol. 85, no. 2 B, pp. 181–184, 2007, https://doi.org/10.1205/psep05002.

13. N. Zhang, L. S. Lin, and D. Gang, "Adsorptive selenite removal from water using iron-coated GAC adsorbents," *Water Res.*, vol. 42, no. 14, pp. 3809–3816, 2008, https://doi.org/10.1016/j.watres.2008.07.025.
14. G. B. Jegadeesan, K. Mondal, and S. B. Lalvani, "Adsorption of Se (IV) and Se (VI) using copper-impregnated activated carbon and fly ash-extracted char carbon," *Water. Air. Soil Pollut.*, vol. 226, no. 8, 2015, https://doi.org/10.1007/s11270-015-2520-5.
15. K. L. Wasewar, B. Prasad, and S. Gulipalli, "Adsorption of selenium using bagasse fly ash," *Clean – Soil, Air, Water*, vol. 37, no. 7, pp. 534–543, 2009, https://doi.org/10.1002/clen.200800183.
16. E. I. El-Shafey, "Removal of Se(IV) from aqueous solution using sulphuric acid-treated peanut shell," *J. Environ. Manage.*, vol. 84, no. 4, pp. 620–627, 2007, https://doi.org/10.1016/j.jenvman.2007.03.021.
17. E. I. El-Shafey, "Sorption of Cd(II) and Se(IV) from aqueous solution using modified rice husk," *J. Hazard. Mater.*, vol. 147, no. 1–2, pp. 546–555, 2007, https://doi.org/10.1016/j.jhazmat.2007.01.051.
18. M. Vohra, A. Basaleh, F. Kabir, M. Al-Suwaiyan, and M. Bilal, "Aqueous bromate (BrO3–) removal using FILTRASORB 400 activated carbon: Effect of process variables and modeling," *Desalin. Water Treat.*, vol. 299, pp. 102–109, 2023.
19. M. Bilal, S. Kaur, and J. Mushtaq, "Adsorption investigation of physico-chemical characteristics of different water samples by coalesce coconut shell," *Int. J. Environ. Eng.*, vol. 12, no. 1, pp. 1–14, 2023.
20. T. K. Patle, K. Shrivas, R. Kurrey, S. Upadhyay, R. Jangde, and R. Chauhan, "Phytochemical screening and determination of phenolics and flavonoids in Dillenia pentagyna using UV–vis and FTIR spectroscopy," *Spectrochim. Acta Part A Mol. Biomol. Spectrosc.*, vol. 242, p. 118717, 2020.
21. M. J. Amiri, M. Bahrami, and F. Dehkhodaie, "Optimization of Hg (II) adsorption on bio-apatite based materials using CCD-RSM design: Characterization and mechanism studies," *J. Water Health*, vol. 17, no. 4, pp. 556–567, 2019.
22. K. M. Moria, H. Khurshid, M. R. U. Mustafa, A. Alhothali, and O. O. Bamasag, "Application of the response surface methodology (RSM) in the optimization of acenaphthene (ACN) removal from wastewater by activated carbon," *Sustainability*, vol. 14, no. 14, p. 8581, 2022.
23. A. I. A. Ibrahim and M. S. Vohra, "Mg-Fe-LDH for aquatic selenium treatment: Adsorption, RSM modeling, and machine learning neural network," *Water, Air, Soil Pollut.*, vol. 234, no. 7, p. 433, 2023.

19 Structural Performance of Sustainable Concrete With Waste-Derived Recycled Aggregates and Pozzolans

A Review

Wisal Ahmed, Muhammad Basit Khan, Jitong Ding and Arslan Akbar

19.1 INTRODUCTION

Globally, concrete is an extensively used construction material owing to its robustness, endurance, flexibility, and low cost. Still, it imposes a substantial environmental burden due to its high intake of natural resources and energy during the fabrication of cement. The cement manufacturing process alone is responsible for about 8% of global carbon emissions (Monteiro et al. 2017; Li et al. 2024; Shivaprasad et al. 2024). This makes the cement manufacturing sector the single most significant industry causing this much carbon emissions. Moreover, continuous extraction of natural resources for cement and concrete production from quarries disrupts the ecosystems and reduces finite geological reserves. Concrete consists of cement, aggregate, and water, as well as some admixtures, with aggregates making 70%–80% of the total concrete volume. Recently, the global production of natural aggregates has been estimated at nearly 40 billion tons/year (Xing et al. 2022). Most importantly, the mining process for extracting natural aggregates involves heavy equipment and makes a significant contribution to energy consumption. Simultaneously, rapid urbanization and construction growth have intensified the issue of construction and demolition waste (C&D), which is about one-third of the global solid debris (Islam et al. 2019; Ahmed et al. 2025a). This dual challenge of resource depletion and waste accumulation has driven the scientific and engineering community toward developing innovative, sustainable alternatives to conventional concrete.

Repurposing C&D waste as crushed concrete for use in new aggregate materials offers a sustainable pathway for resource recovery and greener concrete production (Ossa et al. 2016; Khan et al. 2024). RCA, obtained by crushing and processing

 DOI: 10.1201/9781003664284-19

demolished concrete, offers multiple benefits. For instance, its use in new concrete making can reduce reliance on virgin aggregates, conserve natural materials, and alleviate pressure on ecosystems. Studies suggest that natural aggregates can be substituted by up to 20–50% RCA without significantly altering the concrete integrity (Kou et al. 2011; Elansary et al. 2021; Al-Kheetan et al. 2024; Ahmed et al. 2025b). Furthermore, by diverting C&D waste from landfills, RCA supports waste minimization, aligning with global sustainability goals.

However, the application of RCA in structural concrete causes major problems, particularly at the microstructural level, which affects mechanical and durability characteristics (Xiao et al. 2005; Thomas et al. 2018; Panghal and Kumar 2025). RCA's inherent defects, such as adhered mortar residues, microcracks from crushing, and higher porosity, result in a feebler interfacial transition zone (ITZ), which is critical for load transfer and overall structural integrity. The porosity of the adhered mortar increases moisture uptake and leads to a less dense matrix that is prone to shrinkage and cracking (Ahmed and Lim 2023b; Pereira et al. 2023). This facilitates the ingress of moisture and deleterious ions, compromising the concrete's structural performance over time. It has been well established that the concrete comprising RCA typically exhibits inferior compressive strength (f'_{CS}), abridged tensile strength (f'_{TS}), and a lesser elastic modulus (MOE) compared to traditional concrete, which limits its use in high-load-bearing structures (Kisku et al. 2020; Ahmed and Lim 2021a; Tang et al. 2022; Xiao et al. 2022a). Durability concerns are equally critical, as the porous ITZ and high water absorption accelerate degradation mechanisms such as carbonation, chloride ion penetration, and freeze–thaw damage (Bravo et al. 2015; Jalilifar and Sajedi 2021; Mistri et al. 2023; Peng et al. 2023; Haider et al. 2025). These limitations necessitate innovative strategies to strengthen the performance of RCA-concrete for structural applications.

To address these challenges, various strength-enhancing techniques, including RCA surface treatment via acid washing or carbonation method, fiber reinforcing, and supplementary cementitious materials (SCMs), have been employed (Chen et al. 2017; Khushnood et al. 2020; Tam et al. 2021; Ma et al. 2023). However, the latter offers one of the most effective, economical, and practically viable approaches. Recently, fly ash (FA), volcanic ash (VA), iron-ore tailings (IOT), and glass powder (GP) have emerged as effective strength-enhancing additives. These materials improve RCA-concrete through pozzolanic activity, where they interact with calcium hydroxide to formulate calcium silicate hydrate (C-S-H) gel and densify the microstructure, refine pore structures, and enhance the structural properties (Kou et al. 2011; Xie et al. 2019a; Ahmed et al. 2025b).

The growing academic interest in sustainable green concrete incorporating RCA and SCMs is evident in Figure 19.1, which illustrates the annual and cumulative number of publications on this topic (Almohana et al. 2022; Bhat et al. 2024; Firoozi and Firoozi 2025). The graph reveals a marked increase in research activity, particularly over the last five years, with cumulative totals exceeding those of the previous five years, reaching more than 1300 articles by mid-2025. Additionally, the co-occurrence map in Figure 19.2 represents the interconnected keywords and concepts related to sustainable concrete, explicitly focusing on the employment of RCAs and SCMs. By mapping the frequency and relationships of terms extracted from relevant Scopus

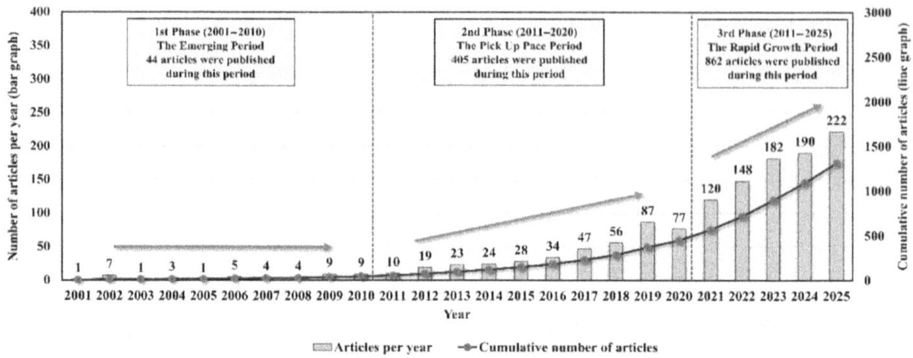

FIGURE 19.1 Annual publications trend on sustainable concrete incorporating waste-derived recycled aggregates and SCMs, based on data from Scopus (2001–2025).

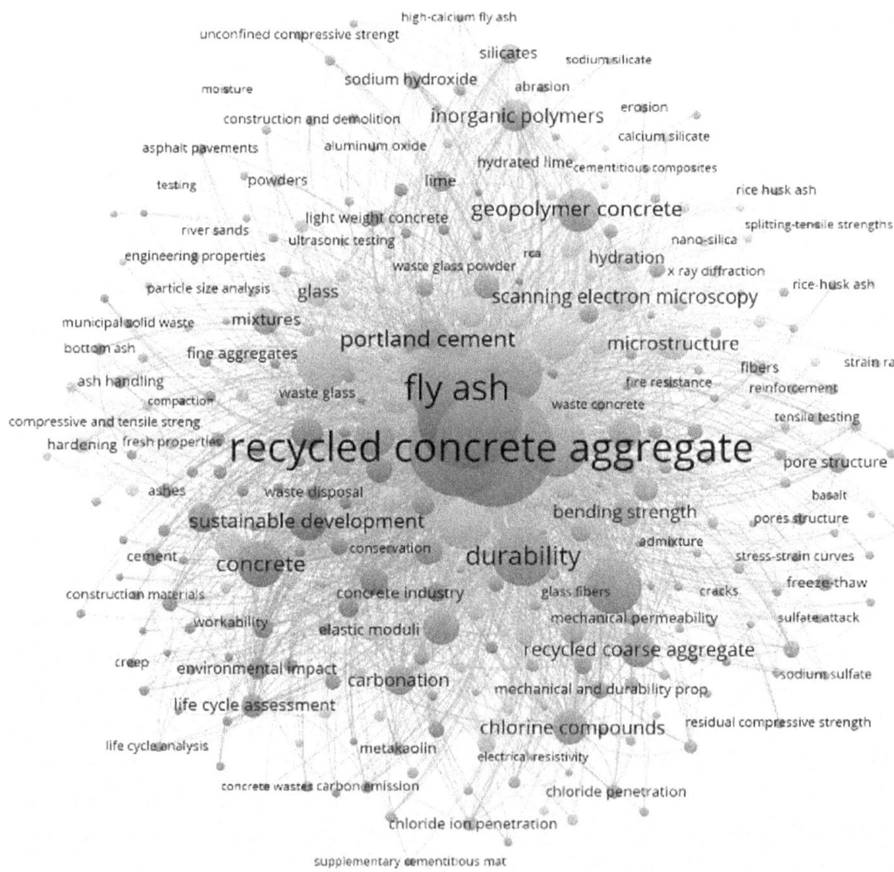

FIGURE 19.2 Co-occurrence map of keywords in sustainable concrete comprising RCAs and SCMs, generated using VOSviewer.

data-based literature, the figure highlights critical research terms such as "recycled concrete aggregate," "recycled concrete," "compressive strength," "fly ash," "durability," and "ground granulated blast furnace slag," highlighting the focus on structural properties and environmental benefits. The subsequent sections of this chapter provide a critical review that synthesizes recent advancements in RCA-SCM-based concrete systems, with particular emphasis on evaluating structural performance under both normal and severe environmental conditions. Through an integrated analysis of microstructural mechanisms, mechanical properties, and long-term durability, this work aims to guide future research toward the development of high-performance and environmentally sustainable concrete.

19.2 STRUCTURAL PERFORMANCE OF RCA CONCRETE

The employment of waste-derived RCA as a sustainable substitute for natural gravels and sand represents a pivotal step in lowering the environmental impacts of concrete. Derived from crushed and processed C&D waste, RCA repurposes end-of-life concrete into a valuable resource, aligning with circular economy principles. Moreover, RCA diverts C&D waste streams from landfill, minimizes the unsustainable use of virgin aggregate, and cuts the carbon footprint of concrete (Ali et al. 2020; Xiao et al. 2022b; Ma et al. 2025). Nevertheless, RCA's heterogeneous composition, including adhered mortar, micro-cracks, and residual contaminants, adversely affects the early workability and the strength parameters of concrete after curing. The following subsections discuss these inherent challenges and the mechanistic role of pozzolans in overcoming them and improving the properties of RCA-based concrete.

19.2.1 Microstructure

The adoption of RCA in structural concrete is hindered by several challenges, primarily related to its poor physical and mechanical properties. These negatively influence the overall microstructure quality, strength, and durability of the resulting concrete. RCA derived from demolished concrete typically contains adhered mortar from its parent concrete (as depicted in Figure 19.3a), which results in higher porosity and a weaker ITZ. The adhered mortar, as illustrated in Figure 19.3b, often comprises 20–50% of RCA's volume and exhibits a porous structure and a high water absorption rate that usually varies from 3 to 15% (Dimitriou et al. 2018) (Ahmed and Lim 2021b). This creates a less dense matrix with several defects, including weak ITZs and voids (Figure 19.3c), thus increasing the concrete's susceptibility to microcracking and shrinkage. The ITZ, where the aggregate interfaces with the cement paste, is further compromised by microcracks formed during the crushing process and residual contaminants, such as dust or organic matter, which weaken the bond strength. Several studies revealed that RCA-based concrete contains more voids and less cohesive C-S-H gel compared to natural aggregate concrete, reducing load transfer efficiency and overall structural integrity (Kumar and Mishra 2022; Ahmed and Lim 2022b).

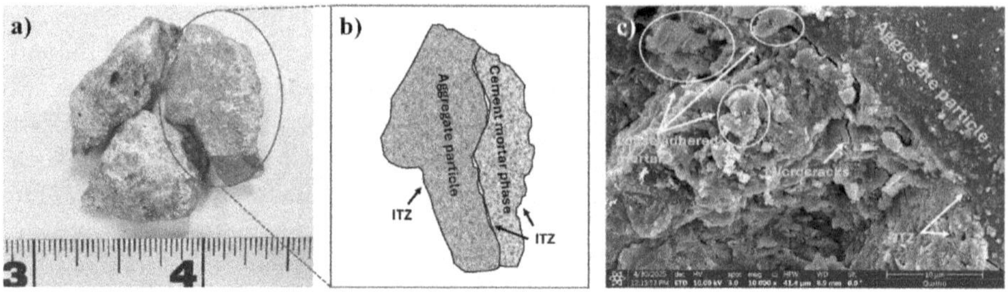

FIGURE 19.3 (a) Sample of RCA with adhered mortar, (b) illustrative figure of a single RCA particle showing the aggregate and cement mortar phase, and (c) Scanning Electron Microscopy (SEM) micrograph of RCA-based concrete highlighting the weak ITZs and microstructural defects.

19.2.2 Mechanical Performance

Table 19.1 summarizes the impact of varying RCA replacement levels on the mechanical and resilience properties of concrete, as reported in multiple studies. It has been well established that the strength performance of recycled concrete declines with increasing RCA content from 0 to 100%. A recent study found that incorporating 20% RCA results in a 12–16% decrease in f'_{CS}, f'_{TS}, and f'_{FS} strengths, along with an increase of up to 5.9% in water absorption (Ahmed et al. 2025b). These reductions in the mechanical characteristics further exacerbate with the higher substitution levels of RCA. Available literature reveals that with the integration of 100% RCA, f'_{CS} of concrete can fall by 16–34%, MOE by 35–46%, and f'_{TS} by 14–26% (Akca et al. 2015; Dimitriou et al. 2018; Thomas et al. 2018; Arora et al. 2019). The f'_{FS} remains comparable to control at 30% RCA but declines by 9.8% and 12.2% at 50% and 100%, respectively (Ali et al. 2020).

19.2.3 Durability

Durability is another critical concern for RCA-based concrete, as its porous microstructure accelerates degradation under environmental exposure. The eminent moisture assimilation of RCA facilitates the ingress of deleterious constituents, such as chloride ions and carbon dioxide, which compromise long-term performance. It has been observed that the water absorption and void content of concrete increase by more than 100% at full substitution of RCA, while its sorptivity rises between 0.021 and 0.036 mm/min (Gonzalez Corominas et al. 2017). Similarly, the permeability of RCA-based concrete reveals up to 90% higher values at 365 days over the reference sample (Mahmood et al. 2022). In cases of carbonation resistance, poor performance in RCA-concrete with a 1.5–2 times greater carbonation depth has been observed (Mahmood et al. 2022). Several other studies have reported higher chloride conductivity and poor freeze–thaw resistance, particularly at RCA replacement levels exceeding 40% (Olorunsogo and Padayachee 2002; Kox et al. 2019). This is mainly associated with the porous and weakly adherent mortar to the RCA, which

TABLE 19.1
Performance of Recycled Concrete Comprising Varying Percentages of Rca

RCA content (%)	Effect on Concrete Performance	Reference
25, 50, and 100	Concrete comprising up to 25% RCA revealed little or no difference in strength performance as compared to NCA-concrete. However, in the case of 100% RCA, a maximal drop of about 16% in f'_{CS}, 35% in MOE, and 14% in f'_{TS} was noticed. Durability of concrete was also adversely affected with the gradual increase of RCA from 25 to 100%.	Thomas et al. (2018)
30, 50, and 100	Replacement levels of RCA up to 50% produced concrete with tensile strength and fracture energy comparable to that of NCA-based concrete. However, beyond 50%, there was a significant decrease in the peak load, toughness, and maximum displacement.	Elansary et al. (2021)
0 – 100 (in increments of 25%)	By adding RCA in the range of 25–100%, the f'_{CS} reduced in the range of 8–24%.	Arora et al. (2019)
20	The structural properties of RCA-concrete, including f'_{CS}, f'_{TS}, $f'_{FS,}$ and UPV reduced by 12%, 9%, 16%, and 13%, respectively. The water absorption value was increased to 5.9% from the reference value of 3.6% for the reference mix.	Ahmed et al. (2025b)
20, 50, and 100	The utilization of RCA resulted in a reduction of about 7.5% in f'_{CS}. Moreover, the RCA-concrete attained the lowest f'_{TS} of 2.37 MPa and MOE of 30.6 GPa. The abrasion resistance of concrete comprising 50% RCA was the highest (10%) in contrast to the reference sample.	Fonseca et al. (2011)
25, 30, and 55	RCA decreased the workability of the mix (resulting in a slump of only 10 mm) and raised the air void percentage to 1.7%. In mixes with higher RCA proportions, f'_{CS} dropped by 40%, MOE reduced by 46%, and f'_{TS} declined by 26%.	Akca et al. (2015)
20 and 40	The employment of 40% RCA with a 0.5 w/c ratio significantly reduced f'_{CS} and resulted in a higher mass loss after freeze–thaw exposure.	Kox et al. (2019)
30, 50, 70, and 100	With a gradual increase in the RCA content (30–100%), f'_{CS} reduced in the range of 10–26%, while f'_{TS} and f'_{FS} decreased in the range of 10–38%. Additionally, the permeability of the concrete matrix comprising RCA was higher, with the R100 sample revealing the highest 90% difference at 365 days from the reference mix. The concrete resistance to CO_2 penetration was also negatively affected by the amalgamation of RCA. Among all the mixes, the R100 showed the highest carbonation depth, varying from 48 to 98 mm at 28 to 720 days.	Mahmood et al. (2022)
30, 50, and 100	Addition of 30% RCA produced no or little effect on f'_{FS}; however, with 50 and 100% RCA, there was a notable reduction of 9.8% and 12.2%, respectively. The f'_{CS} of RCA-concrete was lower by about 4–18%, while the MOE was lower by 5–23% in association with the control sample.	Ali et al. (2020)

(Continued)

TABLE 19.1 (*Continued*)
Performance of Recycled Concrete Comprising Varying Percentages of Rca

20, 50, and 100	RCA adversely affected the imbibition and void content (amplified by more than 100%) of concrete. The sorptivity values increased from 0.021 to 0.036 mm/min, which were greater than those of traditional concrete.	Gonzalez Corominas et al. (2017)
30, 50, 70, and 100	The f'_{CS} and MOE of concrete made with 100% RCA were lower by approximately 12% and 45%, while its peak strain was 20% higher than the control formulation.	Xiao et al. (2005)
20 and 100	The samples prepared with 100% RCA showed a decrease of 19%, 14%, and 15% in f'_{FS} f'_{TS}, and f'_{CS}. The water absorption values also revealed an increase of about 119% over the reference sample.	Ahmed and Lim (2021b)
100	The f'_{CS} and f'_{FS} of laboratory waste-derived RCA-concrete reduced by about 17% and 20% while those made with RCA of field waste reduced by 34% and 24%, respectively.	Dimitriou et al. (2018)
50 and 100	The incorporation of RCA decreased the oxygen permeability index of concrete by 10% and increased the water sorptivity and chloride conductivity by 38.5% and 73.2%, respectively.	Olorunsogo and Padayachee (2002)

is prone to water saturation and subsequent cracking under aggressive conditions. Consequently, these deficiencies compromise the long-standing strength and endurance of RCA-concrete.

19.3 STRUCTURAL PROPERTIES OF SCM-BASED RECYCLED CONCRETE

The weak micro and macro-scale characteristics of RCA-based concrete can be effectively enhanced via the incorporation of suitable SCMs. SCMs, including FA, VA, GGBS, and GP, have been reported to improve concrete performance through pozzolanic reactions, particle packing, and microstructure densification, which can enhance the suitability of RCA-concrete for structural requirements (Celik et al. 2014; Baikerikar et al. 2023; Hamada et al. 2023; Ahmed et al. 2025b). By reacting with $Ca(OH)_2$, a by-product of cement hydration, these materials increase the CSH gel production and thereby upgrade the strength of concrete. FA, a by-product of coal combustion, contributes silica and alumina that enhance CSH formation, reducing porosity and strengthening the ITZ through improved particle packing. GGBS, derived from iron production, promotes long-term strength gains by forming a denser matrix, with studies showing 15–25% improvements in compressive strength at 30% cement replacement levels (Berndt 2009; Kou et al. 2011). VA and GP are rich in reactive silica, which accelerates pozzolanic activity, thereby reducing permeability and enhancing homogeneity. This section evaluates the structural performance of SCM-modified recycled concrete, focusing on f'_{CS}, f'_{ST}, and MOE, supported by past studies that highlight advancements in achieving structural adequacy.

19.3.1 Compressive Strength

In structural design, f'_{CS} is a key parameter related to concrete's ability to bear axial loads. RCA-based concrete typically exhibits 10–30% lower f'_{CS} than normal concrete because of the porous adhered mortar and inferior ITZ (Etxeberria et al. 2007; Malesev et al. 2010; Alexandridou et al. 2018; Ahmed and Lim 2021a). However, several studies confirmed that the addition of suitable SCMs overcomes these limitations by forming additional C-S-H gel, which helps in the concrete matrix densification and strengthening of ITZs. A study by Kou et al. (2011) investigated the performance of RCA-concrete manufactured with various mineral intermixtures and found that samples comprising 35% FA and 55% GGBS demonstrated the highest compressive strength gains of 71% and 69% between 4 and 90 days (Kou et al. 2011). Similarly, a recent experimental study concluded that by adding 30–45% cementitious materials such as GP, VA, GGBS, and pulverized fuel ash (PFA), the mechanical characteristics of concrete can be significantly enhanced to meet the structural concrete requirements (Ahmed et al. 2025b). These pozzolan-based mixes also enhance microstructural characteristics through the formation of additional CSH and CASH gels, while reducing carbon emissions by up to 39% and production costs by up to 16%, making them a sustainable alternative to Ordinary Portland Cement (OPC)-based concrete (Ahmed et al. 2025b). Additionally, GP-RAC with 10–40% GP replacement exhibited significant increases in compressive strength at 90 days (up to 40%), which was attributed to the activated pozzolanic effect of GP at late curing ages (Peng et al. 2023). Table 19.2 presents details regarding the structural properties of RCA-based concrete prepared with various SCMs.

19.3.2 Tensile Strength

f'_{TS} reflects the capacity of concrete against tensile forces, and it is crucial for structural members under bending or shear, such as beams and slabs. RCA-concrete typically exhibits 15–25% lower f'_{TS} associated with traditional concrete because of microcracks in the adhered mortar and ITZs (Etxeberria et al. 2007; Elhakam et al. 2012; Çakir 2014; Jamil et al. 2025; Liang et al. 2025). To overcome these obstacles, several studies examined the accumulation of SCMs in RCA-concrete (Kou et al. 2011; Ahmed and Lim 2021b). They uncovered that SCMs significantly improve the weak ITZs of RAC by promoting secondary hydration reactions and refining the microstructure. For instance, the findings of a recent study reveal that RAC incorporating 20% PFA and 5% SF achieved an f'_{TS} of 3.2 MPa at 28 days, representing a 16% improvement over RAC without SCMs (2.75 MPa) (Ahmed and Lim 2021b). Similarly, Kou et al. (2011) found that mixes with 35% PFA and 55% GGBS exhibited substantial tensile strength gains at 7 and 90 days, attributed to the pozzolanic responses that densify the ITZ and enhance bonding of the cement matrix with the aggregate granules (Kou et al. 2011). Additional study by Ahmed et al. (2025b) showed that combining 10% GP with 30% VA, GGBS, or PFA enhanced the f'_{TS} of recycled concrete by 5%, 8%, and 10%, respectively, compared to non-treated RCA-concrete (Ahmed et al. 2025b). These improvements indicate the effectiveness of SCMs to fill microcracks and reduce porosity, which enables the concrete matrix to sustain greater tensile loads.

Furthermore, the integration of SCMs with advanced treatments, such as carbonation of recycled aggregates, has shown promising results. Carbonation-treated RAC combined with SCMs can achieve tensile strengths of 4–5 MPa, meeting the requirements for structural applications like slabs and precast elements (Jamil et al. 2025). The use of fine SCM particles is particularly effective, as their small size facilitates crack-bridging and improves the homogeneity of the concrete matrix (Ibrahim 2021; Ahmed and Lim 2022a, 2023b; Amin et al. 2023). Additionally, SCMs contribute to long-term tensile strength development in recycled concrete by the pozzolanic activity beyond 28 days, which is beneficial for the long-term performance of buildings (Shaban et al. 2019, 2021; Omrane and Rabehi 2020; Jamil et al. 2025). These findings highlight the pivotal role of SCMs in overcoming the inherent limitations of RAC, enabling its broader adoption in structural concrete applications.

19.3.3 Modulus of Elasticity

The modulus of elasticity (MOE) measures concrete's stiffness and deformation resistance under load, a key factor in structural design that ensures minimal deflection in load-bearing elements such as beams and columns. Like other mechanical characteristics, the MOE of recycled concrete is negatively affected due to RCA inclusion. This is majorly due to the reduced stiffness of recycled concrete, which results from the

TABLE 19.2
Performance of Scm-Modified Rca-Concrete

SCMs	Structural Properties of Rca-Concrete	Reference
SF (10%), FA (35%), MK (15%), and GGBS (55%)	The amalgamation of 10% silica fume (SF) or 15% metakaolin (MK) in RCA-concrete improves both short- as well as long-term f'_{CS} and f'_{ST}, as well as endurance properties like chloride ion penetration resistance and drying shrinkage. In contrast, 35% FA or 55% GGBS reduces f'_{CS} but improves durability after longer curing, with greater performance improvements in recycled concrete.	Kou et al. (2011)
VA (30%), GGBS (30%), PFA (30%), and GP (5 – 15%)	The use of 30–45% pozzolanic constituents in RCA-concrete significantly enhances mechanical properties (f'_{CS}, f'_{TS}, f'_{FS}), with the 10% GP and 30% GGBS combination showing the highest improvements, alongside 5–14% increases in ultrasonic pulse velocity and reduced water absorption (2.9–3.6%).	Ahmed et al. (2025b)
GP (0 – 40%)	The incorporation of GP in RCA-concrete at replacement ratios of 10–40% significantly enhances compressive strength at 90 days by up to 40% due to the pozzolanic activity and denser particle packing, with 10% and 20% GP mixes surpassing the reference RAC strength. Additionally, GP improves chloride penetration resistance, particularly at 90 days, with a significant 40% reduction in chloride migration coefficient.	Peng et al. (2023)
SF, FA, and nano-silica	Treatment of RCA with pozzolan slurry demonstrated superior strength and durability performance as compared to the non-treated RCA-based concrete. The samples made with modified RCA revealed 4.5–18% lower f'_{CS}, 3–18.5% lower f'_{FS}, and 8–22% lower elastic modulus compared to normal concrete at 28 days, while improving chloride ion migration resistance.	Shaban et al. (2021)
SF (10%), zeolite powder (5%), and limestone powder (10%)	The addition of SCMs (SF, zeolite powder, and limestone powder) in carbonated recycled aggregate concrete resulted in reduced f'_{CS} and f'_{TS}, with decreases of up to 40% and 24% compared to CRAC with OPC at water-to-binder ratios of 0.55 and 0.42, respectively. The accelerated carbonation of RCA significantly enhanced concrete performance, increasing f'_{CS} by 30% and f'_{TS} by 42%, while reducing absorption to 6.06% and improving resistance to acid and sulfate attacks.	Jamil et al. (2025)
PFA and GGBS (20%)	With the combined use of SCMs (PFA and GGBS) and hybrid fibers, recycled concrete revealed significant enhancement in durability, with SCM-based hybrid fiber mixes showing a minimal 0.055% change in length, a lower mass loss of 0.39%, and a high 4137 m/s UPV subsequently completing 300 cycles of freeze–thaw.	Ahmed and Lim (2023b)
GGBS and FA	The combination of GGBS and FA in recycled concrete enhances workability and mechanical properties, with FA primarily improving consistency and GGBS significantly boosting f'_{CS} and MOE. A 50% GGBS and 50% FA mix with a 0.5 water-binder ratio optimizes the synergistic effects, resulting in superior performance for GRAC made with 100% RCA.	Xie et al. (2019a)
FA (20%) and SF (5%)	The synergistic use of modified RCA (treated with pozzolans slurry) and basalt fiber significantly enhances the mechanical performance of RAC. The 20%- and 100%-treated RCA formulations showed maximum improvements of 16–57% and 8–37% in f'_{CS}, f'_{TS}, and f'_{FS}, respectively.	Ahmed and Lim (2021b)

superior porosity and feeble ITZs in the matrix (Silva et al. 2015; Wang et al. 2019). The integration of suitable pozzolans enhances the MOE by making a more compact and denser microstructure with better ITZs through pozzolanic reactions. Shaban et al. (2021) found that modifying RCA via pozzolana slurry resulted in improvements of approximately 16–30% and 22–28% in the MOE of concrete prepared with water-to-cement ratios of 0.48 and 0.39, respectively (Shaban et al. 2021). Furthermore, the synergistic use of SCMs and fiber reinforcement has shown significant promise in mitigating the MOE deficiencies of RAC. Ahmed and Lim (2023a) found that incorporating basalt fiber with SCM (20% FA and 5% SF) -treated recycled concrete aggregate (TRCA) increased the MOE from 27.97 GPa to 30.46 GPa (9.07% improvement) for 20% TRCA and from 24.76 GPa to 29.08 GPa (4.1% improvement) for 100% TRCA with 1% BF dosage, compared to reference specimens (Ahmed and Lim 2023a). These improvements highlight the synergistic effect of BF and SCMs in enhancing the matrix quality and stress distribution in recycled concrete. The improvement of concrete matrix due to SCMs, coupled with the crack-bridging capacity of fibers, contributes to a more homogeneous and stiffer material and, thus, makes such approaches highly effective for refining the structural performance of green concrete. Consistent with these findings, other studies have reported similar positive effects of PFA and GGBS on the MOE of recycled concrete, confirming their efficacy in enhancing stiffness and structural reliability (Xie et al. 2019b; Kumar and Mishra 2022).

19.3.4 Durability

Durability is a crucial factor in determining the long-term performance and service life of RCA-based concrete, particularly under harsh conditions. The inherent weaknesses of RCA make concrete more susceptible to degradation mechanisms like carbonation, freeze–thaw cycles, sulfate attack, and chloride ion penetration (Shaban et al. 2021; Ahmed and Lim 2023b; Jamil et al. 2025). Numerous studies have demonstrated that SCMs, such as FA, GGBS, GP, and VA, significantly enhance the durability of RAC by addressing these weaknesses. The densified microstructure resulting from the pozzolanic reactions and the filling effect of SCMs produces additional CSH gel, which reduces concrete permeability and limits the ingress of aggressive agents. For instance, test results from a previous experimental study reveal that treating RCA with FA and SF slurry resulted in a 20–46% enhancement in chloride resistance of modified concrete compared to non-treated RCA-concrete (Shaban et al. 2021). The authors attributed it to the enhanced pore structure and strengthened ITZ in the SCM-modified RCA matrix.

Similarly, Kou et al. (2011) found that RAC incorporating 35% PFA or 55% GGBS exhibited superior chloride ion penetration resistance due to reduced porosity and enhanced matrix density (Kou et al. 2011). The use of suitable pozzolans is also valuable for improving the freeze–thaw resistance of concrete. A recent study demonstrated that combining 30% of VA, GGBS, or PFA with 10% GP lowered the imbibition of concrete manufactured with 20% RCA from 4.6% to 2.9% (Ahmed et al. 2025b). Furthermore, the incorporation of SCMs (GGBS and FA) along with hybrid basalt, polypropylene, and glass fibers can enhance freeze–thaw durability by providing multi-scale reinforcement (Ahmed and Lim 2023b). For example,

SCM-modified hybrid fiber-reinforced RAC with 0.75% BF and 0.25% PPF or 0.5% BF and 0.5% GF exhibited minimal mass losses of 0.49% and 0.39%, respectively, following 300 freeze–thaw cycles. This synergistic effect of SCMs and hybrid fibers not only enhances durability but also improves the overall structural integrity of waste-derived sustainable concrete.

Beyond performance enhancements, pozzolans can directly contribute to the sustainability improvement of RCA-based concrete by lowering both cement quantity and CO_2 emissions by approximately 20–40% per 1 m^3 of concrete (Ahmed et al. 2025b). The utilization of waste-derived pozzolans, such as FA and GP, further supports circular economy principles by reusing industrial by-products. Economically, SCMs reduce material costs, as they are often cheaper than cement, and their improved workability can decrease construction time and labor expenses.

19.4 SUMMARY

The integration of waste-derived RCA and pozzolanic SCMs into structural concrete represents a highly effective and environmentally friendly approach to addressing the ecological challenges of the construction industry. By repurposing C&D waste as functional construction materials, RCA-concrete has a great potential to reduce landfill reliance and conserve virgin aggregates. SCMs, such as PFA, GGBS, VA, and GP, mitigate RCA's inherent weaknesses through pozzolanic activity that densifies the microstructure and strengthens ITZs. This chapter has demonstrated that SCM-enhanced RCA-concrete achieves superior f'_{CS} values, along with f'_{TS} gains of up to 16% and MOE improvements of up to 30%, making it suitable for engineering applications such as beams, slabs, and precast elements. The utilization of hybrid fiber reinforcement can further enhance the mechanical properties (up to a 57% increase in f'_{FS}) and durability, with the least mass loss (0.39–0.49%) after 300 freeze–thaw cycles. These advancements, together with a 40% reduction in CO_2 emissions and 20–30% cost savings, emphasize the ability of SCM-modified RCA-concrete as a green substitute for conventional concrete.

Despite these advancements, challenges related to RCA's material variability, limited scalability due to fragmented C&D waste supply chains, and a lack of universal standards hinder widespread adoption. Solutions such as advanced sorting and treatments, as well as harmonized standards specifying RCA quality for various structural applications, can address these barriers and increase the global utilization of sustainable concrete. Henceforward, the study should focus on long-standing endurance under extreme conditions, scalable treatments like pozzolana slurry coating, and machine learning for mix design optimization and reducing testing costs and time. By addressing these challenges and bridging the current research gaps, excellent SCM-based RCA-concrete has great potential to make modern-day construction a more resource-efficient and eco-friendly endeavour.

REFERENCES

Ahmed W, Lim CW (2021a) Production of sustainable and structural fiber reinforced recycled aggregate concrete with improved fracture properties: A review. *J Clean Prod* 279:123832

Ahmed W, Lim CW (2021b) Coupling effect assessment of vacuum based pozzolana slurry encrusted recycled aggregate and basalt fiber on mechanical performance of fiber reinforced concrete. *Constr Build Mater* 300:124032

Ahmed W, Lim CW (2022a) Effective recycling of disposable medical face masks for sustainable green concrete via a new fiber hybridization technique. *Constr Build Mater* 344:128245. https://doi.org/10.1016/J.CONBUILDMAT.2022.128245

Ahmed W, Lim CW (2022b) Effective utilization of chopped basalt fiber and pozzolana slurry TRCA for sustainable recycled structural concrete with improved fire resistance. *J Mater Civ Eng* 35:04022442. https://doi.org/10.1061/(ASCE)MT.1943-5533.0004613

Ahmed W, Lim CW (2023a) Evaluating fracture parameters of basalt fiber reinforced and pozzolana slurry modified recycled concrete produced from waste. *Structures* 50:1476–1492. https://doi.org/10.1016/j.istruc.2023.02.111

Ahmed W, Lim CW (2023b) Multicriteria performance assessment of sustainable recycled concrete produced via hybrid usage of basalt, polypropylene and glass fiber. *Constr Build Mater* 397:132462

Ahmed W, Lu G, Ng ST, Liu G (2025a) Innovative valorization of solid waste materials for production of sustainable low-carbon pavement: A systematic review and scientometric analysis. *Case Stud Constr Mater* 22. https://doi.org/10.1016/j.cscm.2025.e04541

Ahmed W, Ye C, Lu G, et al (2025b) Low-carbon concrete comprising high-volume pozzolan and recycled aggregate: Evaluating mechanical performance, microstructure, environmental impact, and cost efficiency. *J Clean Prod* 518:145796. https://doi.org/10.1016/j.jclepro.2025.145796

Akca KIR, Cakir O, Ipek M (2015) Properties of polypropylene fiber reinforced concrete using recycled aggregates. *Constr Build Mater* 98:620–630. https://doi.org/10.1016/j.conbuildmat.2015.08.133

Al-Kheetan MJ, Jweihan YS, Rabi M, Ghaffar SH (2024) Durability enhancement of concrete with recycled concrete aggregate: The role of Nano-ZnO. *Buildings* 14:353

Alexandridou C, Angelopoulos GN, Coutelieris FA (2018) Mechanical and durability performance of concrete produced with recycled aggregates from Greek construction and demolition waste plants. *J Clean Prod* 176:745–757. https://doi.org/10.1016/j.jclepro.2017.12.081

Ali B, Qureshi LA, Khan SU (2020) Flexural behavior of glass fiber-reinforced recycled aggregate concrete and its impact on the cost and carbon footprint of concrete pavement. *Constr Build Mater* 262:120820. https://doi.org/10.1016/j.conbuildmat.2020.120820

Almohana AI, Abdulwahid MY, Galobardes I, et al (2022) Producing sustainable concrete with plastic waste: A review. *Environ Challenges* 9:100626

Amin M, Agwa IS, Mashaan N, et al (2023) Investigation of the physical mechanical properties and durability of sustainable ultra-high performance concrete with recycled waste glass. *Sustain* 15. https://doi.org/10.3390/su15043085

Arora S, Singh B, Bhardwaj B (2019) Strength performance of recycled aggregate concretes containing mineral admixtures and their performance prediction through various modeling techniques. *J Build Eng* 24:100741. https://doi.org/10.1016/J.JOBE.2019.100741

Baikerikar A, Mudalgi S, Ram VV (2023) Utilization of waste glass powder and waste glass sand in the production of eco-friendly concrete. *Constr Build Mater* 377:131078. https://doi.org/10.1016/j.conbuildmat.2023.131078

Berndt ML (2009) Properties of sustainable concrete containing fly ash, slag and recycled concrete aggregate. *Constr Build Mater* 23:2606–2613

Bhat M, Gilani TA, Mushtaq J, et al (2024) Novel Use of Waste Rubber Tyres in Pavement Materials. In: International Conference on Advances in Concrete, Structural, & Geotechnical Engineering. Springer, pp 265–277

Bravo M, De Brito J, Pontes J, Evangelista L (2015) Mechanical performance of concrete made with aggregates from construction and demolition waste recycling plants. *J Clean Prod* 99:59–74. https://doi.org/10.1016/j.jclepro.2015.03.012

Çakir O (2014) Experimental analysis of properties of recycled coarse aggregate (RCA) concrete with mineral additives. *Constr Build Mater* 68:17–25. https://doi.org/10.1016/j.conbuildmat.2014.06.032

Celik K, Jackson MD, Mancio M, et al (2014) High-volume natural volcanic pozzolan and limestone powder as partial replacements for portland cement in self-compacting and sustainable concrete. *Cem Concr Compos* 45:136–147. https://doi.org/10.1016/j.cemconcomp.2013.09.003

Chen P, Wang J, Wang L, et al (2017) Producing vaterite by CO_2 sequestration in the waste solution of chemical treatment of recycled concrete aggregates. *J Clean Prod* 149:735–742. https://doi.org/10.1016/j.jclepro.2017.02.148

Dimitriou G, Savva P, Petrou MF (2018) Enhancing mechanical and durability properties of recycled aggregate concrete. *Constr Build Mater* 158:228–235. https://doi.org/10.1016/j.conbuildmat.2017.09.137

Elansary AA, Ashmawy MM, Abdalla HA (2021) Effect of recycled coarse aggregate on physical and mechanical properties of concrete. *Adv Struct Eng* 24:583–595. https://doi.org/10.1177/1369433220963792

Elhakam AA, Mohamed AE, Awad E (2012) Influence of self-healing, mixing method and adding silica fume on mechanical properties of recycled aggregates concrete. *Constr Build Mater* 35:421–427. https://doi.org/10.1016/j.conbuildmat.2012.04.013

Etxeberria M, Vázquez E, Marí A, Barra M (2007) Influence of amount of recycled coarse aggregates and production process on properties of recycled aggregate concrete. *Cem Concr Res* 37:735–742. https://doi.org/10.1016/j.cemconres.2007.02.002

Firoozi AA, Firoozi AA (2025) Recycled aggregate concrete: A sustainable approach to concrete production. *Recent Dev Innov Sustain Prod Concr* 415–459

Fonseca N, De Brito J, Evangelista L (2011) The influence of curing conditions on the mechanical performance of concrete made with recycled concrete waste. *Cem Concr Compos* 33:637–643. https://doi.org/10.1016/j.cemconcomp.2011.04.002

Gonzalez Corominas A, Etxeberria M, Poon C sun (2017) Influence of the quality of recycled aggregates on the mechanical and durability properties of high performance concrete. *Waste and Biomass Valorization* 8:1421–1432. https://doi.org/10.1007/s12649-016-9637-7

Haider I, Yaqub M, Inam I, et al (2025) Investigating the synergistic effects of Metakaolin and silica fume on the strength and durability of recycled aggregate concrete at elevated temperatures. *Sci Rep* 15:29510

Hamada HM, Abed F, Beddu S, et al (2023) Effect of Volcanic Ash and Natural Pozzolana on mechanical properties of sustainable cement concrete: A comprehensive review. *Case Stud Constr Mater* 19:e02425. https://doi.org/10.1016/j.cscm.2023.e02425

Ibrahim KIM (2021) Recycled waste glass powder as a partial replacement of cement in concrete containing silica fume and fly ash. *Case Stud Constr Mater* 15:e00630. https://doi.org/10.1016/j.cscm.2021.e00630

Islam R, Nazifa TH, Yuniarto A, et al (2019) An empirical study of construction and demolition waste generation and implication of recycling. *Waste Manag* 95:10–21

Jalilifar H, Sajedi F (2021) Micro-structural analysis of recycled concretes made with recycled coarse concrete aggregates. *Constr Build Mater* 267:121041. https://doi.org/10.1016/j.conbuildmat.2020.121041

Jamil S, Idrees M, Akbar A, Ahmed W (2025) Investigating the mechanical and durability properties of carbonated recycled aggregate concrete and its performance with SCMs. *Buildings* 15:1–23. https://doi.org/10.3390/buildings15020201

Khan ZA, Balunaini U, Costa S, Nguyen NHT (2024) A review on sustainable use of recycled construction and demolition waste aggregates in pavement base and subbase layers. *Clean Mater* 13:100266

Khushnood RA, Qureshi ZA, Shaheen N, Ali S (2020) Bio-mineralized self-healing recycled aggregate concrete for sustainable infrastructure. *Sci Total Environ* 703:135007. https://doi.org/10.1016/j.scitotenv.2019.135007

Kisku N, Rajhans P, Panda SK, et al (2020) Microstructural investigation of recycled aggregate concrete produced by adopting equal mortar volume method along with two stage mixing approach. *Structures* 24:742–753. https://doi.org/10.1016/j.istruc.2020.01.044

Kou S, Poon C, Agrela F (2011) Comparisons of natural and recycled aggregate concretes prepared with the addition of different mineral admixtures. *Cem Concr Compos* 33:788–795

Kox S, Vanroelen G, Van Herck J, et al (2019) Experimental evaluation of the high-grade properties of recycled concrete aggregates and their application in concrete road pavement construction. *Case Stud Constr Mater* 11:e00282. https://doi.org/10.1016/j.cscm.2019.e00282

Kumar G, Mishra SS (2022) Effect of recycled concrete aggregate on mechanical, physical and durability properties of GGBS–fly ash-based geopolymer concrete. *Innov Infrastruct Solut* 7:1–18. https://doi.org/10.1007/s41062-022-00832-w

Li L, Joseph P, Zhang X, Zhang L (2024) A study of some relevant properties of concrete incorporating waste ceramic powder as a cement replacement agent. *J Build Eng* 87:109106

Liang T, Ahmed W, Duan Z, et al (2025) A critical review on the applications of modified recycled coarse aggregate in conventional concrete and high-performance concrete. *J Build Eng* 112601

Ma X, Hu H, Luo Y, et al (2025) A carbon footprint assessment for usage of recycled aggregate and supplementary cementitious materials for sustainable concrete: A life-cycle perspective in China. *J Clean Prod* 490:144772. https://doi.org/10.1016/j.jclepro.2025.144772

Ma Z, Hu R, Yao P, Wang C (2023) Utilizing heat-mechanical synergistic treatment for separating concrete waste into high-quality recycled aggregate, active recycled powder and new concrete. *J Build Eng* 68:106161. https://doi.org/10.1016/j.jobe.2023.106161

Mahmood W, Khan A ur R, Ayub T (2022) Mechanical and durability properties of concrete containing recycled concrete aggregates. *Iran J Sci Technol – Trans Civ Eng* 46:2111–2130. https://doi.org/10.1007/s40996-021-00692-x

Malesev M, Radonjanin V, Marinković S (2010) Recycled concrete as aggregate for structural concrete production. *Sustainability* 2:1204–1225. https://doi.org/10.3390/su2051204

Mistri A, Dhami N, Bhattacharyya SK, et al (2023) Performance of biocement treatment in improving the interfacial properties of recycled aggregate concrete. *Constr Build Mater* 369:130509. https://doi.org/10.1016/j.conbuildmat.2023.130509

Monteiro PJM, Miller SA, Horvath A (2017) Towards sustainable concrete. *Nat Mater* 16:698–699. https://doi.org/10.1038/nmat4930

Olorunsogo FT, Padayachee N (2002) Performance of recycled aggregate concrete monitored by durability indexes. *Cem Concr Res* 32:179–185. https://doi.org/10.1016/S0008-8846(01)00653-6

Omrane M, Rabehi M (2020) Effect of natural pozzolan and recycled concrete aggregates on thermal and physico-mechanical characteristics of self-compacting concrete. *Constr Build Mater* 247:118576

Ossa A, García JL, Botero E (2016) Use of recycled construction and demolition waste (CDW) aggregates: A sustainable alternative for the pavement construction industry. *J Clean Prod* 135:379–386

Panghal H, Kumar A (2025) Enhancing durability and strength of concrete through an innovative abrasion and cement slurry treatment of recycled concrete aggregates. *Miner Eng* 220:109109

Peng L, Zhao Y, Ban J, et al (2023) Enhancing the corrosion resistance of recycled aggregate concrete by incorporating waste glass powder. *Cem Concr Compos* 137:104909. https://doi.org/10.1016/j.cemconcomp.2022.104909

Pereira MML, Capuzzo VMS, de Brito J (2023) Concrete produced with recycled concrete aggregate exposed to treatment methods. *Case Stud Constr Mater* 18:e01938

Shaban WM, Elbaz K, Yang J, et al (2021) Effect of pozzolan slurries on recycled aggregate concrete: Mechanical and durability performance. *Constr Build Mater* 276:121940. https://doi.org/10.1016/j.conbuildmat.2020.121940

Shaban WM, Yang J, Su H, et al (2019) Properties of recycled concrete aggregates strengthened by different types of pozzolan slurry. *Constr Build Mater* 216:632–647. https://doi.org/10.1016/j.conbuildmat.2019.04.231

Shivaprasad KN, Yang H-M, Singh JK (2024) A path to carbon neutrality in construction: An overview of recent progress in recycled cement usage. *J* CO_2 *Util* 83:102816

Silva RV, Brito J De, Dhir RK (2015) Establishing a relationship between modulus of elasticity and compressive strength of recycled aggregate concrete. *J Clean Prod*. https://doi.org/10.1016/j.jclepro.2015.10.064

Tam VWY, Wattage H, Le KN, et al (2021) Methods to improve microstructural properties of recycled concrete aggregate: A critical review. *Constr Build Mater* 270:121490. https://doi.org/10.1016/j.conbuildmat.2020.121490

Tang Y, Xiao J, Zhang H, et al (2022) Mechanical properties and uniaxial compressive stress-strain behavior of fully recycled aggregate concrete. *Constr Build Mater* 323:126546. https://doi.org/10.1016/j.conbuildmat.2022.126546

Thomas J, Thaickavil NN, Wilson PM (2018) Strength and durability of concrete containing recycled concrete aggregates. *J Build Eng* 19:349–365. https://doi.org/10.1016/J.JOBE.2018.05.007

Wang Y, Hughes P, Niu H, Fan Y (2019) A new method to improve the properties of recycled aggregate concrete: Composite addition of basalt fiber and nano-silica. *J Clean Prod* 236:117602. https://doi.org/10.1016/j.jclepro.2019.07.077

Xiao J, Li J, Zhang C (2005) Mechanical properties of recycled aggregate concrete under uniaxial loading. *Cem Concr Res* 35:1187–1194. https://doi.org/10.1016/j.cemconres.2004.09.020

Xiao J, Tang Y, Chen H, et al (2022a) Effects of recycled aggregate combinations and recycled powder contents on fracture behavior of fully recycled aggregate concrete. *J Clean Prod* 366:132895. https://doi.org/10.1016/j.jclepro.2022.132895

Xiao J, Zhang H, Tang Y, et al (2022b) Fully utilizing carbonated recycled aggregates in concrete: Strength, drying shrinkage and carbon emissions analysis. *J Clean Prod* 377:134520. https://doi.org/10.1016/j.jclepro.2022.134520

Xie J, Wang J, Rao R, et al (2019a) Effects of combined usage of GGBS and fly ash on workability and mechanical properties of alkali activated geopolymer concrete with recycled aggregate. *Compos Part B Eng* 164:179–190. https://doi.org/10.1016/j.compositesb.2018.11.067

Xie J, Wang J, Zhang B, et al (2019b) Physicochemical properties of alkali activated GGBS and fly ash geopolymeric recycled concrete. *Constr Build Mater* 204:384–398

Xing W, Tam VW, Le KN, et al (2022) Life cycle assessment of recycled aggregate concrete on its environmental impacts: A critical review. *Constr Build Mater* 317:125950. https://doi.org/10.1016/j.conbuildmat.2021.125950

20 Laboratory Performance of Composite-Modified Binders and Asphalt Mixes Containing Waste Plastic

Jasim Nisar, Muneeb Farooq and Mohammad Younis

20.1 INTRODUCTION

Asphalt binder functions as the fundamental adhesive material in flexible pavement construction worldwide. Increasing traffic volumes, along with heavier and larger trucks, have posed significant difficulties to conventional asphalt pavements (Gautam et al., 2018). Additionally, harsh weather conditions adversely affect pavement performance. Usually, two primary distress types impact asphalt pavement performance: rutting and fatigue cracking (Alsheyab et al., 2024). Rutting is described by surface deformation due to traffic under high temperatures, and fatigue cracking is caused by repeated loading (Albayati, 2023). These distresses are closely related to asphalt binder characteristics. To address these issues, modified asphalt has attracted substantial interest in the pavement sector over the past decade (Albayati, 2023). Two main categories of modifiers are used: polymers and nanomaterials (Madkour et al., 2019). Polymer-modified asphalt (PMA) is a widely adopted and effective method for improving asphalt properties. Polymer-modified binders have been successfully applied in high-stress areas such as busy street intersections, airports, weigh stations, and race tracks.

The use of natural and synthetic polymers for asphalt modification has its roots in early 1900s, with a patent for a bituminous composition containing polyisobutylene granted as early as 1940 (Feldman and Akovali, 2005; Lewandowski, 1994). Research on various polymers, including different polyethylene (PE) forms, gained momentum in the late 1990s and early 2000s (Nizamuddin et al., 2021). Presently, polymers used for bitumen modification fall into two groups: plastomers and thermoplastic elastomers (Albayati et al., 2025). According to studies, plastomers have minimal elastic behavior, leading to not only rapid early strength but also permanent deformation or brittle failure. Thermoplastic elastomers soften when heated, harden

 DOI: 10.1201/9781003664284-20

upon cooling, and resist permanent deformation by stretching under load and recovering elastically afterward, making them more successful than plastomers as bitumen modifiers (Stroup-Gardiner, 1997).

Plastomeric polymers like polyethylene (PE) have given rise to increased interest as asphalt modifiers. Studies indicate that uniting PE enhances binder and mixture performance at elevated temperatures (Tahmoorian et al., 2023; Abdy et al., 2023). Beyond performance improvements, reusing PE offers significant environmental benefits. Utilizing recycled plastic waste in road construction also helps mitigate the environmental problem of plastic accumulation in landfills (Nisar et al., 2024). However, low density polyethylene (LDPE)-modified asphalt has drawbacks, including reduced elasticity, phase separation, and decreased fatigue cracking resistance (Costa et al., 2019). Previous work has shown that blending multiple polymers to create composites can improve high- or low-temperature performance, elasticity, and toughness (Chen et al., 2024; Zheng et al., 2021). One of the most commonly used elastomers in asphalt modification is styrene butadiene styrene (SBS), which has been demonstrated to enhance both the physical and rheological properties across intermediate and high temperature ranges (Behnood and Modiri Gharehveran, 2019; Liu et al., 2020). Contrasting single-polymer modified binders and composite-modified binders are designed to resist various distresses and offset individual drawbacks. Therefore, composite-modified binder pavements tend to have a longer and more consistent service life due to improved performance. In composite modification, SBS improves elasticity, increasing fatigue cracking resistance, while LDPE significantly boosts stiffness and high-temperature performance (Aurilio et al., 2021; Behnood and Modiri, 2019). Given these challenges, exploring innovative methods for asphalt binder modification is necessary. This study focuses on evaluating the possible use of LDPE/SBS as a composite modifier. LDPE/SBS was chosen as it has proven effectiveness as an individual modifier. The effect of composite-modified binders on asphalt mixture properties is highly unexplored, making LDPE/SBS modified mixtures a novel option for this research. The objective of the study is examining various performances of binder as well as mix of binder modified with LDPE/SBS through rheological tests and mechanical tests.

20.2 MATERIALS AND METHODOLOGY

20.2.1 Raw Materials

The base binder used in this investigation was VG-10 grade binder, which was provided by a local vendor in Srinagar, India. SBS elastomer and LDPE plastomer were among the modifiers employed in the investigation. SBS and LDPE were acquired from RishiChem Distributors Private Limited, Gujarat. Table 20.1 presents the physical characteristics of the base asphalt binder and modifiers.

20.2.2 Preparation of Composite-Modified Binder

A high shear mixer mixes the base binder and modifiers. The VG-10 base was heated to 165°C for fluidity. Meanwhile, the proportions of LDPE are taken as 2%, 4%, and

TABLE 20.8
General Characteristics of Base Binder and Modifiers

Specifications	Standards	Test Temperature (°C)	Value
Penetration (0.1 mm)	IS-1203	25°C	88
Softening point (°C)	IS-1205	-	46
Ductility (cm)	IS-1208	25°C	>100
Rotational viscosity (Pa.s)	IS-1206	135°C	0.27
Density @ 25°C (g/cm^3)			
LDPE			0.92
SBS			0.94
Color			
LDPE			Black
SBS			White
Size (mm)			
LDPE			2–3
SBS			2–3
Melting Point (°C)			
LDPE			125
SBS			102

6%, and the proportion of SBS remains constant at 3% (by the weight of the base binder), which was chosen for preparation of composite-modified asphalt binders and mixtures. As reported in previous literature, higher concentration of polymer leads to phase separation; therefore, the modifier content used in the study was restricted to lower concentrations (Joohari et al., 2022). Mixing occurred for 1.5 hours at 180 ± 2°C using a high-shear mixer at 2500 ± 10 rpm, following protocols from prior research (Nisar et al., 2023). The complete experimental flowchart is depicted in Figure 20.1.

20.2.3 Preparation of Composite-Modified Mixtures

All mixtures were produced for semi-dense graded bitumen macadam gradation suitable for binder courses. The samples were prepared at the optimum bitumen percent according to the Asphalt Institute's MS-2 specification. The determined bitumen proportion was 5.5% by mixture weight. Asphalt specimens were created for evaluating stability, flow, and indirect tensile strength (ITS). The binder used was a composite-modified bitumen consisting of LDPE/SBS.

20.2.4 Rheological Testing

Dynamic shear rheometer (DSR) was employed to evaluate the rheological properties of asphalt binder. Short-term aged specimens underwent testing for high-temperature rutting resistance with 25 mm parallel plates spaced 1 mm apart.

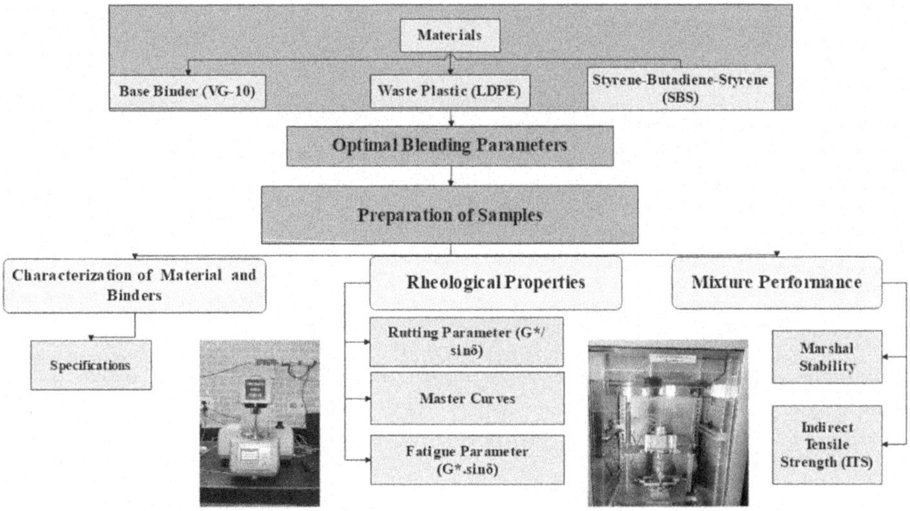

FIGURE 20.1 Experimental flowchart.

Intermediate-temperature behavior was examined using different arrangement of 8 mm plates and a 2 mm gap. These setups enabled precise characterization of viscoelastic responses across the relevant service temperature range. Factors such as complex modulus (G*) and phase angle (δ) were examined at temperatures ranging from 46°C to 88°C. The increase in the temperature should be 6°C after every run.

20.2.4.1 Fatigue Test (Intermediate Temperature)

G*.sinδ was employed to evaluate the fatigue resistance of asphalt binder. The theory suggests that reduced dissipated energy per cycle results in less distress accumulation, indicating that materials with lower G*.sinδ values exhibit superior fatigue resistance. SHRP establishes an upper limit of 5,000 kPa for G*.sinδ.

20.2.5 Marshall Stability and Indirect Tensile Strength

The Marshall stability test was carried out on specimens prepared with composite-modified binders. The samples were submerged in a water bath maintained at 60°C for 30 minutes before testing. Tensile strength is a critical characteristic of pavement mixtures and can be assessed using the ITS test. Typically, the ITS test evaluates the susceptibility of specimens to water damage. Following ASTM D6931–12 standards, the ITS test was conducted at 20 °C.

20.3 RESULTS AND DISCUSSION

20.3.1 Rheological Property

Complex modulus (G*) values were determined from DSR at different temperature variations from 46°C to 88°C. As seen in Figure 20.2, there is a clear increase in

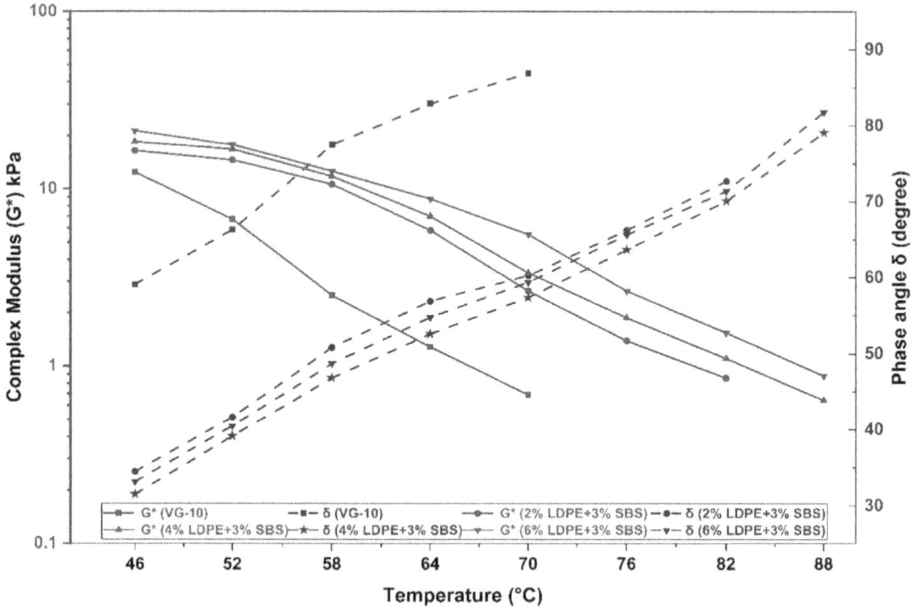

FIGURE 20.2 G* and δ for different types of samples.

G* value with an increase in polymer content, either LDPE or SBS. This is due to the viscoelastic characteristics shared by both the polymers and the asphalt binder. As shown in Figure 20.2, both modifiers significantly enhance the complex modulus, with LDPE exerting a greater influence than SBS. Of all composite modifications, 6% LDPE/3% SBS exhibits the highest value for complex modulus among all tested combinations, then followed by the 4% LDPE/3% SBS composite test formulation. The complex modulus value increased due to interactions between asphalt and additives at their interface. A higher G* indicates increased binder stiffness, correlating with enhanced resistance to permanent deformation (Zaumanis et al., 2014). Remarkably, the phase angle (δ) of composite binders decreases signifying improved recovery potential compared to unmodified variants. Desirable high-temperature properties include elevated G* and reduced δ (Zaumanis et al., 2014). Matching all samples at a specific temperature, it can be said that 4% LDPE/3% SBS contributes in achieving a more stable modified binder.

20.3.2 Master Curve

Master curve for G* for virgin binder and modified binders is given in Figure 20.3. The graph demonstrates that as LDPE content increases, so does the complex modulus, indicating enhanced rut resistance in the composite-modified asphalt binders. This improvement is particularly visible in low-frequency ranges, which is equivalent to high temperatures. However, even at higher frequencies, the 6L3S modified asphalt binders exhibit a higher complex modulus. This characteristic of composite-modified

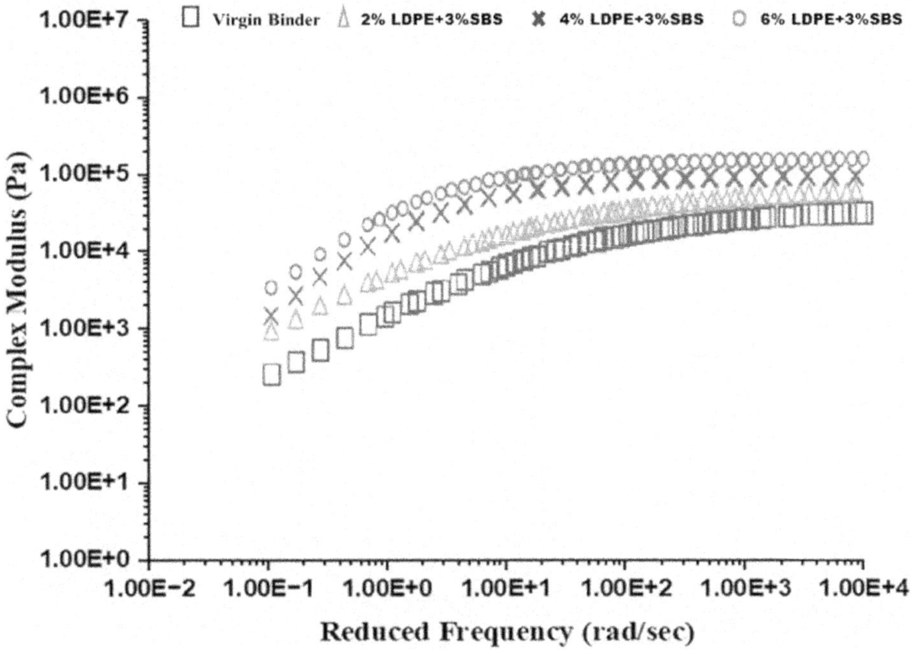

FIGURE 20.3 Master curves of composite-modified asphalt samples.

asphalt binders may be attributed to the varying roles of LDPE/SBS in stress controlling across different frequency ranges. At higher frequencies, LDPE/SBS might have a reduced impact on stress absorption, whereas at lower frequencies, the LDPE/SBS network becomes activated and starts to resist the applied stresses.

20.3.3 Fatigue Property

Figure 20.4 shows the variation of G*.sinδ for different asphalt binder samples with varying LDPE content and a fixed SBS concentration. It is evident from Figure 20.4 that adding LDPE increases the stiffness of the asphalt binder, resulting in a higher G*.sinδ value. Specifically, 4% LDPE combined with 3% SBS raised G*.sinδ by 29.25%, while 6% LDPE with 3% SBS led to an increase of 48.39%. Thus, it can be inferred that incorporating LDPE decreases the fatigue life of the control asphalt binder. On the other hand, SBS contributes to lowering the G*.sinδ values in the composite binder. As mentioned before, the presence of SBS softens the asphalt binder by reducing its stiffness, which in turn decreases G*.sinδ values and enhances fatigue performance. This improvement is attributed to the fact that SBS affects the phase angle more significantly than the complex modulus, creating the binder more elastic and less rigid, thereby improving fatigue resistance. The combined use of LDPE and SBS in modifying the asphalt binder results in G*.sinδ values below the Superpave threshold of 5,000 kPa.

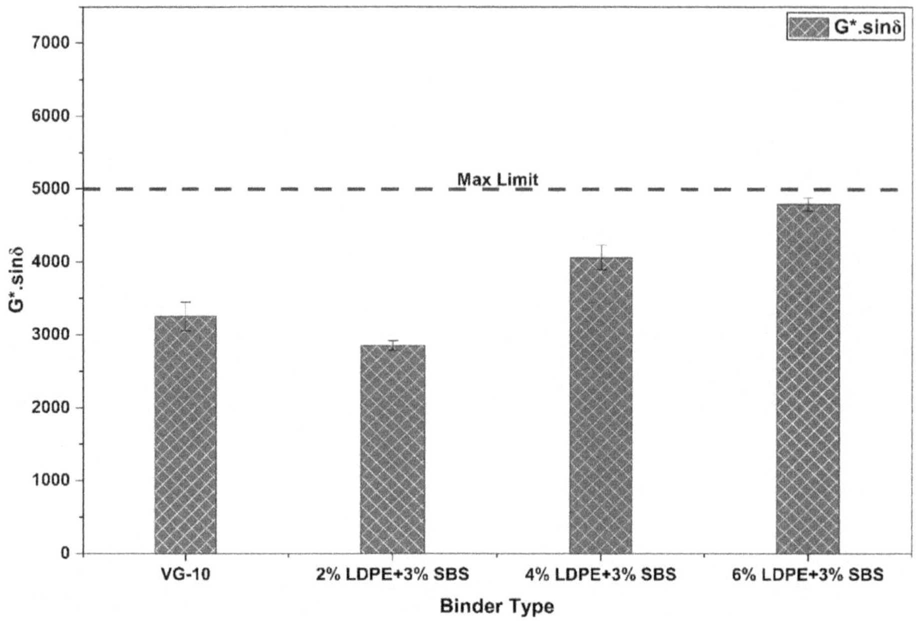

FIGURE 20.4 Fatigue property of composite-modified asphalt binders.

20.3.4 Mechanical Properties Such As Marshall Stability and Indirect Tensile Strength

The results of the Marshall test on hot mix samples containing virgin materials and various proportions of LDPE/SBS composite-modified binder are presented in Figure 20.5. The highest stability value, 14.18 kN, was observed in the asphalt mix prepared with a binder containing 6% LDPE and 3% SBS. Stability remained relatively steady when the LDPE content increased from 4% to 6%, and potential phase separation at higher polymer concentrations could be the reason for it. Nonetheless, all samples surpassed the minimum stability requirement of 9 kN. Additionally, the addition of the modified binder in the mix led to an improvement in stability, possibly due to the stiffer nature of the binder. Indirect tensile strength (ITS) measures the resistance of asphalt mixes along the specimen's diametric plane. Failure in ITS occurs mainly due to weakened adhesion between aggregates and bitumen and improper aggregate orientation caused by unstable mixes. Figure 20.6 shows the ITS results (wet and dry) for samples made with virgin materials and those with LDPE/SBS modified binder. The highest ITS values were found in mixes with 4% LDPE and 3% SBS composite-modified binder, reaching 806.6 kPa (wet) and 942.5 kPa (dry). This improvement is accredited to the elastic properties of SBS and the stiffness of LDPE. Increasing LDPE content more than this percentage led to a decrease in ITS values to 723.6 kPa (wet) and 807.4 kPa (dry), likely due to reduced binder-aggregate adhesion caused by phase separation.

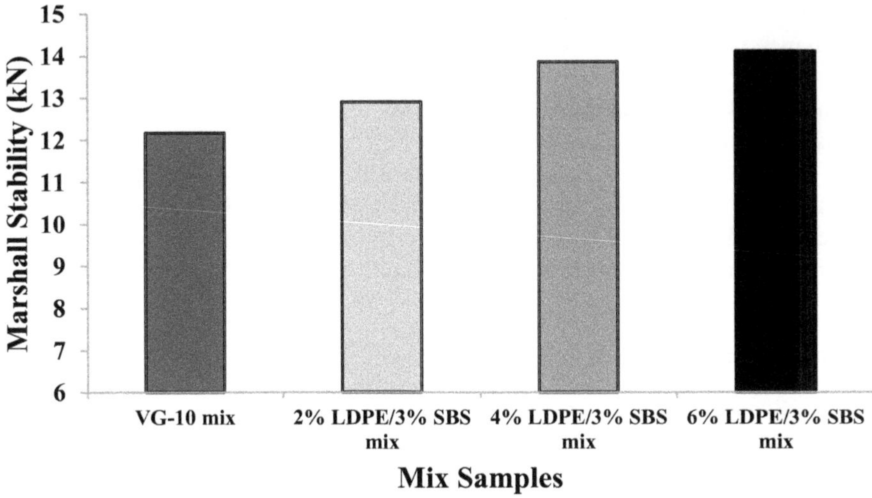

FIGURE 20.5 Variation of Marshall stability.

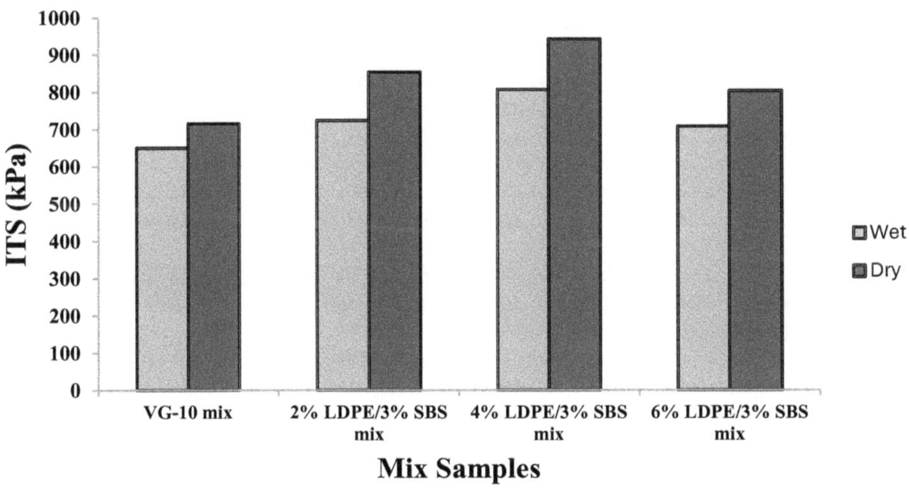

FIGURE 20.6 Variation of ITS.

20.4 CONCLUSIONS

The study demonstrated that LDPE/SBS can serve as an efficient and sustainable composite-modifying agent for asphalt binders. Incorporation of LDPE significantly enhanced the stiffness characteristics of asphalt binder and SBS and significantly helped in maintaining the elastic characteristics of asphalt, which are equally important for long-term performance of the viscoelastic material. As LDPE and SBS modifiers produce favorable outcomes, the utilization of LDPE/SBS as composite modifier is recommended. The optimum proportion of LDPE/SBS modifier, which has enhanced rutting resistance performance and comparable fatigue performance, is chosen as 4% LDPE/3% SBS. However, other values of 6% LDPE/3% SBS can be chosen as workable but at higher temperatures. The LDPE/SBS composite not only increases the polymer content but also provides environmental advantages by facilitating the recycling of waste LDPE. This research has investigated the rheological performance and mixture characteristics of LDPE and SBS modifiers at different dosages as composite modifiers for binder. Future investigations will aim to study the low-temperature performance and storage stability to fully understand its potential.

REFERENCES

Abdy, C., Zhang, Y., Wang, J., Cheng, Y., Artamendi, I., & Allen, B. (2023). Investigation of high-density polyethylene pyrolyzed wax for asphalt binder modification: Mechanism, thermal properties, and ageing performance. *Journal of Cleaner Production*, *405*, 136960.

Albayati, A. H. (2023). A review of rutting in asphalt concrete pavement. *Open Engineering*, *13*(1), 20220463.

Albayati, A. H., Al-Ani, A. F., Mohammed, A. M., Al-Kheetan, M. J., Moudhafar, M. M., & Jweihan, Y. S. (2025). Understanding the effectiveness of elastomeric and plastomeric polymers on the high-temperature performance of asphalt binders. *Innovative Infrastructure Solutions*, *10*(9), 388.

Alsheyab, M. A., Khasawneh, M. A., Abualia, A., & Sawalha, A. (2024). A critical review of fatigue cracking in asphalt concrete pavement: A challenge to pavement durability. *Innovative Infrastructure Solutions*, *9*(10), 386.

Aurilio, M., Tavassoti, P., Elwardany, M., & Baaj, H. (2021). Impact of styrene-butadiene-styrene (SBS) content on asphalt Binder's fatigue resistance at various aging levels using viscoelastic continuum damage and fracture mechanics. *Construction and Building Materials*, *305*, 124627.

Behnood, A., & Gharehveran, M. M. (2019). Morphology, rheology, and physical properties of polymer-modified asphalt binders. *European Polymer Journal*, *112*, 766–791.

Chen, S., Zhuo, S., Xu, G., Chen, X., Yu, L., & Xu, Q. (2024, March). Rheological and chemical indices to characterize long-term oxidative aging of SBS/rubber composite-modified asphalt binders. *Frontiers in Materials*, *11*, 1–10. https://doi.org/10.3389/fmats.2024.1346754.

Costa, L. M. B., Silva, H. M. R. D., Peralta, J., & Oliveira, J. R. M. (2019). Using waste polymers as a reliable alternative for asphalt binder modification – performance and morphological assessment. *Construction and Building Materials*, *198*, 237–244. https://doi.org/10.1016/j.conbuildmat.2018.11.279.

Feldman, D., & Akovali, G. (2005). The Use of Polymers in Construction: Past and Future Trends. In *Handbook of Polymers in Construction* (pp. 13–34). Shrewsbury: GBR.

Gautam, P. K., Kalla, P., Jethoo, A. S., Agrawal, R., & Singh, H. (2018). Sustainable use of waste in flexible pavement: A review. *Construction and Building Materials*, *180*, 239–253.

Joohari, I. B., Maniam, S., & Giustozzi, F. (2022, January). Enhancing the storage stability of SBS-Plastic waste modified bitumen using reactive elastomeric terpolymer. *International Journal of Pavement Research and Technology*. https://doi.org/10.1007/s42947-021-00132-z

Lewandowski, L. H. (1994). Polymer modification of paving asphalt binders. *Rubber Chemistry and Technology*, 67(3), 447–480.

Liu, G., Nielsen, E., Komacka, J., Greet, L., & Ven, M. Van De. (2014). Rheological and chemical evaluation on the ageing properties of SBS polymer modified bitumen: From the laboratory to the field. *Construction and Building Materials*, *51*, 244–248. https://doi.org/10.1016/j.conbuildmat.2013.10.043.

Madkour, M., Bumajdad, A., & Al-Sagheer, F. (2019). To what extent do polymeric stabilizers affect nanoparticles characteristics? *Advances in Colloid and Interface Science*, *270*, 38–53.

Nisar, J., Mir, M. S., & Vivek. (2024). Exploring the potential of waste plastic–modified asphalt: A systematic review of blending ratios, mixing conditions, and rheological properties. *Environmental Science and Pollution Research*, *31*(8), 11507–11528.

Nisar, J., Shafi Mir, M., & Vivek. (2023, August). Study on optimal preparation and rheological characteristics of waste low density polyethylene (LDPE)/styrene butadiene styrene (SBS) composite modified asphalt binder. *Construction and Building Materials*, *407*, 133459. https://doi.org/10.1016/j.conbuildmat.2023.133459.

Nizamuddin, S., Boom, Y. J., & Giustozzi, F. (2021). Sustainable polymers from recycled waste plastics and their virgin counterparts as bitumen modifiers: A comprehensive review. *Polymers*, *13*(19), 3242.

Stroup-Gardiner, M. A. (1997). *Relationships between Rheological Properties and Morphology for Polymer-modified Asphalts*. University of Minnesota.

Tahmoorian, F., Liyanapathirana, S., Yeaman, J., & Egwurube, J. (2023). Performance of hot-mix asphalt and modified binders containing polyethylene. *Journal of Transportation Engineering, Part B: Pavements*, *149*(1), 04023001.

Zaumanis, M., Mallick, R. B., & Frank, R. (2014). Determining optimum rejuvenator dose for asphalt recycling based on Superpave performance grade specifications. *Construction and Building Materials*, *C*(69), 159–166. https://doi.org/10.1016/J.CONBUILDMAT.2014.07.035.

Zheng, X., Xu, W., & Xie, S. (2021). Study on ultraviolet aging mechanism of carbon nanotubes/sbs composite-modified asphalt in two-dimensional infrared correlation spectroscopy. *Materials*, *14*(19). https://doi.org/10.3390/ma14195672.

Index

For Product Safety Concerns and Information please contact our EU representative GPSR@taylorandfrancis.com
Taylor & Francis Verlag GmbH, Kaufingerstraße 24, 80331 München, Germany

www.ingramcontent.com/pod-product-compliance
Ingram Content Group UK Ltd.
Pitfield, Milton Keynes, MK11 3LW, UK
UKHW022217190926
475702UK00003B/406